CRC

Chemoprophylaxis and Virus Infections of the Respiratory Tract

Volume II

Editor:

J. S. Oxford, Ph.D.

Member of the Scientific Staff
National Institute for Biological
Standards and Control
Holly Hill, Hampstead
London, England

Published by

CRC PRESS

CRC PRESS, Inc.
18901 Cranwood Parkway · Cleveland, Ohio 44128

First published 1977 by CRC Press
Taylor & Francis Group
6000 Broken Sound Parkway NW, Suite 300
Boca Raton, FL 33487-2742

Reissued 2018 by CRC Press

Library of Congress Cataloging in Publication Data

Main entry under title:

Chemoprophylaxis and virus infections of the respiratory
tract.

Bibliography: p.
Includes index.
1. Respiratory organs — Diseases — Prevention.
2. Antiviral agents. I. Oxford, John Sidney. [DNLM:
1. Influenza vaccine. 2. Virus diseases — Prevention and
control. 3. Respiratory tract infections — Prevention
and control. WC505 C517]
RC732.C44 616.2 76-45674
ISBN 0-8493-5136-7

A Library of Congress record exists under LC control number: 76045674

Publisher's Note
The publisher has gone to great lengths to ensure the quality of this reprint but points out that some imperfections in the original copies may be apparent.

Disclaimer
The publisher has made every effort to trace copyright holders and welcomes correspondence from those they have been unable to contact.

ISBN 13: 978-1-315-89152-1 (hbk)
ISBN 13: 978-1-351-07062-1 (ebk)

Visit the Taylor & Francis Web site at http://www.taylorandfrancis.com and the
CRC Press Web site at http://www.crcpress.com

PREFACE

Some of the problems associated with any future chemoprophylactic control of the most serious respiratory virus infection of man (namely influenza) and also relevant in some respects to the control of parainfluenza and picornavirus infections have been highlighted recently with the isolation of A/New Jersey/8/76 ($H_{swine\,1}N_1$) virus from a military recruit who died of influenzal pneumonia this winter at Fort Dix, New Jersey. The total. number of cases of infection by the swine-like virus at Fort Dix was serologically estimated to be 500 in a camp of 12,000 persons. The virus A/NJ/8/76 is antigenically related to the swine influenza virus of Shope (A/swine/30 $H_{swi\,1}N_1$), implicated as the causative agent of the influenza pandemic of 1918 which resulted in 20 million deaths (548,000 of these in the United States) and perhaps 200 million clinical cases of influenza world-wide.[1] Will the A/NJ/76 virus spread and, if so, are we able to prevent a new pandemic by vaccination alone? If we had a very effective inhibitor of the virus, how could we use it in practice? Amantadine has antiviral activity against influenza A viruses, but is it sufficiently effective to play a role in a preventative campaign? Should we look more carefully at the possibility of combined chemo- and immunoprophylaxis? It must be realized that the chances of eliminating the virus from the pig population will be very low;[1] thus, the specter of this influenza A virus crossing the interspecies barrier to man will be with us for a long time to come, especially as the potentially susceptible antibody-free group in the population increases each year while older, presumably immune people die.

To date, influenza has remained completely uncontrolled as an epidemic disease. The policy in most countries has been to offer inactivated vaccine to groups at special risk, such as chronic bronchitics, asthmatics, and persons with heart disease of any etiology. More widespread vaccination with live, attenuated virus has been carried out in the Soviet Union and China but with rather variable results. For the first time, therefore, a country (the U.S.) is attempting to prevent a possible influenza outbreak by a mass immunization campaign. Canada will also produce enough vaccine to immunize half of its population. This brings us to a consideration of one of the main obstacles which to date has not been overcome with influenza vaccines. Influenza viruses are antigenically variable: faced with immunological pressure from an immune or partly immune population, antigenic mutants with a survival value are selected. In the face of a mass immunization campaign, the A/NJ/8/76 virus may simply "drift" antigenically and thus still be able to infect and spread in the community. A mutant thus selected may have an even greater capacity to spread than the original parent.

Large-scale trials conducted by the Medical Research Council in the U.K. over the last 20 years have shown extremely variable protection resulting from vaccination against influenza subtypes, varying from negative results to a maximum of 70% protection.[2] More recent trials in the public service in England[3] have highlighted another difficulty of large-scale influenza immunization – low initial acceptance of the vaccine (around 42%) and even lower acceptance in succeeding years (14% in one factory). In these industrial trials, it was concluded that the use of inactivated influenza vaccine in healthy adults was unlikely to produce any great benefit in winters where the incidence of influenza infection is low. Two contributing factors comprised the incomplete protective effect of the vaccine, the low attack rate and the poor acceptance of the vaccine. The saving in sickness absence was approximately 7 days per 100 employees in each winter influenza period, during which about 270 days were lost by each 100 employees. A similar situation is likely to occur in the U.S., especially if there is not an outbreak of A/NJ/76 in the winter of 1977. Will the whole population be vaccinated in the following and succeeding years as well? An alternative plan would be to stockpile vaccine and rely on close surveillance to indicate any spread of virus, whereupon groups of the population could be vaccinated. Given an effective anti-influenza drug, this could be used during the first 10 to 15 days following immunization to give protection during a period when vaccine-induced antibody levels would be low; it could also be used at times of peak exposure to virus to augment any vaccine-induced immunity. Certainly the economic, social, and health gains which would accrue from preventing influenza outbreaks would be considerable, as will be seen in different chapters of this book. For example, in England working days lost through influenza sickness outnumber days lost through industrial strikes. In the U.S. when the Asian (H2N2) influenza arrived in 1957 to 1958, nearly 70,000 deaths occurred, and in 1968 to 1969 the Hong Kong (H3N2) variant caused 33,000 deaths and was costed at 3.8 billion dollars.[4]

Would this supplementation be the main role of influenza chemoprophylaxis? Amantadine has been shown to prevent infection in humans with influenza A subtypes and to resolve clinical symptoms in persons already infected and showing signs of illness. The main problem is the practical question of when to administer the drug. Obviously, a whole population cannot take amantadine daily. The question is partially answered in this book with the description of trials of the compound in general practice in the U.K. Most persons are made aware by the media of the advent of influenza in the community, either at home or at work. Under these conditions, a general practitioner could prescribe a course of compound for 10 or 15 days during a high-risk period. The ideal situation would arise with an antiviral compound which was not effective in completely suppressing viral infection, but rather reduced replication and abrogated clinical illness while allowing subsequent development of protective antibody. As mentioned above, further consideration should be given to the possibility of combined chemo- and immunoprophylaxis. Even with a highly mutable and variable biological entity like influenza, the chances of simultaneous development of drug resistance and antigenic drift would be low.

Even if we find a really effective drug against upper respiratory tract infections, there are strategic problems of administration, particularly if the compound is considered a prophylactic. The tremendous practical advance of recent years resulting from the detailed studies of amantadine both in clinical practice and in the laboratory indicates that the compound has therapeutic activity. Of course the details of the immunopathology of influenza are not clear at present, particularly in molecular terms. Why should a virus infection, predominantly of the upper respiratory tract in man, produce such a manifest clinical response? Respiratory infections have always been difficult to treat. Even a physician presented with a sore throat caused by a β-hemolytic streptococcus will not observe a dramatic effect through treatment with penicillin to which the organism is highly sensitive. Thus, the prevention and, in particular, cure of infections of the upper respiratory tract are fraught with difficulties. However, the whole concept merits more intensive efforts. Respiratory infections are global in impact and, with the exception of influenza, occur throughout the year in all races and classes of people. Although the clinical manifestations are not as dramatic as smallpox, for example, this family of respiratory viruses causes untold misery, morbidity, and mortality. They are worthy of considerable respect and research.

At present, there are few compounds that are known to be effective against influenza and common cold viruses; therefore, the main function of this book is to lay the groundwork for future work and to encourage more active studies in this area. The book has been organized so as to interrelate all aspects of the problem – clinical, epidemiological, pharmacological, molecular biological, etc. Some chapters offer practical advice and techniques, while others are purely theoretical and speculative. Some chapters have been written by specialists in fields other than virology. Thus, fresh ideas can be introduced into an area which to date has been characterized by hard work and no dramatic breakthroughs. Therefore, a wide range of authors have contributed, many of whom are not convinced that these respiratory infections can be controlled by interfering with virus replication through the use of specific inhibitors. However, we are all convinced of the importance of bringing this large group of viruses under control.

There is also controversy over the strategy to be used in the search for new virus-inhibitory molecules. Should we rely on random screening which has been so successful in finding antibacteria, antiprotozoa and antihelminth compounds or should we try to use more contrived and specific approaches? I feel that active searchers have not examined closely enough the explosive increase in our knowledge of the biochemistry or molecular biology of the viruses in question. However, alternative considerations, are given in other chapters of this book particularly in the context of the immense cost of producing a new drug.

Originally, it was our intention to produce a single-volume book covering all aspects and approaches to the problem of specific inhibitors of respiratory viruses. However, as the work progressed, it became obvious that certain chapters, because of the research interests of the authors, concentrated particularly on influenza viruses. It seemed logical, therefore, to divide the book into two volumes, the first emphasizing influenza and the second concentrating on other viruses as well as discussing important general aspects of drug screening and clinical testing, although the second volume does have some chapters which deal mainly with influenza. The two volumes obviously interrelate very closely. For example, Professor Sir Charles Stuart-Harris' chapter on clinical aspects of respiratory infections (Chapter 1, Volume I) is immediately relevant to both volumes, as is Professor Albert's contribution on selective toxicity (Chapter 2, Volume I). Clinical testing of compounds, either in volunteers or in general practice, should be viewed in light of applicability to either influenza or other respiratory virus infections. The two volumes together, then, form the multidisciplinary approach to the problem which we desired.

I hope readers will find the book useful, controversial, and stimulating. It is not meant to be an exhaustive summary of all past work on inhibitors of these virus groups; rather, it is intended to help us define where we are and to offer some suggestions for the future. There is certainly plenty of room for future antiviral compounds outside the respiratory virus group. Work has already been started to search for inhibitors of the cancer-producing oncornaviruses, and a few clinical trials are under way with antiviral compounds and hepatitis B infection, for example.

Obviously, past experience has shown that some viral and bacterial infections, such as smallpox, can be controlled or even eliminated by effective immunization programs. Since this latter approach takes advantage of a natural protective mechanism rather than introducing foreign low molecular weight compounds with all the associated difficulties of toxicity, it would seem to be preferable. Thus, research into more effective influenza vaccines is being intensified, and some work with rhinovirus vaccines is being continued. Inactivated influenza virus vaccines are now composed of relatively pure preparations of whole or partially disrupted virions prepared by continuous-flow rate zonal centrifugation. However, we still are unclear as to whether they induce a high-quality antibody of the correct specificity and why the immunity is incomplete and short-lived. There is an increasing number of experiments with attenuated mutant, recombinants, and low-temperature adapted viruses as candidate live vaccine strains, but the virus is elusive and success seems a long way off. The fascination of the possible intervention with Ehrlich's "magic bullet" continues, and a glance at the bacterial and protozoal world indicates the tremendous changes in the pattern of infective diseases which have resulted from the introduction of chemotherapy. Tuberculosis control, for example, takes advantage of both immunization and chemotherapy. Perhaps as virologists we can learn from this broad approach. If this book encourages a dual approach to the problem of prophylaxis for respiratory virus infections, then it will have succeeded in its objective.

J. Oxford
London
July 1976

REFERENCES

1. Editorial, Swine influenza campaign. Should we vaccinate the pigs?, *Science,* 192, 870, 1976.
2. **Belyavin G.,** Experience in the control of influenza by vaccination – a brief review, in *Chemotherapy and Control of Influenza,* Oxford, J. S. and Williams, J. D., Eds., Academic Press, London, 1976.
3. **Smith, J. W. G., Fletcher, W. B., and Wherry, P. J.,** Vaccination in the control of influenza, *Postgrad. Med. J.,* Vol. 52, 1976.
4. Editorial, *JAMA,* 235, 2753, 1976.
5. **Stuart-Harris, C.,** Swine influenza virus in man. Zoonosis or human pandemic, *Lancet,* 2, 31, 1976.
6. Pandemic of Influenza, 1918–1919, Ministry of Health Report No. 4, HMSO, London, 1920.

THE EDITOR

J. S. Oxford, Ph.D., is Member of the Scientific Staff, National Institute for Biological Standards and Control, Division of Virology, London, England.

Dr. Oxford obtained his B.Sc. degree in 1963 from the University of Reading and received his Ph.D. degree in 1966 from the University of Sheffield.

Prior to his present position, Dr. Oxford was Member of the Scientific Staff, Medical Research Council, National Institute for Medical Research, Mill Hill. Dr. Oxford has also served as Research Fellow, Australian National University and Lecturer in Medical Microbiology, The Medical School, University of Sheffield.

Dr. Oxford is a member of the Society for General Microbiology and the British Society for Antimicrobial Chemotherapy.

Dr. Oxford is currently a member of the Editorial Board of the *Journal of Antimicrobial Chemotherapy* and *Experimental Cell Biology*. He has published more than 70 scientific papers, mainly on the immunology and replication of influenza virus. Dr. Oxford's current research interest is in aspects of the control of influenza by immunoprophylaxis and chemoprophylaxis.

CONTRIBUTORS

A. S. Beare, M.D., F.R.C.Path., D.T.M.
Member of the Scientific Staff
Medical Research Council
MRC Common Cold Unit
Harvard Hospital
Salisbury, Wilts
England

R. A. Bucknall, Ph.D.
Virus Research Unit
Imperial Chemical Industries Limited
Pharmaceuticals Division
Macclesfield, Cheshire
England

M. Clemens, D.Phil.
Member of the Scientific Staff
Medical Research Council
National Institute for Medical Research
Mill Hill
London, England

P. Cooper, Ph.D., D.Sc., M.A.S.M.
Senior Fellow
Institute for Advanced Studies
John Curtin School of Medical Research
Australian National University
Canberra, Australia

A. Galbraith, M.B., B.Chir.
Medical Advisor
Geigy Pharmaceuticals
Hurdsfield Industrial Estate
Macclesfield, Cheshire,
England

Albert Z. Kapikian, M.D.
Epidemiology Section
Laboratory of Infectious Diseases
National Institute of Allergy and Infectious Diseases
National Institutes of Health
Bethesda, Maryland

D. D. Perrin, M.Sc., Ph.D., D.Sc., F.R.I.C.
Professorial Fellow
Medical Chemistry Group
John Curtin School of Medical Research
Australian National University
Canberra, Australia

C. W. Potter, Ph.D., M.R.C.Path.
Professor of Virology
University of Sheffield Medical School
Sheffield, England

Sylvia E. Reed, M.B., B.Chir., M.R.C.Path.
Member of the Scientific Staff
Medical Research Council
MRC Common Cold Unit
Harvard Hospital
Salisbury, Wilts
England

D. A. Rutty, B.Vet.Med., M.R.C.V.S., M.R.C.Path.
Project Manager
Safety Evaluation Section
Safety of Medicines Department
Imperial Chemical Industries Limited
Pharmaceuticals Division
Macclesfield, Cheshire
England

D. A. J. Tyrrell, M.D., F.R.C.P., F.R.C.Path.
Head, Division of Communicable Diseases
Deputy Director and Consultant Physician
Clinical Research Centre
Harrow, Middlesex
England

G. I. Watson, (O.B.E.), M.A., M.D., F.R.C.G.P.
Founder and Honorary Consultant Director
Epidemic Observation Unit
Royal College of General Practioners
London, England

ACKNOWLEDGMENTS

I would like to thank Gillian Oxford for assistance with many aspects of the book and Alan Hodge for the line drawings. The photographic departments of the Medical Research Council Laboratories at Mill Hill and Hampstead were of considerable assistance. The staff of CRC Press has provided assistance at every stage in the preparation of the book.

TABLE OF CONTENTS

Chapter 1

ANIMAL MODELS OF INFLUENZA VIRUS INFECTION AS APPLIED TO THE INVESTIGATION OF ANTIVIRAL COMPOUNDS

C. W. Potter and J. S. Oxford

TABLE OF CONTENTS

I. INTRODUCTION

The initial screening of compounds for antiviral activity has usually been carried out in tissue cultures. In some cases the in vitro antiviral activity shown by these techniques has been subsequently demonstrated in both animal and clinical studies; thus, the antiviral activity of the compound N-methylisatin-β-thiosemicarbazone against vaccinia virus infection has been demonstrated in tissue culture systems,[1] experimental animals,[2] and clinical studies.[3] However, studies with other compounds have not shown this agreement; isoprinosine was reported to be active against influenza virus infection in tissue cultures and mice,[4,5] but tests in human volunteers failed to show significant antiviral activity.[6] Although in clinical studies these results fail to demonstrate the results obtained in laboratory experiments, it is possible that the reverse may occur, and compounds potentially active in clinical tests are discarded because of a lack of in vitro activity. For these reasons, in vitro and in vivo studies should both be conducted to ensure that useful compounds are not overlooked. Since it is not ethically possible to simultaneously carry out in vitro and clinical studies, there is a need for studies in suitable animal model(s); the choice of animals depends principally on the virus infection for which compounds are to be evaluated.

The evaluation of antiviral compounds against influenza could be carried out in a number of animal models, since mice,[7] hamsters,[8,9] dogs,[10] guinea pigs,[11] ferrets,[12,13] monkeys,[14] and pigs[15] are all susceptible to infection by human influenza viruses. Monkeys are too expensive to maintain for initial studies of antiviral compounds, and the same is true of dogs and pigs. Ferrets are highly susceptible to infection with unadapted influenza virus; infection produces a clinical response similar to that observed in man, with symptoms including a febrile reaction,[12,13,16,17] nasal obstruction,[18,19] and an increase in nasal wash protein.[20] However, ferrets are relatively large laboratory animals that are difficult and expensive to maintain; they are an outbred species that show individual variation of susceptibility to influenza virus infection. In addition, ferrets do not produce local 11S secretory IgA antibody,[21] and in this respect the ferret is not an exact model for influenza in man.

Mice are small and relatively easy to maintain; they are available as inbred strains that would be expected to give a reproducible response to influenza virus infection. A considerable body of experience and information on antiviral compounds has been accumulated by experimentation in mice which allows the relative values of many different compounds to be assessed. On the other hand, there are disadvantages in using mice. These animals are difficult to infect with many strains of influenza virus; although these viruses can be adapted to produce lung lesions and death in mice, this can be criticized as an artificial model system quite distinct from influenza in man. Furthermore, influenza virus infection of mice does not produce the febrile response seen in man and ferrets, while infection does induce a heterotypic immunity[22,23] which has not been reported in other species.[9] Both of these reactions suggest that the mouse is not an ideal model for the study of influenza virus infection.

Studies of influenza virus infection in guinea pigs has shown these animals to be susceptible to infection with unadapted influenza viruses; however, guinea pigs do not produce a clinical response to infection, and the sera of these animals contain high titers of inhibitors.[11]

Hamsters are very sensitive to infection with human influenza viruses, and this species may provide a valuable model system.[9,24] Hamsters breed easily and are relatively inexpensive to maintain. They are closely related, since all the hamsters in England originate from a single litter; this would suggest that results obtained in this species would be reproducible. Influenza virus infection in hamsters is principally an upper respiratory tract infection, and relatively high titers of virus can be recovered from nasal washings.[9]

It is likely that no single animal species will prove to be a completely adequate model for human influenza; the extremes of convenience supplied by mice and the similarity to influenza infection of man which is shown by ferrets may be two models that should be used in parallel. However, results obtained for chemotherapeutic compounds against influenza virus infection in mice and/or ferrets cannot be translated to man without qualification; although animal studies are necessary and can be of considerable value, the effect of antiviral compounds on influenza in man can only be accurately assessed by clinical studies.

However, the rapid antigenic changes detected in influenza A virus (Chapter 4, Volume I) raise the important question of whether clinical trials have to be repeated each year with any new influenza variant and an antiviral compound licensed for use against influenza H3N2 strains. In particular, amantadine has been shown to be an active inhibitor of influenza A (H2N2) and (H3N2) subtypes in the laboratory, in animal models, and in man. We consider that an in vitro or in vivo demonstration of efficacy against any new influenza A variant would provide sufficient data to decide its potential usefulness without the delaying requirement of new clinical trials. This is particularly pertinent at present as regards the possible emergence of A/New Jersey/76 ($H_{swine}1$ N1) as a possible epidemic virus. Chemoprophylaxis in influenza could be used as a stopgap measure until sufficient vaccine has been accumulated. If, on the other hand, lengthy clinical trials have to be performed, then the potential usefulness of the compound at a perhaps critical stage in the spread of the virus will have been lost.

II. THE FERRET MODEL

Ferrets are susceptible to infection by all human influenza A and B strains that have been tested.[12,13,17,18,25,26] Virus strains do not require adaptation to grow to high titer in this species, and virus infection can easily spread to other ferrets[27] or to man.[28] The results of infection can be studied using a number of parameters, and the effect of chemotherapeutic compounds on influenza virus infection in ferrets can be assessed both quantitatively and qualitatively by measuring one or more of these responses in control and drug-treated ferrets.

Virus infection can be given to ferrets by nose drops; anesthetics can be used for this procedure, but this is not always necessary for ferrets that have become used to handling. The animals are simply suspended by holding them by the head and neck with one hand, and the inoculum is applied dropwise directly into the nose (Figure 1). The inoculum must be given to nonanesthetized animals slowly in small drops or the animal will blow back. Various techniques have been developed by different workers to infect ferrets and other animals with influenza viruses in aerosols; these can be given either directly, from nebulizers through a face mask, or directly to animals held in a closed environment to which aerolized virus is added. An example of the latter procedure is shown in Figure 2. The authors of this technique place the animal in airtight container(s) and introduce the virus or antiviral compound into the chamber through a Wright nebulizer, producing an aerosol of 1- to 2-μm particle size which reaches the pulmonary alveoli.[29] Although this method of virus infection has some advantages over other methods for some studies, there are problems in standardizing the dose. This is particularly important if an antiviral compound is to be administered in aerosol chambers; a standard amount of com-

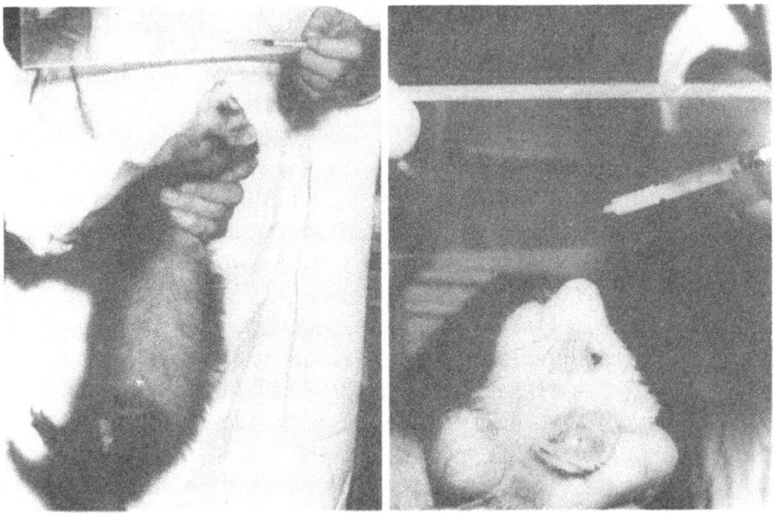

FIGURE 1. Inoculation of ferrets. The ferret is held firmly with one hand, and the inoculum is given dropwise directly into the nose.

FIGURE 2. Inoculation of animals with aerosols. Apparatus for exposing animals to aerosols of virus or antiviral compounds. (Left) Multicompartmental chamber for exposing many animals to the same aerosol. (Right) Multiple chambers for comparison of different aerosols. (From Boyd, M. R. and Beeson, M. F., *J. Antimicrob. Chemother.*, in press. With permission.)

pound may reach the lungs in aerosol form, but much may adhere to the fur to be licked off and ingested in an uncontrolled manner.

A. The Response of Ferrets to Influenza Virus Infection

1. Temperature

Unlike any other animal species, ferrets infected with influenza A virus develop a sharp, febrile response; this is a measurable reaction similar to that found in man and is one of the most important reasons why the ferret is considered a good model system for the study of chemotherapeutic drugs against influenza. Ferrets should be housed in individual cages for at least 1 week prior to any critical assessment of the temperature response to influenza virus infection. During this time, a routine of feeding and cleaning should be used, and the temperature should be taken by a single operator at a fixed time or times each day. This is necessary since ferrets are highly excitable, and their temperature can be made to rise by banging on the cage or otherwise exciting the animals. During the period of 1 week, however, ferrets become tolerant to the routine of feeding, cleaning, and handling.

Two procedures can be used for measuring ferret temperatures. The temperature can be taken per rectum using a clinical thermometer or a thermocouple probe; the temperatures are recorded daily for 4 days and subsequently once daily for 3 days after virus infection.[17] The temperature response of ferrets to intranasal infection with influenza virus A/Port Chalmers/73 (H3N2) using this method is shown in Figure 3. The criteria by which a febrile response is measured in virus-infected animals given antiviral compounds are arbitrary, depending on the response seen in control ferrets; either the elevation of the temperature response above the mean preinfection level and/or the maximum temperature can be used. Two temperature readings of over 40°C measured in the period 24 to 72 hr after infection can be taken to indicate a significant temperature response, and two temperature readings of 1°C or more above the mean preinfec-

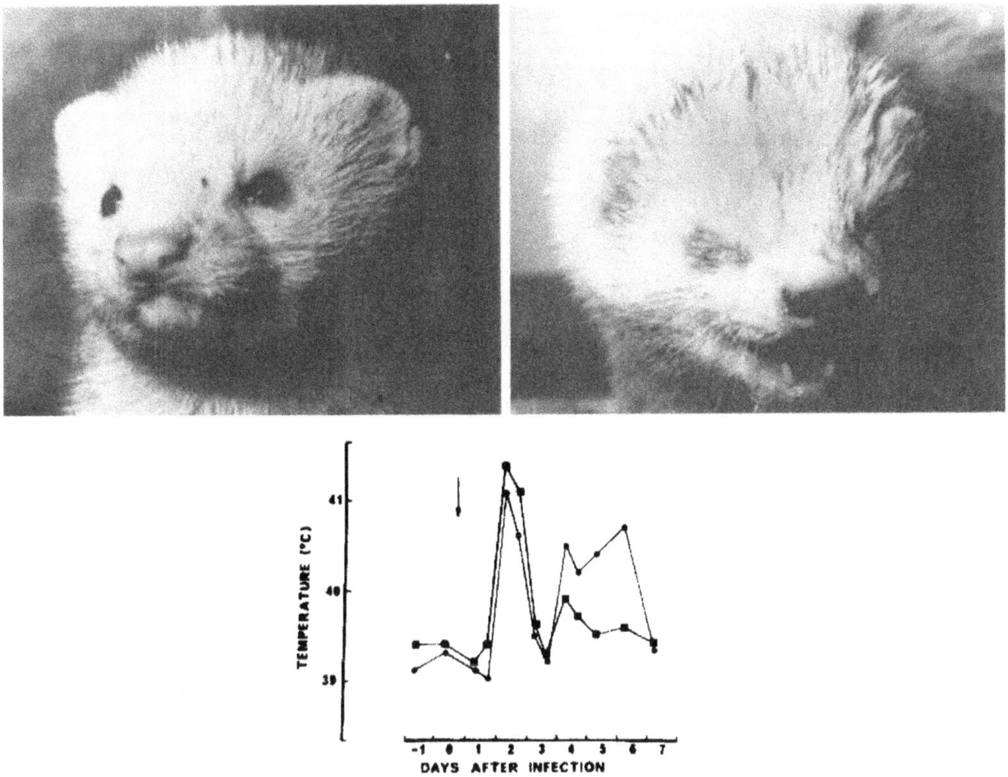

FIGURE 3. Temperature response of ferrets to influenza virus A/Port Chalmers/73 infection. (Top left) Normal ferret. (Top right) Ferret showing febrile response to influenza infection. The animal shows damp, ruffled fur; running nose and, to a lesser extent, eyes; and excessive moisture around the mouth. (Bottom) The fever is maximal 24 to 48 hr postinfection, when the temperature has risen more than 1°C above the base line to a level of 40°C.

tion temperature can be taken to indicate a significant rise in temperature.[17,30]

An alternative method of assessing the temperature response of ferrets to influenza virus infection is to plot a graph of the temperature response and calculate the area enclosed by the lines marking the rise and fall in the temperature. However, for accuracy this method would require more than two temperature readings a day. To achieve this and to eliminate both the time taken and the variations that may be introduced by handling the animals, Squires[31] has developed an electronic method of measuring the temperature of ferrets. In this technique a transmitter is surgically placed under the skin on the back of the ferret, and the skin fold is stitched over; the operation is simple, and the transmitter is too small to irritate or discomfort the animal. The transmitter sends out a variable radio signal which can be detected with a short-wave receiver and translated by a data logger to a graph of the temperature of the animal (Figure 4). In this way, temperatures can be registered as often as required, or even continuously, simply by tuning into the transmitter. The apparatus can be multiplied so that several transmitters can be monitored at the same time using the same receiver. Using this method, the febrile response of a ferret to influenza virus infection can be measured by observing the maximum height of the febrile response, the duration of the febrile reaction, or the area under the temperature profile which takes into account both of these factors (Figure 4).

2. *Virus Multiplication*

Unadapted human influenza A viruses replicate to high titers in ferrets. Virus replication occurs chiefly in the nasal turbinates;[26,32] extracts of nasal turbinates or single nasal washings taken into 10 ml of diluent can contain $10^{5.0}$ to $10^{7.0}$ EID_{50}/ml of virus at 2 to 3 days after virus infection, depending on the strain of virus used.[17,32,33] In an analysis of tissues and organs taken from ferrets 2 to 3 days following intranasal

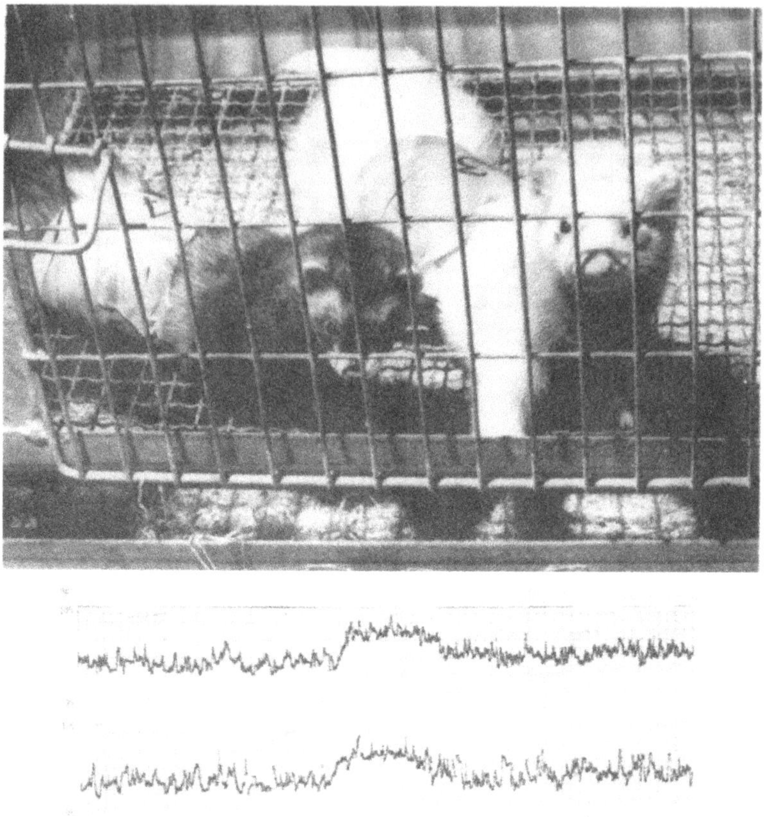

FIGURE 4. Automated method of recording ferret temperatures. (Top) Two ferrets with transmitters in place. (Bottom) Graph of the temperature response to influenza virus A infection. The signal from the transmitter is detected by a short-wave receiver and translated by a data logger to a graphic reading. (From Squires, S. L., *Ann. N.Y. Acad. Sci.*, 173, 229, 1970. With permission.)

infection, virus was recovered in high titer from nasal turbinates, but was also recovered from conjunctiva, trachea, pharynx, esophagus, lung, bladder, uterus, and oviduct.[26] However, Sugg and Nagaki,[34] who also found virus in the nasal turbinates of ferrets infected with influenza virus, failed to detect virus in lung extracts and did not test bladder, uterus, or oviduct. The main tissue for virus replication is, therefore, the nasal turbinates, where virus can be detected from 1 to 6 days after infection.[13,17] The titers of virus present in nasal washings taken at different times from ferrets infected with influenza virus A/Port Chalmers/73 are shown in Figure 5.

3. Local Symptoms

The pathological changes which occur in the nasal turbinates of ferrets infected with virulent influenza virus have been fully described.[35,36] Following virus inoculation by the intranasal route, there is little pathological change during the first 24 hr; however, at 48 hr the turbinates show necrosis of the respiratory epithelium with congestion, edema, and cellular infiltration. At 2 to 4 days postinfection, the ciliated epithelium in the respiratory area of the turbinates has completely disappeared, leaving the basal cells and basement membrane intact. By the 6th day, reparative processes are seen, and a transitional epithelium 2 to 4 cells thick has formed over the whole respiratory area of the turbinates. In the week following this, the transitional epithelium is replaced by normal ciliated epithelium (Figure 6). In further experiments the transitional epithelium was found to be resistant to iontophoresis or prolonged irrigation with a 1.0% solution of zinc sulfate; both of these treatments cause extensive damage to the ciliated epithelium of normal ferrets.[37] At the 7th to 8th day postinfection, the ferrets were also resistant to reinfection with

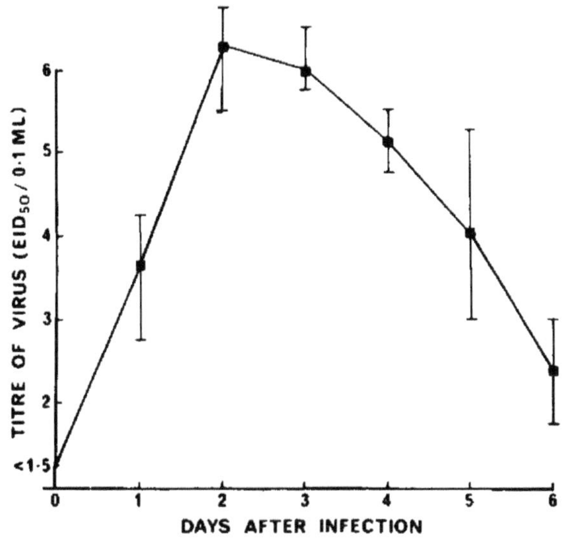

FIGURE 5. Titers of virus in nasal washings from ferrets infected with influenza virus A/Port Chalmers/73. Nasal washings collected in 10 ml of PBS were titrated in 10-day embryonated eggs. (■ Mean titer (EID_{50}/ml) of virus (four ferrets). (I) Range of results

homologous influenza virus. This may not be due to the presence of a resistant, transitional epithelium, since serum and local antibody can be demonstrated at this time,[21] and 7 to 8 days later ferrets previously infected with influenza A were susceptible to influenza B and vice versa.[38]

In a detailed study of the rhinitis induced in ferrets by influenza virus infection, Haff and his colleagues[13] showed that nasal washings from influenza virus-infected ferrets contained increased numbers of leucocytes; this had been described previously.[25] The increase in cells was variable, however, and because of the wide variation from animal to animal, it was difficult to quantitate the increased number of leucocytes or the increase in numbers of individual cell types.

An alternative method of assessing the rhinitis which occurs in ferrets infected with influenza virus is to measure the nasal patency; in this technique the nasal airway resistance is measured by passing air through the nasal passages of anesthetized ferrets and relating the transnasal pressure to the rate of air flow.[18,39] A similar method has been described by Boyd and Beeson,[19] and this technique and the results obtained in ferrets infected with influenza virus A/Port Chalmers/73 are shown in Figure 7 and in Table 1.

Ferrets infected with influenza virus develop

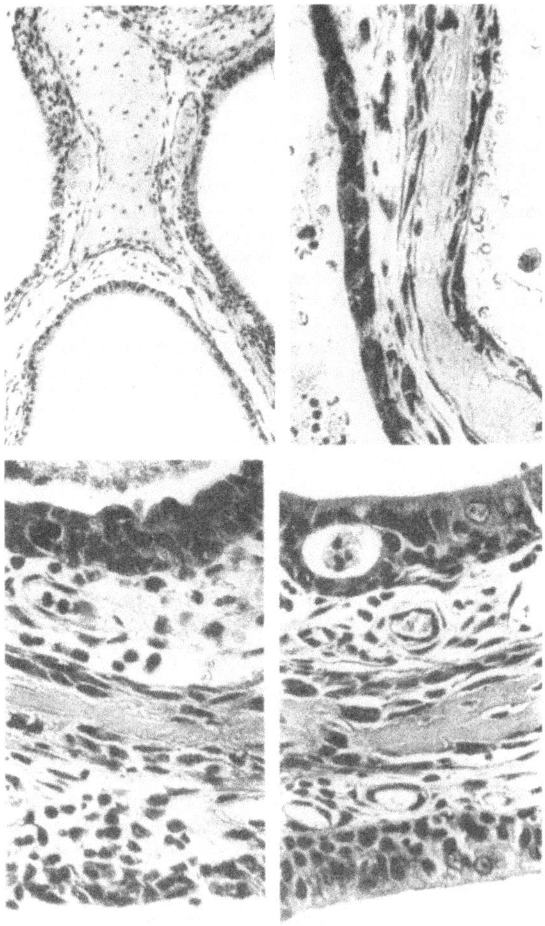

FIGURE 6. Pathology of ferret turbinates during influenza virus A/PR/8/34 infection. (Top left) Normal appearance. (Top right) Destruction of ciliated epithelium at 48 hr after infection. (Bottom left) Transitional stratified columnar epithelium, three-cells thick covering the engorged submucosa at 6 days after infection. (Bottom right) Regenerated, normal mucosa, numerous goblet cells, and normal epithelium from which cilia are appearing at 14 days after infection. (From Stuart-Harris, C., *Influenza and Other Viruses of the Respiratory Tract*, Edward Arnold, London, 1953. With permission.)

increased levels of protein in nasal washings. The peak increase in nasal wash leucocyte content has been shown to occur 4 to 7 days after virus infection,[13] and the peak levels of nasal wash protein occur at approximately the same time.[20] Nasal washings can be obtained by instilling 0.3 to 0.4 ml of phosphate-buffered saline (PBS) into one nostril. After approximately 1 to 2 sec, the ferret will forcibly express the PBS, which is collected in a petri dish; this is due to a natural sneeze reflex.[16] The nostrils are washed out alternately

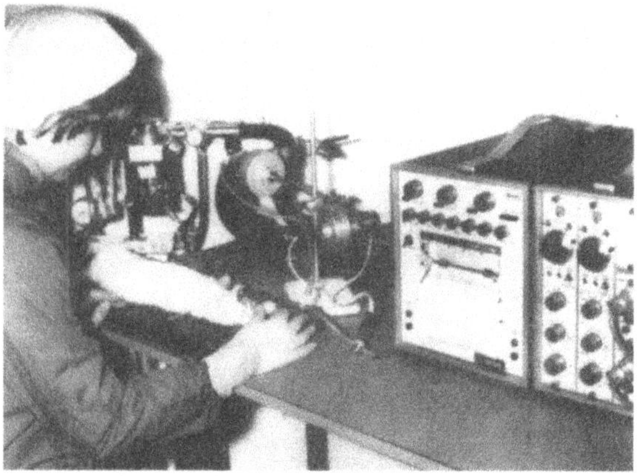

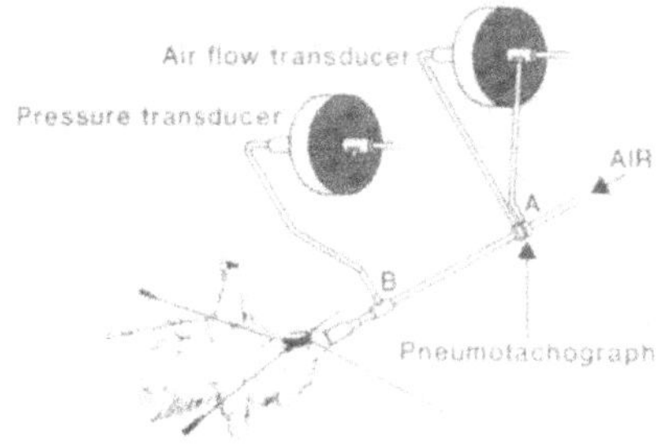

FIGURE 7. Measurement of nasal airway resistance of ferrets. (From Boyd, M. R. and Beeson, M. F., *J. Antimicrob. Chemother.*, in press. With permission.)

with small volumes using a total of 10 ml of PBS. The total specimen is then shaken with glass beads and centrifuged at 3000 rpm for 10 min; the supernatant fluids are separated and concentrated tenfold by dialysis against 30% Carbowax®. The protein content of concentrated washings from a ferret infected with influenza virus A/Hong Kong/68 is shown in Figure 8. In this example, and for other influenza H3N2 viruses tested, the peak concentration was found 7 to 9 days after virus infection, at which time the protein levels were three- to fivefold greater than that of specimens taken prior to infection.[20,30]

4. Nasal Wash Antibody

Nasal washings from ferrets infected with influenza virus have been shown to contain specific antibody.[17,21,39] In a study of nasal washings from ferrets infected with influenza virus A/Hong Kong/68, neutralizing and hemagglutination-inhibiting antibody was first detected in specimens collected 7 days after virus infection (Figure 8); antibody was also found 9 and 11 days postinfection, but not in specimens collected 13 or more days after virus infection.

TABLE 1

Nasal Airway Resistance of Ferrets Before and After Intranasal Infection with A/Port Chalmers/73 (H3N2) virus

	Nasal airway resistance (mm H_2O/l/min)		
		Postinfection	
Ferret	Preinfection	Day + 2	Day + 3
1	45;44	85	121
2	30;47	101	130
3	48;54	40	98
4	24;22	84	59
5	34;31	28	82

From Boyd, M. R. and Beeson, M. F., *J. Antimicrob. Chemother. Suppl.*, in press.

The appearance of antibody in nasal washings coincided with the rise and fall in nasal wash protein levels (Figure 8). An examination of pooled, concentrated nasal washings from influenza virus-infected ferrets in linear sucrose gradients indicated that the antibody in these specimens was a 7S immunoglobulin; in addition, the absence of IgA in ferret nasal washings and the presence of relatively high concentrations of IgG suggested that the nasal wash antibody was IgG.[21] Quantitative studies of the proteins in nasal washings from virus-infected ferrets indicated that the antibody in these specimens was largely, but not necessarily entirely, transudated serum antibody.[21] If this finding is confirmed, the measurement of nasal wash antibody may indirectly indicate the virus-induced damage to the nasal respiratory epithelium and thus serve as a marker of a pathological effect of influenza virus infection. The titer of neutralizing antibody found in the nasal washings of virus-infected ferrets has been found to vary considerably for different influenza virus strains. Thus, relatively high titers were found in washings from ferrets infected with unadapted influenza virus A/Hong Kong/68[17] or A/PR/8/34 (H0N1),[33] but significantly lower titers were found in specimens from ferrets infected with A/Port Chalmers/73,[40] although the serum antibody titers measured 21 days after infection were comparable for the three viruses. In view of the previous findings for nasal wash antibody mentioned above, this could reflect differences in virulence of these viruses for ferrets.

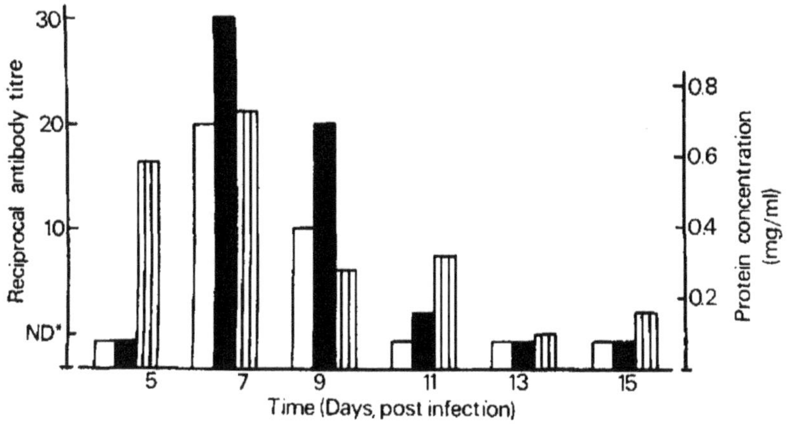

FIGURE 8. Protein and antibody levels in serial nasal washings from a ferret infected with influenza virus A/Hong Kong/68. (*) Antibody not detected. (□) HI antibody titer. (■) Neutralizing antibody titer. (▥) Protein concentration. (From Potter, C. W., Oxford, J. S., Shore, S. L., McLaren, C., and Stuart-Harris, C. H., *Br. J. Exp. Pathol.*, 53, 153, 1972. With permission.)

5. Serum Antibody Response

Sera collected from ferrets 14 to 21 days after influenza virus infection contain high titers of hemagglutination-inhibiting (HI) antibody. These specimens have relatively low titers of serum inhibitors and a relatively high degree of specificity which make them valuable in the serological differentiation of influenza viruses. Serum HI antibody is first detected at approximately 4 days postinfection;[13] relatively high titers are detected at 8 days after virus infection, when most of the antibody is IgM;[21] and peak titers are found 17 to 21 days postinfection, when the antibody is predominantly IgG.[21] It has been suggested that the serum antibody response may be a useful measurement of the effect of antiviral compounds on influenza virus infection in animals.[41] Antiviral compounds which limit virus replication would reduce the antigenic stimulus, and serum antibody titers lower than those seen in controls would develop. Provided that adequate controls are used to insure that the drug does not directly affect the immune mechanisms, serum antibody measurements could provide a marker of antiviral activity in both animals and man.[42-44]

6. Interferon Response

Extracts of nasal turbinates from ferrets infected with influenza virus have been shown to contain interferon;[13,45] interferon could not be demonstrated in nasal washings or blood taken at the same time.[13] The titer of interferon in the nasal turbinates was maximal 2 to 4 days after virus infection, but was not detectable in specimens collected 7 or more days postinfection.[13,45] This result closely parallels the changes in virus titer present in the nasal washings,[13,17] and interferon titers may be directly related to the titer of virus present.

7. Delayed Hypersensitivity Response

Ferrets infected with influenza virus A/Hong Kong/68 or A/PR/8/34 develop a delayed hypersensitivity response to inactivated influenza virus; this can be demonstrated by skin tests, macrophage migration inhibition tests,[46] or by measuring foot pad swelling after intradermal inoculation.[45] The hypersensitivity reaction could be demonstrated 14 and 35 days after virus infection, but disappeared after this time.[46] Histological examination of hypersensitivity lesions induced by the intradermal inoculation of 40 international units of homologous, inactivated virus vaccine into sensitized ferrets showed a diffuse inflammatory reaction with edema; the deeper layers exhibited some mononuclear infiltration with perivascular cuffing. Although the response was clearly a hypersensitivity reaction, the appearance was not as striking as in guinea pigs or man.[11]

B. The Study of Antiviral Compounds in Ferrets

Ferrets have been used extensively for the study of influenza virus infection and for the evaluation of vaccines.[30,47] The clinical response of ferrets to influenza virus, together with the

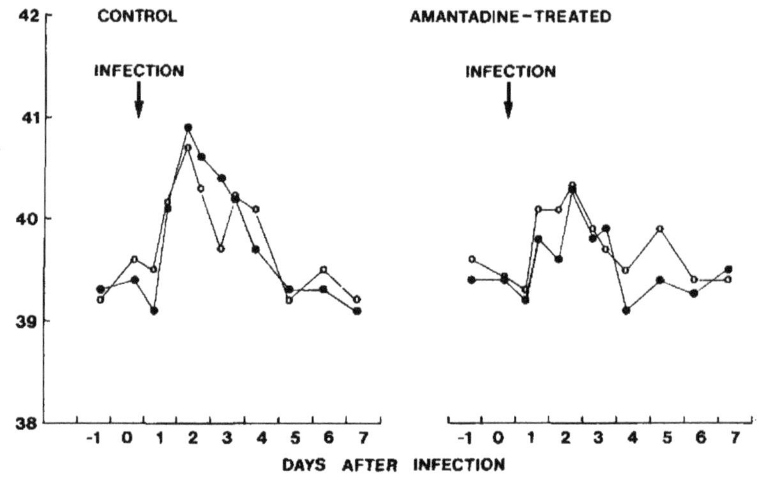

FIGURE 9. Temperature response of ferrets infected with influenza virus A/Port Chalmers/73.

other measurements of infection, has meant that a number of parameters can be used to assess the value of a particular vaccine. Indeed, some of the parameters show degrees of reaction; thus, degrees of immunity can be measured.[47] In theory, all or any combination of the responses to infection mentioned above can be used to assess the response of normal or immunized animals to influenza virus inoculation, and these have been used in the study of influenza vaccines for many years. However, the use of ferrets for the study of antiviral compounds is a comparatively recent development.

1. Studies of Amantadine

The compound 1-adamantanamine hydrochloride (amantadine) has exhibited antiviral activity against some human influenza viruses (Chapter 5, Volume I). Thus, using a variety of tissue culture systems, influenza A virus H2N2 and H3N2 strains have been shown to be sensitive to amantadine; HON1 and H1N1 strains of influenza A virus and influenza B virus were less sensitive or insensitive to this compound.[48-51] Similar results have been obtained in mice.[48,49,52]

In animal experiments a dose of 50 mg/kg/day of amantadine, given subcutaneously for 12 days from the time of infection, was found to have no antiviral effect on influenza virus A/PR/8/34 (HON1) infection of ferrets; indeed, the authors reported that the treated ferrets were more ill than untreated control ferrets.[53] These results were based on mortality and lung pathology; deaths occurred in the drug-treated group, whereas none was found in the control group. The inability of these experiments to demonstrate antiviral activity for amantadine was not surprising, since later studies indicated that A/PR8 virus was insensitive to amantadine by in vitro tests and in mice.[49]

Squires[31] studied the action of amantadine given by the oral route to ferrets infected with influenza virus A/Singapore/57 (H2N2); the author was able to show that the compound did not modify the temperature response of ferrets to virus infection or reduce the clinical signs of influenza in these animals. The virus used in this study was more sensitive to amantadine than the A/PR/8/34 virus used by Cochran and his colleagues,[53] but the strain was not as sensitive to amantadine as the more recent H3N2 strains of influenza A virus.[51] In a study in ferrets of the effects of amantadine on influenza virus A/Port Chalmers/73 (H3N2) infection, the compound was found to have antiviral activity.[54] Ferrets were inoculated subcutaneously with 100 mg/kg/day of amantadine from 1 day before to 7 days after intranasal virus infection with $10^{3.0}$ ferret-infective doses of A/Port Chalmers/73 virus; the temperature response, titers of virus present in nasal washings, increase in nasal wash protein, the changes in serum and nasal wash antibody were measured in both drug-treated and control animals. The temperature response of two of four ferrets treated with amantadine is shown in Figure 9; compared to control animals, amantadine caused a modification of the temperature response to infection. In addition, the titers of virus recovered from nasal washings collected 1 to 6

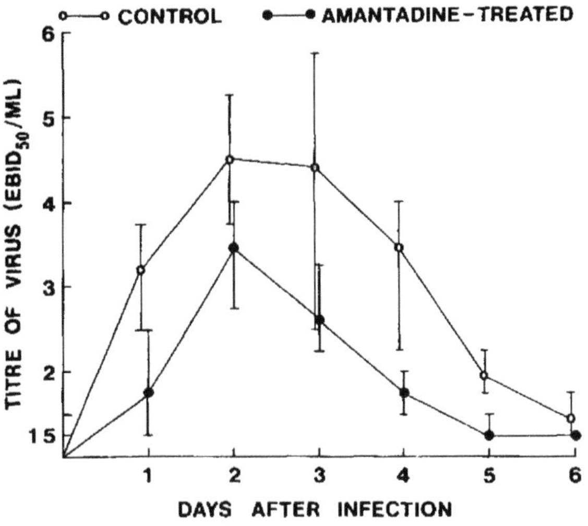

FIGURE 10. Titers of virus in nasal washings of ferrets infected with influenza virus A/Port Chalmers/73. Nasal washings collected in 10 ml of PBS were titrated by the allantois-on-shell method. (o and •) Mean results of four ferrets. (Ɪ) Range of titers.

days after infection were less than those found in control ferrets (Figure 10). Although amantadine treatment reduced both the febrile response and virus production in influenza virus-infected animals, the compound did not reduce the nasal symptoms, as measured by the increase in nasal wash protein concentration, or the titers of serum or nasal wash antibody. Other studies have shown that amantadine also modified influenza virus A/England/72 (H3N2) infection of ferrets; this was most marked when the compound was given as an aerosol.[55] The increased antiviral activity of amantadine when given by aerosol, as compared with oral administration, has also been demonstrated in influenza virus-infected mice.[56]

There are several reports of the antiviral activity of amantadine in natural or experimental influenza A virus infections of man;[57-62] although the doses used to treat patients and the strain of infecting virus vary, the results obtained for the majority of studies indicate that amantadine limits fever and the duration of symptoms of influenza A virus infection (Chapters 2 and 3). This finding is in agreement with the results of in vitro and in vivo tests in mice, but does not agree as well with the results obtained in ferrets. Thus, amantadine was found to be active against influenza A (H2N2) virus infection of tissue cultures, mice, and man,[48,49,60,63] but had no effect on H2N2 virus infection of ferrets in the single published experiment.[31] The result suggests that the ferret may not be an ideal animal model for the study of antiviral compounds for man.

In most studies of antiviral agents where there has been lack of correlation between in vitro and in vivo results, the findings have usually indicated a failure to confirm in vivo activity demonstrable in vitro, rather than the converse. This has probably been due to a failure of the compound under test to reach the site of virus infection in a form or at a concentration comparable to that used in in vitro experiments. In the present context, amantadine was found to have antiviral activity against influenza virus A (H2N2) infection of ferrets when given by aerosol, but when given orally,[55] this may be due to the absence of an adequate concentration of the compound at the site of virus infection following oral administration. By extrapolation, the administration of other antiviral components in aerosol form to influenza virus-infected ferrets may show either the same or greater activity than that demonstrable when compounds are given by other routes. Moreover, before any animal can be considered a suitable experimental model, compounds must be shown to reach the site of virus infection in an animal in an active form comparable to that found in man.

2. Studies with Ribavirin

The compound 1-β-ribofuranosyl-1,2,3-triazole-3-carboxamide (ribavirin) has been reported to inhibit the replication of a number of DNA and RNA viruses;[64-67] specifically, this compound was reported to inhibit influenza virus infection in vitro[64,66,68,69,73] (see Chapter 5, Volume I).

The antiviral activity of ribavirin against influenza virus A/Port Chalmers/73 infection has been studied in ferrets. In one experiment, virus-infected ferrets were given 100 mg/kg/day intraperitoneally for 3 days from 1 day before infection. In a second experiment virus-infected ferrets were given the same daily dose of compound for 8 days from 1 day before infection. The two studies were carried out independently, and in each experiment the results were compared with those obtained in nontreated ferrets.[71] The temperature response to infection for two ferrets from each of the two drug-treated groups and for untreated control ferrets is shown in Figure 11. The control ferrets showed both a significant temperature (>40°C) and a significant rise in temperature (>1°C rise above the mean, preinfection temperature) at 24 to 72 hr post-

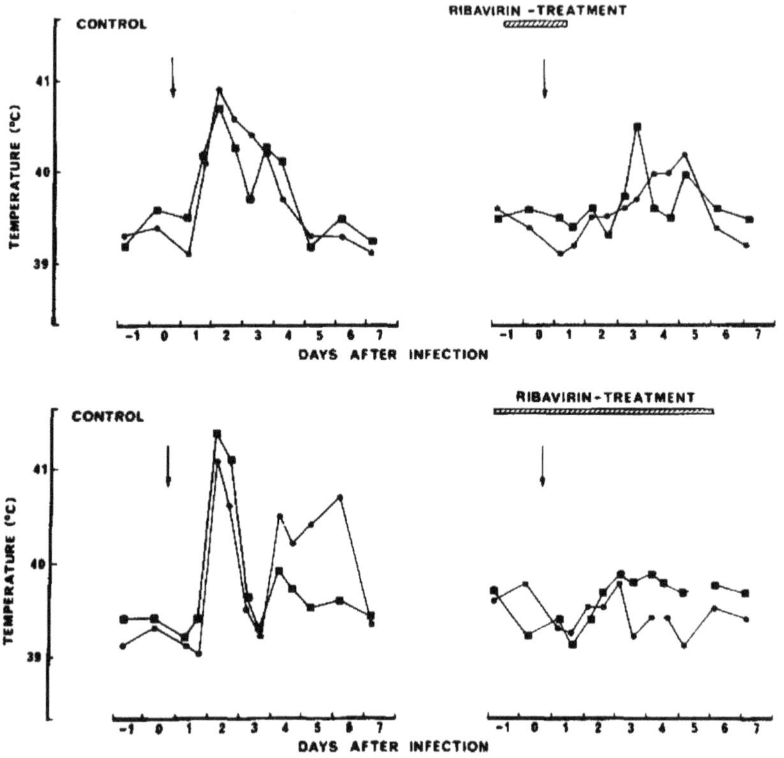

FIGURE 11. Temperature response of ferrets treated with ribavirin to infection with influenza virus A/Port Chalmers/73. (From Schofield, K. P., Potter, C. W., Edey, D., Jennings, R., and Oxford, J. S., *J. Antimicrob. Chemother.*, in press. With permission.)

infection. In contrast, virus-infected ferrets given three doses of ribavirin showed a modified temperature response, and the peak temperature, which for three of the four animals did not reach a significant level, occurred at 3 to 5 days postinfection. Ferrets treated for 7 days with ribavirin showed no febrile response to infection (Figure 11). The titers of virus in nasal washings collected 1 to 7 days after influenza virus infection in drug-treated and control ferrets are shown in Figure 12. Compared to the titers of virus found in nasal washings from control ferrets, animals treated with three doses of ribavirin had reduced titers of virus on days 1 to 2 after infection, but the titers of virus present after this time were as high or higher than those found in control ferrets. The titers of virus present in nasal washings from ferrets given seven doses of ribavirin were 10- to 50-fold less on each day than those found in control animals.

Ribavirin had a significant effect on other responses of ferrets to influenza virus infection. Ferrets that received three doses of ribavarin showed an increase in the concentration of nasal wash protein, but the increase was delayed. In control animals, the peak levels of protein were seen at 5 to 7 days, whereas for drug-treated ferrets peak levels occurred at 9 to 11 days after virus infection (Figure 13). Virus-infected animals that received seven doses of ribavirin showed no increase in nasal wash protein (Figure 13). Ribavirin treatment also affected the antibody response to infection. Ferrets that received three doses of ribavirin showed a delayed serum and nasal wash antibody response, while ferrets given seven doses of the compound did not produce demonstrable serum or nasal wash antibody. Further investigation indicated that the absence of an antibody response to virus infection was probably not entirely due to the antiviral effect of ribavirin which reduced the antigenic stimulus, but was due to the immunosuppressive activity of the compound. Thus, ribavirin reduced or abrogated the serum antibody response of mice and guinea pigs to intramuscular inoculation with inactivated influenza virus or purified virus hemagglutinin in Freund's complete adjuvant.[71] The above results suggest that ribavirin is a potent antiviral com-

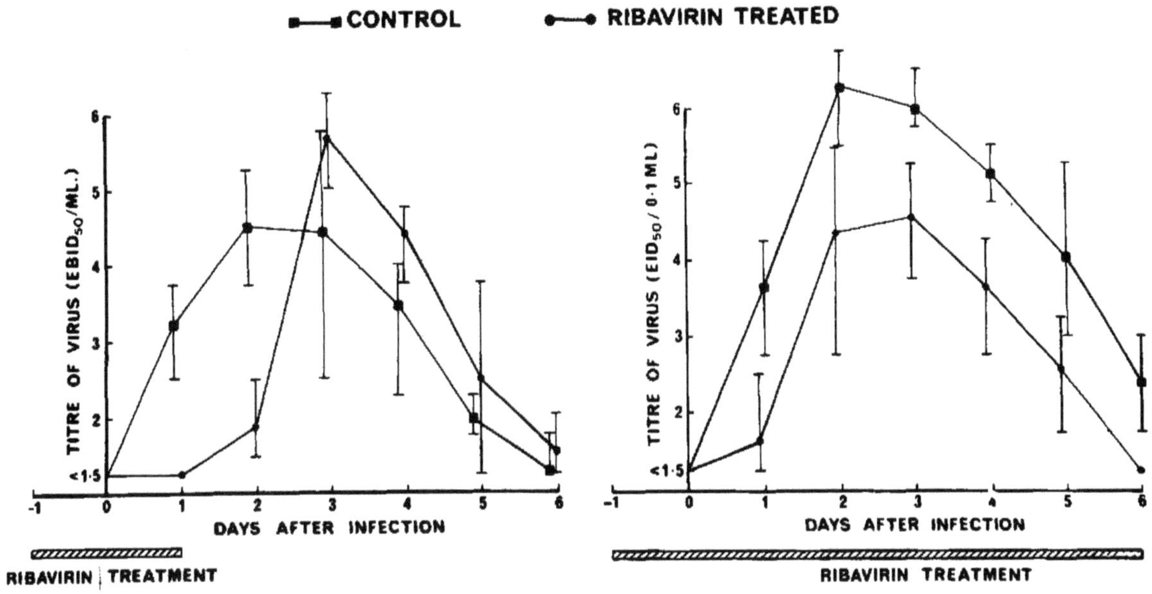

FIGURE 12. Titers of virus in nasal washings of ferrets treated with ribavirin and infected with influenza virus A/Port Chalmers/73. (From Schofield, K. P., Potter, C. W., Edey, D., Jennings, R., and Oxford, J. S., *J. Antimicrob. Chemother.*, in press. With permission.)

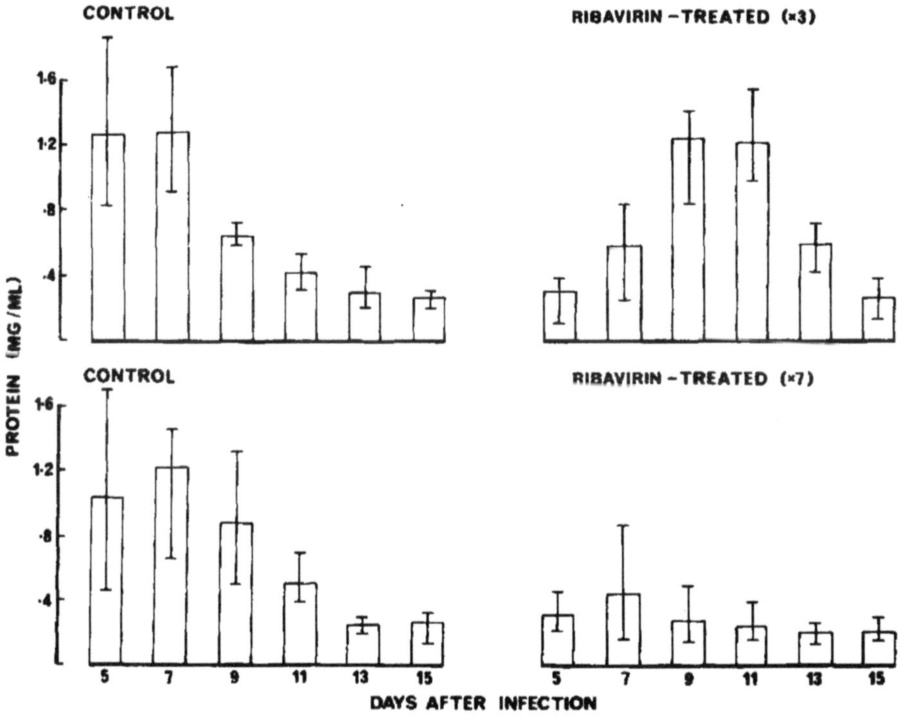

FIGURE 13. Concentration of protein in nasal washings from ferrets infected with influenza virus A/Port Chalmers/73. (□) Mean concentrations for four ferrets. ( ) Range of concentrations. (From Schofield, K. P., Potter, C. W., Edey, D., Jennings, R., and Oxford, J. S., *J. Antimicrob. Chemother.*, in press. With permission.)

pound against influenza infection of ferrets. For man, the use of this compound parenterally, but not necessarily by other routes, is limited by the immunosuppressive effect; not only is this a direct disadvantage, but ideally an antiviral compound should modify or abrogate illness while allowing antibody formation that would protect against subsequent infection.[72]

3. Studies with Cyclooctylamine

The compound cyclooctylamine hydrochloride has been reported to have antiviral activity against influenza A and B virus, both in tissue culture and in mice;[75] the effect of this compound on three different influenza A virus infections of ferrets was reported.[74] In this study cyclooctylamine at a concentration of 50 mg/kg was given subcutaneously 1.5 hr before, 1 to 5 hr after, and twice daily to groups of ferrets infected with influenza A virus H2N2, H1N1, or A/swine. The effects of drug treatment were evaluated by measuring the temperature response, titers of virus in nasal turbinates, leucocyte counts in nasal washings, and neutrophil/lymphocyte ratio in the blood. Cyclooctylamine was found to significantly diminish the temperature response to infection by all three influenza viruses. In addition, the drug-treated ferrets showed significantly less leucocytes in the nasal washings and a lower blood neutrophil/lymphocyte ratio. Treated ferrets also had significantly less virus in nasal turbinates than control animals at 24 hr postinfection; after this time, however, the differences decreased. The compound had the most significant effect on influenza virus H2N2 infection of ferrets, but antiviral activity was also demonstrated in ferrets infected with the other viruses tested. The spread of influenza virus H2N2 from infected ferrets to drug-treated and control ferrets was also studied; infection was determined by virus isolation from nasal washings and by sero-conversion of control animals. In these studies the spread of the virus to control cage contacts occurred more frequently than to drug-treated contacts.[74]

A study of the chemotherapeutic activity of cyclooctylamine on experimental influenza virus infection of man showed results similar to those obtained in ferrets (Chapter 2); thus, the compound modified the clinical response to virus infection.[44] The febrile response to infection was significantly reduced, compared to that of placebo-treated patients, and the symptom score was also reduced in treated patients. Virus shedding was more easily demonstrated, and virus was recovered for longer periods after infection of control patients than for drug-treated patients. Lastly, the titers of serum antibody were significantly less at 14 days postinfection in patients treated with cyclooctylamine than in control patients; however, at 21 to 24 days after infection the antibody titers for the two groups were comparable. The lowered serum antibody titer of treated subjects may reflect a reduction of virus replication and a reduced antigenic stimulus; this has been suggested as a valuable marker of activity for an antiviral compound.[42,44] However, before this conclusion can be drawn, the compound under test must be shown not to have direct, immunosuppressive activity. Thus, in tests of ribavirin in influenza virus-infected ferrets, lower titers of serum and nasal wash antibody were found in drug-treated animals than in control animals; this is due, at least in part, to the immunosuppressive activity of this compound.[40,75]

4. Studies with Aspirin

The value of ferrets as an animal model for the study of antiviral compounds against influenza virus infection is found in the clinical reaction of this animal to virus infection. Using this model it is possible, therefore, to study not only compounds which directly affect virus infection, but also compounds which may ameliorate symptoms. In this respect, Haff and Pinto reported the results of aspirin treatment on the nasal congestion induced by influenza virus infection in ferrets.[18] Animals were given different doses of aspirin either orally or intranasally using a variety of regimes. The authors were able to show that repeated intranasal application of aspirin at concentrations of 0.001% or more had no effect on nasal potency of normal ferrets. When given 1 hour prior to virus infection, the compound had a significant decongestant activity on virus-infected animals, as measured by quantitating the nasal airway resistance to a controlled air flow. This compound was also effective when given orally.

5. Comparative Studies

Because of the variety of both clinical and virological responses, the ferret has a number of advantages over other animals in the study of antiviral compounds against influenza virus infection. In addition, by measuring a number of these responses, the relative activity of different

TABLE 2

Effects of Inhibitors on Influenza A Virus Infection of Ferrets

Compound tested	Dosage	Response to infection with influenza virus A/Port Chalmers/73				
		Temperature	Virus isolation	n/w protein	n/w antibody	Serum HI antibody
Nil	b	d	d	d	d	d
Ribavirin	100 mg/kg from day −1 to 7 (once daily; i.p.)	b	c	b	b	a,b
Amantadine	100 mg/kg from day −1 to 7 (once daily; s.c.)	c	c	d	d	d

[a]This compound was shown to have an immunosuppressive action at the concentration used.
[b]No response to challenge infection.
[c]Modified response to challenge infection.
[d]Unmodified response to challenge infection.

compounds can be assessed. In Table 2 the activity of two compounds is compared; other compounds tested which showed no chemotherapeutic activity are not included.[76] The compounds were given to groups of ferrets by different routes and using different dosage schedules, but in each case the animals were infected intranasally with $10^{3.0}$ ferret-infective doses of A/Port Chalmers/73 virus. By monitoring a group of responses of the animals to infection, it can be seen that the compound ribavirin had the most dramatic effect in virus infection, since four of five of the responses measured were abrogated; however, the absence of serum antibody was probably due to the immunosuppressive activity of the compound.

Amantadine modified two responses. In this way the relative values of compounds can be assessed. By expanding the number of responses of the ferrets to virus infection to include others mentioned above, a more detailed assessment can be made. The results shown in Table 2 are not intended to indicate that one compound is better than another, except under the conditions in which the tests have been carried out. Thus, by varying the concentration of compound and, in particular, the route and times of administration, a totally different table of relative values may be obtained.

III. THE MOUSE MODEL

Usually, laboratories involved in the search for inhibitors of influenza virus use virus-infected mice as a primary or secondary screening procedure. Compared to ferrets, for example, mice are inexpensive, and large numbers can be handled with relative ease to give statistically significant results; usually, groups of 10 to 20 mice are used with a single concentration of inhibitor under test. However, influenza in mice is not comparable to the infection in man. Most mouse models of influenza infection use highly mouse-adapted viruses which cause acute infection of the lower respiratory tract; the animals die of virus pneumonia (Figure 14), while human influenza is predominantly an acute infection of the upper respiratory tract, only rarely producing pneumonia and death (Chapter 1, Volume I). The mouse in the lower half of Figure 14 is acutely ill with influenza pneumonia and has ruffled fur with arched back and impeded respiration. It might be anticipated that such mouse models present a vigorous test to the efficacy of any anti-influenza compound. We shall consider the responses of mice to infection with influenza virus and how the measurement of these responses can be used in the study of two extensively investigated antiviral compounds, amantadine and ribavirin.

A. Infection and Adaptation of Influenza A Viruses to Mice

In practical terms mice are lightly anesthetized in a glass jar containing a cotton pad soaked in anesthetic ether. They are infected by addition of three drops of virus dilution to the nostrils. Any mice blowing back and producing frothing at the nose are rejected. If mice are inoculated intranasally under ether anesthesia with egg-adapted influenza viruses of high titer, severe lung lesions may be produced at the first passage due to the

FIGURE 14. Clinical effects of influenza infection of CBA mice.

toxic effects of the inoculum. Mice may die, although on further serial passage, the lung lesions disappear. The lesions usually appear again after six to ten passages and increase in severity, with the mice finally dying. Usually adaptation of recent influenza A isolates to mice is a very variable procedure, some strains adapting easily and other strains not at all. However, regardless of whether the virus kills mice, relatively high infectivity titers of virus can be demonstrated in the lungs 48 hr following inoculation.

In experiments with mice it should be remembered that great variation can occur between different strains of mice and their susceptibility to infection; in our experience variation also exists between inbred colonies of the same mouse strain kept in different laboratories. Considerable variation, of course, is noted in the susceptibility of young and old mice to infection with influenza. In our laboratory very reproducible results have been achieved using male inbred and specific pathogen-free CBA mice at 10 weeks of age. A further source of experimental variation is derived from the inoculation method, particularly if animals are anesthetized with ether and infected by dropwise addition of virus to the nose. More reproducible results are achieved if animals are infected by aerosol (Figure 2). This method is particularly convenient since relatively large numbers of mice can be infected simultaneously without anesthetic. In the apparatus illustrated, the particle size obtained is 1 to 2 μm, which ensures that the virus reaches the pulmonary alveoli to initiate the characteristic pneumonitis.[29]

B. Pathology of Influenza in the Mouse

The acute stage of influenza virus pneumonia in the mouse is characterized by a hemorrhagic, edematous exudate in the alveolar air spaces and a hyperemia with cellular infiltration in the thickened alveolar septa.[77] The cellular infiltrate is composed of polymorphonuclear leucocytes, small mononuclear cells, and macrophages. Using immunofluorescence, Hers[78] demonstrated conclusively that the experimental lung lesions in mice were a true viral pneumonia and that the tissues affected were the alveolar cell lining. During adaptation of influenza A virus to mice by serial passage, the virus causes severe damage to the epithelium of the respiratory tree prior to the development of any of the macroscopic or microscopic indication of influenza virus pneumonia. With a low dose of virus, rapid adaptation to the airway epithelium of mice can be obtained without evidence of virus-induced lesions of the lung.[77] Thus, adaptation of a human influenza virus A strain to the mouse lung may be a selection of a variant capable of multiplication in the alveolar cells.

As early as 24 hr after intranasal inoculation of mice with influenza A virus, changes can be detected in the tracheal epithelium; these include cell ballooning, loss of cilia, and vacuolation. By 48 hr only remnants of cilia are present; the basal cells are degenerate and show characteristic shrinkage and nuclear fragmentation. By 72 hr changes can be detected in the bronchus, including disintegration of the epithelial layer (Figure 15); at this stage, degeneration of the alveolar cells with fragmentation of the nuclei is detectable. Immunofluorescence studies reveal the presence of virus antigen in the nucleus of lung alveolar cells as early as 24 hr postinfection. Of considerable interest is the observation that if the whole trachea and bronchus are systematically examined, patches of virus-infected cells and degenerate cells alternate with completely normal cells even late after virus

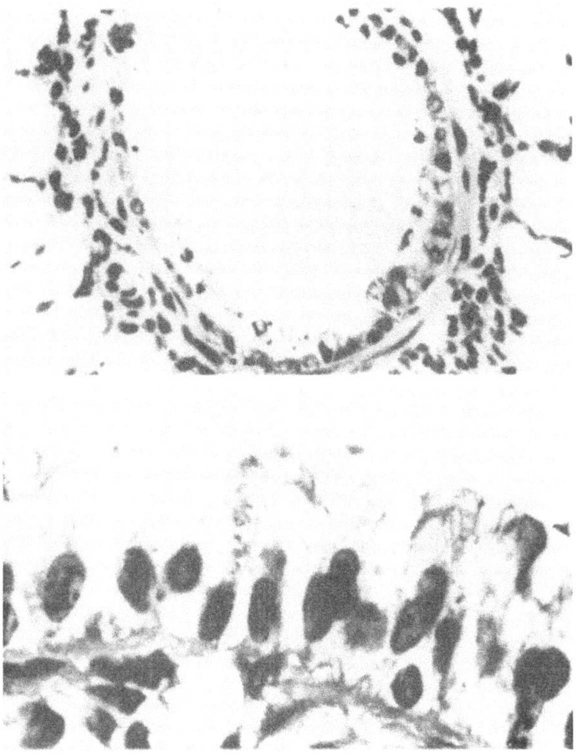

FIGURE 15. Pathology of influenza virus infection in the mouse. (Top) Mouse trachea 24 hr postinfection shows ciliated epithelium with loss of cilia and ballooning. (Bottom) Mouse bronchus 72 hr postinfection shows desquamation, and nuclear fragmentation. (From Mulder, J. and Hers, J. F., *Influenza,* Walters-Noordhoff, Groningen, 1972. With permission.)

infection; this has also been described in ferrets and man. This could be the reason for the therapeutic action of amantadine when given late after virus infection of man or mice.[79] Thus, even at a late stage in infection, the influenza virus is still spreading to previously unaffected cells, and the inhibition of this virus transfer by compounds could resolve the infection more quickly.

C. Response to Infection

1. Death

Since influenza virus strains adapted to grow in mouse lungs frequently produce death, this has been one of the most commonly used parameters of assaying the effects of virus infection in this species. Influenza virus-infected animals are examined for 14 days for sickness, and the number of daily deaths is recorded. The results are expressed both as the percentage of mice surviving the infection at the end of the experiment and the mean day of death of the nonsurvivors. In studies of influenza virus inhibitors, only low doses of virus, in the region 5 to 10 LD_{50}/mouse, are commonly used to infect mice. Under these conditions, most mice would appear ill, with rapid breathing and ruffled fur (Figure 14) about 3 to 4 days postinfection; they would die at about 7 to 9 days after virus infection.

In the cases of most influenza virus infections of mice, the response is quantal; the animals either die or survive. Bauer and Sadler[76] have developed an analysis in which the response is more or less a continuous variable. In this method the mean reciprocal survival time of groups of mice is plotted against the dose of compound on a log scale, and a linear relationship is obtained. The relative activity of an antiviral compound can be calculated by standard methods from the displacement of its dose-response line to the left of that of the standard obtained for nontreated animals. In the same way, such dose-response lines can be used to measure the activity of a single compound against different influenza virus subtypes. The point at which the line intersects the ordinates corresponding to the mean reciprocal survival time of the untreated control animals is a constant and is independent of the dose of virus used for infection. The dose corresponding to the point of intersection is E_0, the zero effect dose. When the dose of virus is varied over 1000-fold range, the E_0 varies less than 2-fold. Therefore, E_0 is essentially an invariant and can be used as an absolute numerical measure of the activity of a particular compound against a particular virus.

The initial criteria for evaluation of anti-influenza virus activity of 156 drugs[80] were increases in the time of mean survival and in the number of survivors 21 days after virus inoculation. The results are statistically evaluated by comparing (by means of the t test) the mean survival time of drug-treated, virus-infected animals dying on, or before day 21 with the mean survival time of the virus control animals. If the P value obtained is < 0.05 but > 0.001, the antiviral activity is considered questionable; however, if the P value is < 0.001, the drug is considered to have possible antiviral activity. The data are questionable if a significant increase in mean survival time occurred in animals treated with higher nontoxic concentrations. The same consideration applies if activity is observed in mice infected with a high level of virus, but not in those infected with lower virus doses. The number of surviving infected,

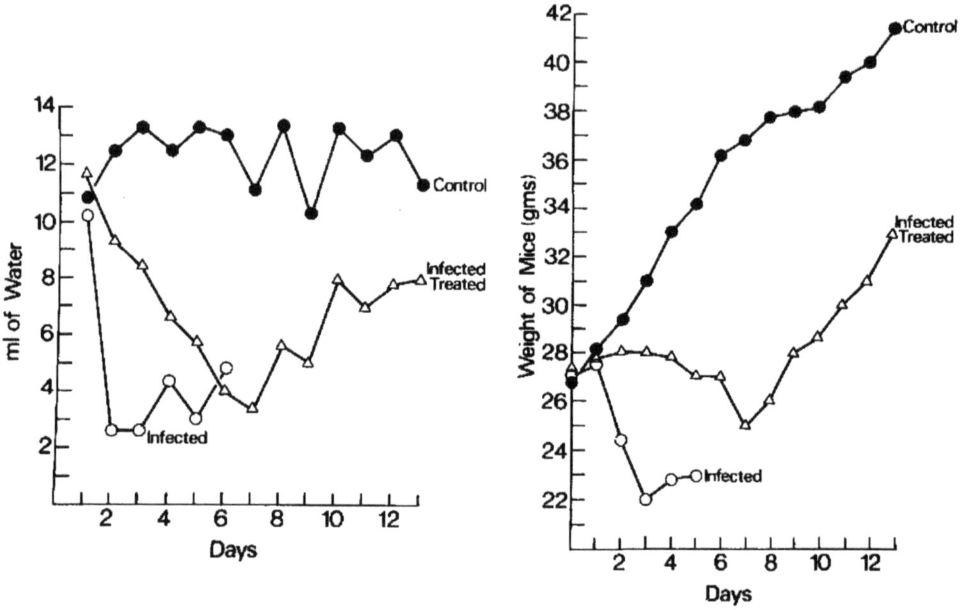

FIGURE 16. Weight and water intake of mice infected with 30 LD_{50} of influenza virus A/Japan/305/57. (Right) Weight of mice following virus infection. In contrast to control mice, virus infection results in a dramatic loss of weight; successful treatment corrects the weight loss. (Left) Water intake of virus-infected mice. There is a rapid decline in the water intake of virus-infected mice which is corrected by treatment with antiviral compounds. (From Rabinovich, S., *Antimicrob. Agents Chemother.*, 1, 408, 1972. With permission.)

treated animals is compared with the number of surviving virus-control mice by means of the chi-square analysis technique. If the P value calculated by this method is <0.5 but >0.05 (approximate standard error), the antiviral activity is questionable, whereas values of <0.05 are indicative of possible antiviral activity.

2. Lung Consolidation

A further method of evaluating influenza virus infection of mice is based on the degree of lung consolidation; this is carried out by scoring each lung on a "blind" basis, i.e., the scorers do not know the history of the donors of the lungs. The lungs are graded according to the following scale: 5, death with consolidation; 4, 100% consolidation; 3, 75% consolidation; 2, 50% consolidation; 1, 25% consolidation; 0, no consolidation. An average lung-consolidation score is calculated by dividing the total grade of consolidation by the number of lungs graded. The data are statistically evaluated using White's modification of the Wilcoxon test. A P value of <0.05 calculated by this method indicates questionable antiviral activity; the same considerations of dose response and virus concentrations described above are also used in this method. Any compound having questionable or possible antiviral activity when tested by the above procedure is retested for confirmation. If possible activity is again seen, additional experiments with other virus strains, different treatment schedules, or different methods for evaluation are then carried out. Mice which have been exposed only to virus diluent are treated with identical drug dosages at the same time as the test animals. These animals are held 30 days after the end of treatment and serve as drug toxicity controls.

3. Virus Hemagglutinin in Mouse Lungs

It is of particular importance to demonstrate that a potential influenza virus inhibitor actually reduces the titer of virus in the lung, rather than exerting some effect on the immunopathology of influenza infection in lung cells, indirectly resulting in animal survival. A useful technique, particularly applicable to mouse-adapted strains of virus, is to remove infected mouse lungs 48 to 72 hr postinfection, grind up the lung in a pestle and mortar with sand, centrifuge at 1500 g (10 min), and titrate the supernatant fluid for virus hemagglutinin (HA) using 0.5% chicken red blood cells. This method can be quantitative and avoids the delay and expense of titrating lung extracts for infective virus.

4. Food and Water Intake and Body Weight

Two interesting developments seen in studies of amantadine in influenza virus-infected mice are the observations of changes in the food and water intake. Thus, virus infection causes a decrease in both water and food intake and a subsequent weight loss; however, the late administration of amantadine and derivatives to infected mice restored their weight gain and water uptake to almost normal levels (Figure 16). These extended models of in vivo infection will undoubtedly prove to be useful for the assessment of other antiviral compounds against influenza virus. In experiments in Swiss white mice infected intranasally with 30 LD_{50} of A/Japan/305/57 (H2N2) virus, the mice appeared ill about 36 hr postinfection; water and food consumption decreased, and weight reduction was easily quantitated. Infected and nontreated mice lost weight progressively until death. Infected mice treated with 24 mg/kg/day of rimantadine (α-methyl-1,-adamantane methylamine) had an initial decrease in weight followed by an increase in weight at a rate parallel to that of the controls. By the end of the experiment, however, they had not reached the same weight as the noninfected animals. Mice given rimantadine and not infected continued to gain weight, drink, and eat the same as the controls.

5. Changes in Lung Weight

A refinement of the method of measuring the effect of virus infection on mice is to measure the lung weight increase in infected animals; these weight increases are a result of pneumonic changes that occur in the lungs of mice infected with lethal virus challenge levels. In a series of experiments in mice infected by aerosol with 1/30 LD_{50} of a mouse-adapted A/Hong Kong/1/68 strain,[29] the lung weights are observed to increase to a maximum 8 days after infection (Figure 17). Prevention of this virus-induced increase in lung weight by compounds under test was indicative of antiviral activity in its broadest sense.

The results of dose-response titrations with some standard antiviral drugs are represented graphically in Figure 18. Six subcutaneous doses of amantadine, spiroamantadine, or tilorone at 10 or 30 mg/kg/day produce substantial reductions in the lung weight increase due to influenza virus. The lung weight method of evaluating influenza infections in mice is, therefore, a sensitive, reproducible technique readily susceptible to statistical analysis.

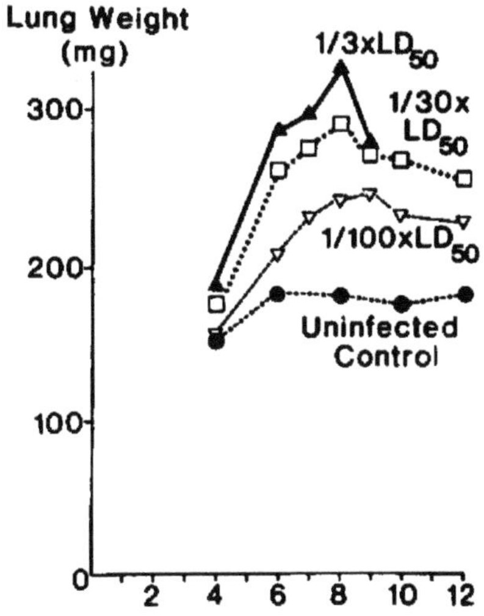

FIGURE 17. Lung weights in influenza virus-infected mice. Lung weight gain in mice infected with different doses of influenza virus A/Hong Kong/68. (From Boyd, M. R. and Beeson, M. F., *J. Antimicrob. Chemother.*, in press. With permission.)

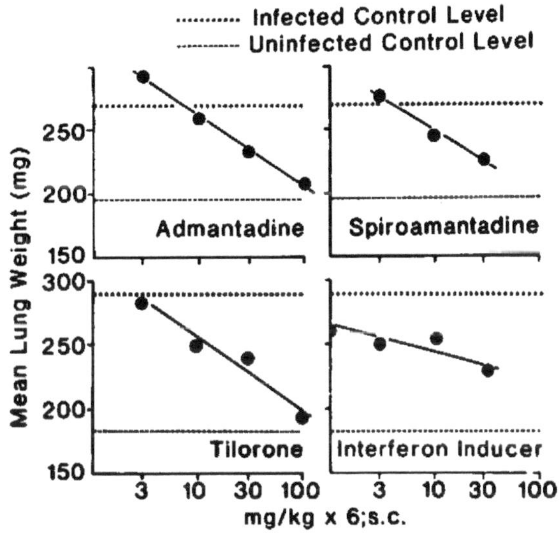

FIGURE 18. Lung weight in virus-infected mice treated with antiviral compounds. Effect of treatment with antiviral drugs of mice infected with influenza virus A/Hong Kong/68. Six doses of the compound were given twice daily from 3 hr before virus infection. Interferon inducer – N, N-dioctadecyl-N',N'-bis(2-hydroxyethyl) propanediamine. (From Boyd, M. R. and Beeson, M. F., *J. Antimicrob. Chemother.*, in press. With permission.)

6. *Contact Infection*

A particularly interesting animal model system has been developed by Schulman.[81] Virus-infected mice are housed with uninfected mice, and these are later examined for infection by transmission. In a series of elegant experiments, the author has demonstrated that if the uninfected mice are given amantadine derivatives, there is a significant reduction in their attack rate from the infector mice. Also, treatment of the infector mice will reduce their ability to transmit virus to uninfected contacts. These results are illustrated in Table 3, from the author's published data. Obviously this model is very close to the natural epidemiology of influenza in man and should be of great interest. The use of ferrets (see above) in place of mice should also improve the model.

TABLE 3

The Effect of Amantadine and Rimantadine on the Ability of Infector Mice to Transmit Influenza A2 Virus Infection

Drug treatment of infector mice	No. of normal contact mice infected	
	No.	%
Saline	35/60	58.3
Amantadine[a]	28/60	48.3
Rimantadine[a]	5/60	8.3

[a]A 20-mg/kg dose, i.p., 0.5 hr before and 12, 24, and 36 hr after infection.

From Schulman, J. L., *Proc. Soc. Exp. Biol. Med.*, 128, 1173, 1968. With permission.

D. Studies of Antiviral Compounds in Mice

1. *Amantadine*

Of particular interest for the future assessment of influenza-inhibitory compounds in the mouse model has been the demonstrated therapeutic activity of amantadine. Thus, amantadine administered by oral intubation 64 hr postinfection, when virus multiplication would have reached a maximum in the mouse lung and when mice were clinically ill, showed prolongation of mouse survival (Table 4).[82] Rabinovich was also able to demonstrate a therapeutic effect of rimantadine against influenza A/Jap/305 (H2N2) in the mouse; when treatment was begun with 24 mg/kg of rimantadine 1 hr postinfection, the survival rate was 90% compared to 20% in the control untreated group. In addition, the survival rates were 60 and 50%, respectively, when treatment was delayed for 24 and 48 hr postinfection; when treatment was delayed for 72 hr there were no differences between treated and control mice.

Of particular interest are the in vivo data showing inhibition of A/swine/30 ($H_{swine}1$ N1) virus by amantadine;[82] a closely related virus, A/New Jersey/75 ($H_{swine}1$ N1), has recently been isolated from an army camp in the U.S. and may spread in the general population since few persons

TABLE 4

Survival and Disease Severity Ratings of Mice Infected with Influenza A/Ann Arbor/2/60 (H_2N_2) and Treated with Amantadine HCl

Dose (mg/kg)[a]	N	Survival day 10(%)	Mean survival day	Mean disease severity rating	Healthy animals (%)
Control	60	15.0	6.2	1.82	3.3
10	60	43.3[c]	8.2[c]	1.22[c]	35.0[c]
5	58	37.9[c]	8.4[c]	1.34[c]	27.6[c]
2.5	60	30.0	6.8	1.57[b]	16.7[b]
1.25	60	26.7	6.8	1.60	16.7[b]

[a]Treatment by oral intubation beginning 64 hr and continuing through 124 hr after infection at 4-hr intervals.
[b]$p = 0.05$.
[c]$p = 0.01$.

From Grunert, R. R., McGahen, J. W., and Davis, W. L., *Virology*, 26, 262, 1965. With permission.

under the age of 50 years have protective antibody to the virus. A virus similar to, or identical with, A/swine/30 ($H_{swine}1$ N1) was the cause of the influenza pandemic in 1918 which resulted in 20 million deaths. Our recent in vitro studies indicate that the A/New Jersey/75 virus is inhibited by amantadine (Chapter 5, Volume I). Also, the earlier studies of Grunert et al.[82] show that the compound inhibited pneumonia in mice infected with A/swine/30 to a similar degree to inhibition of pneumonia induced by A/Ann Arbor/2/60 (H2N2) virus. The data are presented in Figures 19 and 20 which show dose response curves illustrating that the degree of effectiveness of the compound decreases as the virus challenge dose increases (Figure 19) and that the dose of amantadine which increased the percentage of survivors half way from the control value (10 to 30%) to complete survival (the corrected ED_{50}) was 7 mg/kg/4hr.

A qualification of the above results is the demonstrable, but apparently mild, immunosuppressive activity of amantadine.[83] It is known, for example, that administration of antilymphocytic serum to mice after infection reduces lung lesions and deaths.[84] Thus, it is possible that at least part of the beneficial effect of amantadine on animal deaths may be attributed to some immunosuppressive action. Control experiments with influenza B virus, which is not inhibited in vitro by amantadine and thus should not be inhibited in vivo under the above conditions, should be performed to resolve the question.

2. Ribavirin

Most in vivo reports to date of the antiviral activity of ribavirin have used the conventional mouse model with scorings for mouse survival and mean day of death;[70] the more elaborate models of virus transmission, effect on weight gain, etc. have not been used. Apparently, well-tolerated oral doses of ribavirin effectively protected mice against lethal infections produced by strains of both influenza A and B virus.[85] In nontreated animals infected with A/England/42/72 (H3N2), virus titers peaked at $10^{6.0}$ pfu/ml (plaque-forming units) at 24 hr postinfection and remained at this level for 48 hr; most of these mice died between days 5 and 9. In mice treated with a single but relatively high dose of 200 mg/kg ribavirin immediately after infection, the virus titer was reduced 1 to 2 logs, and maximum titers were reached 24 hr later than in nontreated mice;

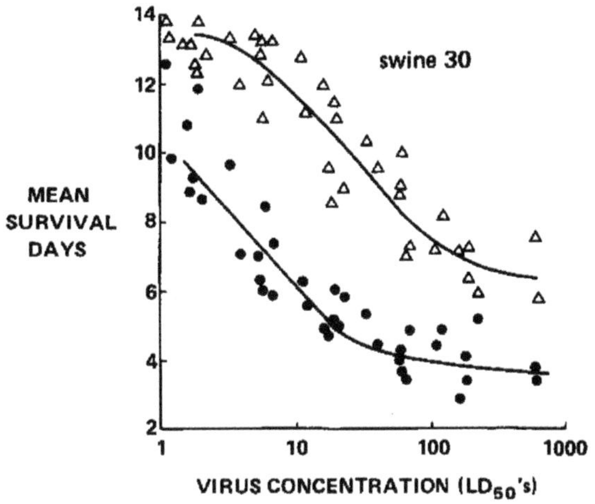

FIGURE 19. Effect of virus concentration on antiviral effect of amantadine. Compound was given 10 min before infection at a dosage of 30 mg/kg. (From Grunert, R. R., McGahen, J. W., and Davis, W. L., *Virology*, 26, 262, 1965. With permission.)

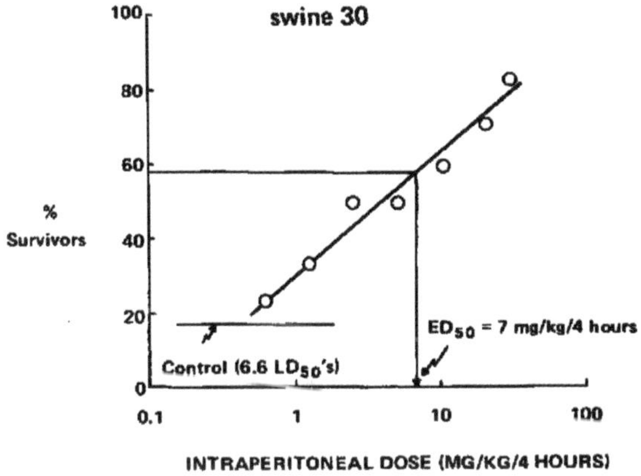

FIGURE 20. Dose response curve of amantadine concentration against infection with A/swine/515. Compound was given i.p. every 4 hr beginning 30 min before infection and continuing to 43½ hr after infection. (From Grunert, R. R., McGahen, J. W., and Davis, W. L., *Virology*, 26, 262, 1965. With permission.)

the results are similar to those observed with the ferret given three doses of ribavirin.[71] Further, treatment with ribavirin resulted in a dose-related suppression of lung consolidation, and 70 to 90% of mice survived. When multiple doses of 75, 37.5, or 18 mg/kg of ribavirin given twice daily for 7 days were tested (Table 5), the greatest activity was found with 75 mg/kg.[86] In separate multiple-dose experiments in mice treated with 100 mg/kg twice daily for 4 days, virus titers remained at levels approximately 1 log pfu/ml lower than in

TABLE 5

Activity of Ribavirin Against Lethal Influenza Virus Infections in Mice

	Drug dosage (mg/kg)[a]	Percent survivors		Mean day of death	
		Ribavirin treated[b]	Virus control[c]	Ribavirin treated	Virus control
Influenza	37.5	90[d]	20	20.1[e]	10.2
A_0 (PR8)	18.8	40	20	12.4[d]	10.2
Influenza	75	100[d]	30	21.0[e]	10.0
A2(Jap/305)	37.5	70	30	16.8[e]	10.0
Influenza	37.5	90[e]	5	20.3[e]	10.6
B (Lee)	18.8	40[d]	0	16[e]	8.4

[a]13–15 g Swiss mice inoculated i.p. at 4 hrs and then 2 × daily for 7 days.
[b]10 mice per group, 21 days post-virus inoculation.
[c]20 mice per group, 21 days post-virus inoculation.
[d]$P = <0.05$
[e]$P = <0.001$

From Sidwell, R., personal communication.

TABLE 6

Effects of Timing or Oral Dosing with Ribavirin on its Efficacy against Influenza A2/England Infections in Mice

Time of dosing relative to viral inoculation	Total no. of doses (100 mg/kg)	Percent survivors/20 mice 14 days postinoculation
Day of inoculation	1	50[a]
1 day before	1	5
10 days before through 1 day before	10	15
9 days before through day of inoculation	10	65[a]
5 days before through 4 days after	10	100[a]
1 day after	1	5
1 day after through 10 days after	10	15
Day of inoculation through 9 days after	10	95[a]
Infected nontreated controls	–	5

[a]$P \leq 0.01$.

From Sidwell, R., personal communication.

nontreated mice for about 96 hr.[86] In another series of experiments, mice were infected with A/England/42/72 (H3N2) virus, and ribavirin was given orally at varying times (Table 6). The results illustrate the importance of time of administration of the compound, since administration of a single dose of compound well before infection or 1 day after infection has no effect on the survival rate. This is in contrast to studies with 1-amino-adamantane, which could be administered as late as 48 hr postinfection and give significant survival rates.[79]

The efficacy of ribavirin against influenza infection in the mouse appears to be directly related to the ability of the compound to inhibit virus growth. However, an important reservation at this stage concerns the possible immunosuppressive activity of ribavirin. Certainly in influenza-infected mice treated with antilymphocyte serum, lung consolidation was reduced and an increase in mouse survival was recorded.[86] However, in the latter group of mice, virus titers remained at a higher level than in the control animals, whereas in ribavirin-treated animals, later in infection, the titers of virus are comparable in the treated and in the control groups.

IV. CONCLUSIONS

As was stated earlier, it is probable that no single animal species is a completely adequate model for human influenza. From the variety of species that could be used for the identification and study of antiviral compounds, the mouse and the ferret complement each other. Primary screening is probably best carried out in mice (which are relatively inexpensive and inbred), while ferrets (which show a clinical response to infection) allow a more detailed study of the effects of antiviral compounds on the individual responses to infection. The recording of mouse death due to virus infection is a long-established and proven measurement; however, the measurement of changes in lung weight may also prove to be valuable, since this is directly related to the known pathology of infection by adapted influenza viruses. Although ferrets are outbred and expensive to maintain, the clinical response of these animals to influenza virus infection makes them a most important model animal for the study of compounds against influenza. The most significant responses of this species in the study of antiviral compounds are probably, therefore, the temperature response and the measurement of rhinitis. Results obtained in any animal system, however, can only be used as an indication of possible activity in man. Thus, there are a number of compounds active in animal systems that have not been reported to be active in clinical tests. In this respect, it must be mentioned that in various tissues and organs, the toxicity, distribution, persistence, metabolism of compounds may well vary for different species. The results of animal experimentation, although necessary, cannot be used to directly predict the findings of human studies.

REFERENCES

1. **Joklik, W. K.,** Poxviruses, *Bacteriol. Rev.,* 30, 33, 1966.
2. **Bauer, D. J., Dumbell, K. R., Fox-Hulme, P., and Sadler, P. W.,** The chemotherapy of variola major infection, *Bull. WHO,* 26, 727, 1962.
3. **Adels, B. R. and Oppé, T. E.,** Treatment of Eczema vaccination with N-methylisatin-β-thiosemicarbazone, *Lancet,* 1, 18, 1966.
4. **Brown, E. R. and Gordon, P.,** Inosine-alkyl-amino alcohol complexes: anti-viral actions, *Fed. Proc.,* 29, 684, 1970.
5. **Brown, E. R. and Gordon, P.,** NPT-10381: suppression of mouse influenza mortality, mortality and replication of virus antigen, *Fed. Proc.,* 30, 242, 1971.
6. **Longley, S., Dunning, R. L., and Waldman, R. H.,** Effect of isoprinosine against challenge with A (H3N2) Hong Kong influenza virus in volunteers, *Antimicrob. Agents Chemother.,* 3, 506, 1973.
7. **Henle, W., Stokes, J., Jr., and Shaw, D. R.,** Passive immunization of mice against human influenza virus by the intranasal route, *J. Immunol.,* 40, 201, 1941.
8. **Taylor, R. M. and Parodi, A. S.,** Use of hamsters (*Cricetus auratus*) for detection of influenza virus in throat washings, *Proc. Soc. Exp. Biol. Med.,* 49, 105, 1942.
9. **Potter, C. W. and Jennings, R.,** The hamster as a model system for the study of influenza vaccines, *Postgrad. Med. J.,* in press.
10. **Todd, J. D. and Cohen, D.,** Study of influenza in dogs. I. Susceptibility of dogs to natural and experimental infection with human A2 and B strains of influenza virus, *Am. J. Epidemiol.,* 87, 426, 1968.
11. **Phair, J. P., Jennings, R., and Potter, C. W.,** Studies of human influenza virus infection in guinea pigs, in preparation.
12. **Smith, W., Andrewes, C. H., and Laidlaw, P. P.,** The virus obtained from influenza patients, *Lancet,* 2, 66, 1933.
13. **Haff, R. F., Schriver, P. W., and Stewart, R. C.,** Pathogenesis of influenza in ferrets: Nasal manifestations of disease, *Br. J. Exp. Pathol.,* 48, 435, 1966.
14. **Berendt, R. F.,** Simian model for the evaluation of immunity to influenza, *Infect. Immunol.,* 9, 101, 1974:
15. **Styk, B., Sabo, A., and Blaskovic, B.,** Experimental infection of pigs with Hong Kong influenza viruses, *Acta Virol.,* 15, 221, 1971.
16. **Shope, R. E.,** The infection of ferrets with swine influenza virus, *J. Exp. Med.,* 60, 49, 1934.
17. **Potter, C. W., Oxford, J. S., Shore, S. L., McLaren, C., and Stuart-Harris, C. H.,** Immunity to influenza in ferrets. I. Response to live and killed virus, *Br. J. Exp. Pathol.,* 53, 153, 1972.

18. **Haff, R. F. and Pinto, C. A.**, The nasal decongestant action of aspirin in influenza infected ferrets, *Life Sci.*, 12, 9, 1973.
19. **Wardell, J. R., Familiar, R. G., and Haff, R. F.**, A technique for measuring nasal airway resistance in ferrets, *J. Allergy*, 40, 100, 1968.
20. **Potter, C. W., Shore, S. L., and McLaren, C.**, Immunity to influenza in ferrets. III. Proteins in nasal secretions, *J. Infect. Dis.*, 126, 387, 1972.
21. **Shore, S. L., Potter, C. W., and McLaren, C.**, Immunity to influenza in ferrets. IV. Antibody in nasal secretions, *J. Infect. Dis.*, 126, 394, 1972.
22. **Schulman, J. L. and Kilbourne, E. D.**, Induction of partial specific heterotypic immunity in mice by a single infection with influenza A virus, *J. Bacteriol.*, 89, 170, 1966.
23. **Werner, G. H.**, Immunite croisée hétérotypique chez la souris entre virus grippaux humans de types A et A2, *C. R. Acad. Sci.*, 264, 1913, 1966.
24. **Mills, J. V., Chanock, R. M., Nusinoff, S. R., Walls, B. E., and Fishburne, I. E.**, Temperature sensitive mutants of influenza virus. I. Behaviour in tissue culture and in experimental animals, *J. Infect. Dis.*, 123, 145, 1971.
25. **Lui, C.**, Studies on influenza infection in ferrets by means of fluorescein-labelled antibody, *J. Exp. Med.*, 101, 665, 1955.
26. **Toms, G. L., Rosztoczy, I., and Smith, H.**, The localization of influenza virus: minimal infections dose determinations and single cycle kinetic studies on organ cultures of respiratory and other ferret tissues, *Br. J. Exp. Pathol.*, 55, 116, 1974.
27. **Squires, S. L.**, The febrile response of ferrets to influenza virus infection, *J. Antimicrob. Chemother. Suppl.*, in press.
28. **Smith, W. and Stuart-Harris, C. H.**, Influenza infection of man from the ferret, *Lancet*, 2, 121, 1936.
29. **Boyd, M. R. and Beeson, M. F.**, Animal models for evaluation of compounds against influenza viruses, *J. Antimicrob. Chemother.*, in press.
30. **Prinzie, A., Delem, A., and Huygelen, C.**, Protective efficacy of "Alice," a live attenuated influenza virus vaccine, *Postgrad. Med. J.*, in press.
31. **Squires, S. L.**, The evaluation of compounds against influenza viruses, *Ann. N.Y. Acad. Sci.*, 173, 229, 1970.
32. **Basarab, O. and Smith, H.**, Quantitative studies on the tissue localization of influenza virus in ferrets after intranasal and intravenous or intracardial inoculation, *Br. J. Exp. Pathol.*, 50, 612, 1969.
33. **McLaren, C., Potter, C. W., and Jennings, R.**, Immunity to influenza in ferrets. XIII. Protection against influenza infection by serum antibody by homologous haemagglutinin and neuraminidase antigens, *Med. Microbiol. Immunol.*, 100, 33, 1974.
34. **Sugg, J. Y. and Nagaki, D.**, Tissue tropism of a strain of influenza virus in ferrets, *J. Immunol.*, 74, 46, 1955.
35. **Francis, T., Jr., and Stuart-Harris, C. H.**, Studies on the nasal histology of epidemic influenza virus infection in the ferret, *J. Exp. Med.*, 68, 789, 1938.
36. **Francis, T., Jr.**, Factors conditioning resistance to epidemic influenza, *The Harvey Lectures*, 1941–42, Vol. 37, Science Press, Lancaster, Pennsylvania, 1942, 69.
37. **Stuart-Harris, C. H. and Francis, T., Jr.**, Studies on the nasal histology of epidemic influenza virus infection in the ferret, *J. Exp. Med.*, 68, 803, 1938.
38. **Sugg, J. Y. and Magill, T. P.**, Susceptibility of convalescent ferrets to reinfection with influenza virus in absence of specific antibodies, *Proc. Soc. Exp. Biol. Med.*, 65, 233, 1947.
39. **Haff, R. F. and Pinto, C. A.**, Comparison of neutralizing activity in nasal secretion and serum of ferrets in response to infections with influenza A viruses, *Arch. Gesamte Virusforsch*, 40, 168, 1973.
40. **Schofield, K. P., Potter, C. W., Edey, D., Jennings, R., and Oxford, J. S.**, Antiviral activity of ribavirin on influenza infection in ferrets, *J. Antimicrob. Chemother*, in press.
41. **Jackson, G. G., Muldoon, R. L., and Akers, L. W.**, Serological evidence for prevention of influenzal infection in volunteers by anti-influenzal drug amantadine hydrochloride, *Antimicrob. Agents Chemother.*, 7, 703, 1963.
42. **Stanley, E. D., Muldoon, R. E., Akers, L. W., and Jackson, G. G.**, Evaluation of antiviral drugs: The effect of amantadine on influenza in volunteers, *Ann. N.Y. Acad. Sci.*, 130, 44, 1965.
43. **Engle, C. G., Frankel, J. W., and Gelzer, J.**, Quantitative assessment of virus inhibition by indirect technique *in vivo*, *Ann. N.Y. Acad. Sci.*, 173, 139, 1970.
44. **Togo, Y., Schwartz, A. R., Tominaga, S., and Hornick, R. B.**, Cyclooceylamine in the prevention of experimental human influenza, *JAMA*, 220, 837, 1972.
45. **Small, P. A.**, International symposium on influenza vaccines for men and horses, *Sym. Ser. Immunobiol. Stand.*, 20, 238, 1973.
46. **Potter, C. W., Rees, R. C., Shore, S. L., Jennings, R., and McLaren, C.**, Immunity to influenza in ferrets. IX. Delayed hypersensitivity following infection or immunization with A/Hong Kong virus, *Microbios*, 10, 7, 1974.
47. **Potter, C. W., Jennings, R., and McLaren, C.**, Immunity to influenza in ferrets. VI. Immunization with adjuvanted vaccines, *Arch. Gesamte Virusforsch.*, 42, 285, 1973.
48. **Davies, W. L., Grunert, R. R., Haff, R. F., McGahen, J. W., Neumayer, E. M., Paulshock, M., Watts, J. C., Wood, T. R., Herrman, E. C., and Hoffman, C. E.**, Antiviral activity of 1-adamantanamine (amantadine), *Science*, 144, 862, 1964.

49. **Schild, G. C. and Sutton, R. N. P.,** Inhibition of influenza viruses *in vitro* and *in vivo* by 1-adamantanamine hydrochloride, *Br. J. Exp. Pathol.*, 46, 403, 1965.
50. **Oxford, J. S. and Schild, G. C.,** Inhibition of the growth of influenza and rubella viruses by amines and ammonium salts, *Br. J. Exp. Pathol.*, 48, 235, 1968.
51. **McLaren, C. and Potter, C. W.,** *In vitro* inhibition of influenza virus A/Mill Hill/1/72 by amantadine, *Lancet*, p. 1157, 1973.
52. **Schulman, J.,** Effect of 1-adamantanamide hydrochloride (amantadine HCl) and methyl-1-adamantethylamine hydrochloride (rimantadine HCl) on transmission of influenza infection in mice, *Proc. Soc. Exp. Biol. Med.*, 108, 1173, 1968.
53. **Cochran, K. W., Maassab, A., Tsunda, A., and Berling, B. S.,** Studies on the antiviral activity of amantadine hydrochloride, *Ann. N. Y. Acad. Sci.*, 130, 432, 1965.
54. **Potter, C. W. and Schofield, K. P.,** The effect of amantadine on the response of ferrets to influenza virus infection, *Ciba-Geigy Symp.*, in press.
55. **Fenton, R. J. and Bessell, C.,** personal communication.
56. **Solovyov, V. N. and Tolmacheva, N. S.,** Therapeutic effect of adamantanamine aerasol in experimental influenza infection of white mice, *Acta Virol.*, 11, 482, 1967.
57. **Finklea, J. F., Hennessy, A. V., and Davenport, F. M.,** A field trial of amantadine prophylaxis in naturally occurring acute respiratory illness, *Am. J. Epidemiol.*, 85, 403, 1967.
58. **Togo, Y., Hornick, R. B., and Dawkins, A. T., Jr.,** Studies on induced influenza in man. I. Double blind studies designed to assess prophylactic efficacy of amantadine hydrochloride against A/Rockville/1/65 strain, *JAMA*, 203, 1089, 1968.
59. **Galbraith, A. W., Oxford, J. S., Schild, G. C., and Watson, G. I.,** Study of 1-adamantanamine hydrochloride used prophylactically during the Hong Kong influenza epidemic in the family environment, *Bull. WHO*, 41, 677, 1969.
60. **Galbraith, A. W., Oxford, J. S., Schild, G. C., and Watson, G. I.,** Protective effect of aminoadamantine on influenza A2 infection in a family environment, *Ann. N.Y. Acad. Sci.*, 173, 29, 1970.
61. **Galbraith, A. W., Oxford, J. S., Schild, G. C., Potter, C. W., and Watson, G. I.,** Therapeutic effect of 1-adamantamine hydrochloride in naturally occurring influenza A2/Hong Kong infection, *Lancet*, 2, 113, 1971.
62. **Iezzoni, D.,** Evaluation of amantadine hydrochloride in the treatment of A2 influenza disease, *Ann. N.Y. Acad. Sci.*, 173, 10, 1970.
63. **Quilligan, J. J., Hirayema, A., and Baernsetin, H. D.,** The suppression of A2 influenza in children by the chemoprophylactic use of amantadine, *J. Pediatr.*, 69, 572, 1966.
64. **Sidwell, R. W., Huffman, J. H., Khare, G. P., Allen, L. B., Witkowski, J. T., and Robins, R. K.,** Broad-spectrum antiviral activity of virazole: 1-β-D-ribofuranosyl-1,2,4-triazole-3-carboxamide, *Science*, 177, 705, 1972.
65. **Witkowski, J. T., Robins, R. K., Sidwell, R. W., and Simon, L. N.,** The design, synthesis and broad spectrum antiviral activity of 1-β-D-ribofuranosyl-1,2,4-triazole-3-carboxamide and related nucleosides, *J. Med. Chem.*, 15, 1150, 1972.
66. **Huffman, J. G., Sidwell, R. W., and Khare, G. P.,** *In vitro* effect of 1-β-D-ribofuranosyl-1,2,4-triazole-3-carboxamide (Virazole, ICN 1229) on deoxyribonucleic acid and ribonucleic acid viruses, *Antimicrob. Agents Chemother.*, 3, 235, 1973.
67. **Sidwell, R. W., Allen, L. B., Huffman, J. H., Witkowski, J. T., and Simon, L. N.,** Effect of 1-β-D-ribofuranosyl-1,-2, 4-triazole-3-carboxamide (Ribavirin) on Friend leukaemia virus in mice, *Proc. Soc. Exp. Biol. Med.*, 148, 854, 1975.
68. **Togo, Y.,** *In vitro* effect of Virazole against influenza viruses, *Antimicrob. Agents Chemother.*, 4, 641, 1973.
69. **Oxford, J. S.,** Inhibition of the replication of influenza A and B viruses by a nucleoside analogue (Ribavirin), *J. Gen. Virol.*, 28,409, 1975.
70. **Khare, G. P., Sidwell, R. W., Witkowski, J. T., Simon, L. N., and Robins, R. K.,** Suppression by 1-β-D-ribofuranosyl-1,2,4-triazole-3-carboxamide (Virazole ICN 1229) of influenza virus-induced infections of mice, *Antimicrob. Agents Chemother.*, 3, 517, 1973.
71. **Schofield, K. P., Potter, C. W., Phair, J. P., Edey, D., Jennings, R., and Oxford, J. S.,** Antiviral activity of ribavirin on influenza virus infection of ferrets, *Antimicrob. Agents Chemother.*, in press.
72. **Oxford, J. S.,** Specific inhibitors of influenza replication as potential chemoprophylactic agents, *J. Antimicrob. Chemother.*, 1, 7, 1975.
73. **Flagg, W. B., Stanfield, F. J., Haff, R. F., Steward, R. C., Stedman, R. J., Gold, J., and Ferlauto, R. J.,** Antiviral activity of cyclooctylamine hydrochloride in cell culture and mouse systems, *Antimicrob. Agents Chemother.*, 8, 194, 1969.
74. **Pinto, C. A. and Haff, R. F.,** Antiviral activity of cyclooctylamine hydrochloride in influenza virus-infected ferrets, *Antimicrob. Agents Chemother.*, 2, 201, 1968.
75. **Potter, C. W.,** unpublished results.
76. **Bauer, S.,** personal communication.
77. **Mulder, J. and Hers, J. F.,** *Influenza*, Walters-Noordhoff, Groningen, 1972.
78. **Hers, J. F. Ph.,** Fluorescent antibody technique in respiratory virus diseases, *Am. Rev. Respir. Dis.*, 88, 316, 1963.
79. **Rabinovich, S.,** Rimantadine therapy of influenza A infection in mice, *Antimicrob. Agents Chemother.*, 1, 408, 1972.

80. **Sidwell, R. W., Dixon, G. J., Sellers, S. M., and Schabel, F. M.,** In vivo antiviral properties of biologically active compounds. II. Studies with influenza and vaccinia viruses, *Appl. Microbiol.,* 16, 370, 1968.
81. **Schulman, J. L.,** Effect of amantadine HCl and rimantadine HCl on transmission of influenza virus infection in mice, *Proc. Soc. Exp. Biol. Med.,* 128, 1173, 1968.
82. **Grunert, R. R., McGahen, J. W., and Davis, W. L.,** The *in vivo* antiviral activity of 1-adamantanamine (amantadine). I. Prophylactic and therapeutic activity against influenza viruses, *Virology,* 26, 262, 1965.
83. **Bredt, A. B. and Mardiney, M. R.,** The effects of amantadine on the reactivity of human lymphocytes stimulated by allogeneic lymphocytes and phytohaemagglutinin, *Transplantation,* 8, 763, 1969.
84. **Suzuki, F., Ohya, J., and Ishida, M.,** Effect of antilymphocyte serum on influenza virus infection in mice, *Proc. Soc. Exp. Biol. Med.,* 146, 78, 1974.
85. **Durr, F. E., Lindh, H. F., and Forbes, M.,** Efficacy of 1-β-D-ribofuranosyl-1,2,4-triazole-3-carboxamide against influenza virus infections in mice, *Antimicrob. Agents Chemother.,* 7, 582, 1975.
86. **Sidwell, R. W.,** personal communication.

Chapter 2

THE STUDY OF ANTIVIRAL COMPOUNDS IN VOLUNTEERS

A. S. Beare and Sylvia E. Reed

TABLE OF CONTENTS

I. INTRODUCTION

A. General

This chapter is solely concerned with respiratory virus infections. Most of these are mild, self-limiting, and clinically ill-defined. It is perhaps fortunate that the infections which are the most difficult to assess under natural conditions are also ethically admissible for human experiments. Common colds, for example, generally have a sporadic incidence, are difficult to assign to particular infecting agents without laboratory help and much delay, and in addition cannot conveniently be studied in animal models. Hence the program of study of a potential antirhinovirus drug is conventionally that of laboratory experiment, toxicological and pharmacological testing in animals and man, and testing in volunteers who are deliberately infected with individual rhinoviruses. Influenza tends to differ somewhat in that it has a wide host range in many animals, some of which can be used in laboratory experiments (Chapter 1, Volume I), and also in the fact that it produces recognizable clinical patterns in man during the course of epidemics (Chapter 1). Nevertheless, there are wide differences in the virulence of influenza viruses for man and animals,[1] and the unpredictability of outbreaks in nature makes the standard epidemiological trial a very hazardous undertaking (Chapter 3).

B. Prerequisites for Studies in Volunteers

A long period of laboratory experimentation intervenes between the synthesis of a new drug and its eventual trial in man. As a rule, trials of a drug in volunteers against experimental virus challenge are a sequel to comprehensive virological studies in the laboratory, a subject outside the scope of this chapter. Many substances found to be active in vitro show no activity in man, but it would seem that the reverse is seldom true and that a high degree of activity in the laboratory is likely to be a prerequisite of any potential therapeutic effect in vivo. Toxicological assessment of a potential new drug includes estimation of the acute toxic dose in small animals and long-term administration to at least two different animal species with careful monitoring of the effects. If the drug is to be used for topical administration to humans, animals must also be used for studies of local irritancy. Experiments are also designed to detect mutagenic or teratogenic effects. During pharmacological testing in animals and subsequently in man, a dose regime is determined that will provide an adequate concentration of the compound at the site of virus replication, will prove nonirritant, and will not produce undesirable side effects. Finally, evaluation of the drug in man against a virus challenge may involve screening by hematological and biochemical means in addition to those of clinical surveillance.

C. Ethical Considerations

Organized volunteer experiments now require a very different approach from the somewhat haphazard methods adopted during the early days of microbiology. Only microorganisms which produce mild disease and from which recovery is complete may be used for human inoculation. While some risk is probably inevitable in all procedures involving the administration of infectious material to man, this risk must be kept to an absolute minimum. Thus, the administration of previously uncharacterized viruses to potentially susceptible volunteers needs to be severely restricted, and legal requirements must, of course, be met. Wherever possible, an incidental benefit, such as immunization against an organism prevalent in nature, should accompany each experiment. The volunteer must give informed consent to his participation in the trial and should be given an opportunity to withdraw at the eleventh hour. In this unit an ethics committee, not itself concerned with the research in progress although sympathetic to the aims of the investigator, does much to help and advise and to restrain any tendencies toward excessive enthusiasm. Nevertheless, very little can really be done by others to help the individual research worker in this field. He must develop a high sense of responsibility toward the people who are trying to help him and not only try to behave humanely at all times, but also to ensure that he has the fullest possible understanding of his subject and his own aims. Andrewes[2] draws attention to the administration without mishap of rhinoviruses, coronaviruses, influenza viruses, and many other agents to 11,000 volunteers at the Common Cold Unit during 27 years of its history.

D. Merits and Defects of Volunteer Trials

Volunteer trials provide a rapid, accurate, and economical means of acquiring data. In their assessment of the efficacy of antiviral compounds, experimenters are made independent of the effects of accidental natural infection, variability in drug administration, seasonal influences on the epidemiology of respiratory virus infection,[3] and the unreliability of a natural virus challenge to which subjects are not uniformly exposed. Dosage of drug, mode of administration, and virus inoculum can be varied at will according to the dictates of safety and convenience. Indeed, the results obtained in preliminary volunteer experiments enable field trials to be more efficiently organized and interpreted at a later date and to give a better return for the immense labor they entail. Because of the small numbers of people involved in the volunteer trial, full advantage must be taken of the opportunities for achieving a high degree of accuracy. Experiments must be carefully designed, clinical data accurately recorded, infection with extraneous agents guarded against, specimens correctly processed, and results carefully analyzed. Nevertheless, the system remains artificial. The people selected are usually young adults free of antibody to the virus to be used, and their responses may differ from those of the very young or old or people with impaired respiratory function. The virus inoculated may be a strain which has undergone laboratory passage, and its dose is accurately measured by a standard assay technique. In order to achieve the highest infection rate in the control group, the virus is delivered directly onto susceptible cells, usually as nose drops. The drug is given so as to afford the best chance of exercising its maximum effect, usually by means of treatment before, during, and after the virus challenge. The results obtained in this way are remarkably precise and can often be obtained with as few as 20 to 30 volunteers.

II. DESCRIPTION OF HARVARD HOSPITAL, SALISBURY

A. Volunteers and their Accommodation

Harvard Hospital seems to be a unique virological unit. It is designed to function on continuous series of trials (about 25 each year). A maximum of 30 people can be accommodated at a time. Trials begin on a Tuesday and end on the Friday of the following week, and the intervening 4 days are used for maintenance and general domestic economy. Buildings, supposedly temporary, have been in use for nearly 30 years and consist of huts (Figure 1) divided into apartments which can each accommodate one, two, or three people. The interiors (Figures 2a and 2b) are plain but comfortable, and, within the limitations imposed by experiments, efforts are made to provide the volunteers with the amenities of a holiday camp, including facilities for indoor and outdoor games. Isolation of individuals or of groups is, however, strictly enforced. Volunteers are allowed to

FIGURE 1. The Common Cold Unit, Salisbury.

FIGURE 2. Accomodation at the Common Cold Unit, Salisbury: (a) living room; (b) bedroom.

FIGURE 2B.

fraternize only with the people sharing their apartments; although they are permitted to walk in the open country, they may not enter towns or any public place. Over the years, remarkably little trouble has been experienced with breaking of bounds, and most people say without prompting that they have enjoyed their stay.

The volunteers themselves are 18 to 50 years old, male and female, and often university students during vacations; there are no particular social groups outside these periods. All expenses are refunded, but the volunteers are otherwise unpaid. Bookings are accepted for trials far in advance (Figure 3), and the summer season is of course the most popular. Women always seem to outnumber men by about two or three to one. There is probably much to be said for the nonpayment principle since it capitalizes on whatever public spirit is available, even though 10 days of free board and lodging are apparently an inducement to the impecunious. The general approach to intending volunteers is illustrated in the printed circulars of the unit (Appendices A and B). The system of enrollment of volunteers, by which the potential participant must take the initiative in making contact with the unit, ensures that those who do decide to apply are generally highly cooperative.

B. Routine of Experimental Trials

The medical superintendent who functions also as clinical observer considers each application form in turn (Appendix C) and, if necessary, may inquire further about the prospective volunteer's medical history. When the volunteer arrives he is given a medical examination. Additional investigations are ordered, if necessary, and people thought to be incubating colds or other infectious illness are sent home or segregated from the other volunteers. A preliminary serum specimen is taken in the laboratory, and is subsequently used for antibody titrations relating to the experiment to be carried out. It is later tested in parallel with a second specimen returned by post from the volunteer's home 2 to 3 weeks after inoculation of virus. The day after his arrival, the volunteer undergoes a routine chest X-ray, and the medical examination is then completed.

On the day of their arrival volunteers are informed about the procedures of the trial, with emphasis on the need for isolation, and are then installed in their own apartments. Throughout the trial, food is delivered to their doors three times daily in heated containers. Visits are made once daily by the medical superintendent, who records appropriate "baseline" observations on the special form (Figure 4). In addition, the resident nurse

FIGURE 3. Recording bookings made for volunteers at the Common Cold Unit, Salisbury.

visits each volunteer twice daily to take temperatures and note symptoms. All visits by the medical and scientific staff are formal, with the staff masked and gowned (perhaps more to impress the volunteers with the importance of maintaining isolation than because of the belief that this procedure is necessary to contain infectious organisms.) At night the volunteer area is patrolled by a night watchman. Any infringement of the isolation regulations is taken very seriously, and the individual is excluded from the trial and sent home. Clinical reactions recorded outside the unit in subjects continuing to perform their normal jobs, have in our experience, never been wholly reliable. However, with the small numbers of people participating in Salisbury trials, and with constant insistence on the need for strict isolation, a high degree of precision in making clinical observations has been reached, and this has enabled us to confidently characterize the virulence of individual viruses.[3]

C. Methods Used in Clinical Surveillance

The study of reactions produced by human viruses of comparatively low virulence has long been an important aspect of the work of the Salisbury group.[3,4] It has been possible to pursue this work continuously in spite of seasonal fluctuations in the incidence of respiratory virus infections in nature. After the completion of the trial,

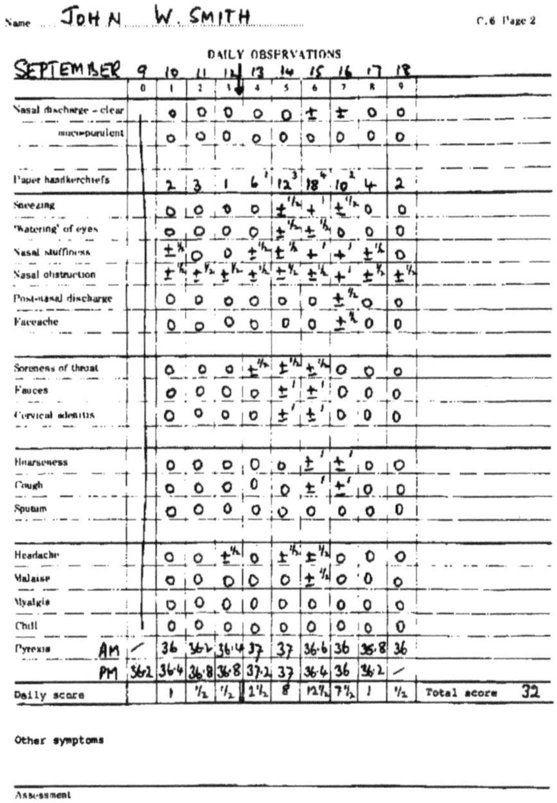

FIGURE 4. Daily clinical record and clinical score for a volunteer developing a mild cold after inoculation on September 12.

subjective symptoms and objective criteria are used to grade reactions according to an arbitrary system. These gradings are very mild (trivial reactions), mild (local symptoms), moderate (both local and constitutional symptoms), and severe (febrile or other unpleasant reactions necessitating bed rest). Many of the simpler observations do in fact readily provide a means of visualizing the type of reaction that might be experienced with a respiratory infection. For example, a baseline of 3 paper handkerchiefs per day will increase to 35 or more when the symptoms of a cold have supervened (Figure 4). A small number of volunteers, commonly those with high antibody titers to the virus used in the trial, function as controls and are given diluent only. These serve as a check on the effect of autosuggestion on subjective symptoms and on technical errors in the subsequent processing of specimens.

After considerable persistence, it has been possible to develop a system for assigning numerical values to individual signs and symptoms (Appendix D). This has enabled each volunteer to be given a "clinical score" daily and at the end of a trial, which can provide a basis for comparison between different volunteer groups. As a rule, a higher grade of reaction will also have a higher score, but anomalies do arise when the groups are small, and the accuracy of the average clinical score calculated for an experimental group depends very much on the number of volunteers involved.

Throughout the trial, the clinical observer and the nurse remain uninformed about the type of inoculum or medication given to individual volunteers.

D. Experimental Procedures

The first 3 days of the trial represent a quarantine period during which each volunteer's fitness to take part is confirmed. Serological screening tests can be conducted and, if appropriate to the protocol, volunteers can be assigned to their respective experimental groups. Antiviral medication may be started on the third day. On the Friday after their arrival, when they have completed 3 days of quarantine, the volunteers are inoculated with virus or control material. The material is prepared in the laboratory in a special vaccine suite and carried on melting ice by the virologist to the individual apartments. As a rule, the virus is diluted in Hanks's balanced salt solution containing 0.2% bovine plasma albumin so as to yield the appropriate dose in 1.0 ml of liquid. Each volunteer is asked to lie on a couch with the shoulders projecting over the edge and the neck fully extended. He is asked to relax and breathe through his mouth. Diluted virus (0.5 ml) is then instilled into each nostril with a calibrated dropper, and the volunteer is kept in this position for 1 min (Figure 5). He is then allowed to rise and is asked not to blow his nose for 1 hr, but simply to dab it if necessary. This method of virus administration is a little uncomfortable for the volunteer, but it has proved very effective and gives reproducible results. Administration of virus as a coarse spray is convenient, but the dose of virus actually reaching susceptible nasal mucosa seems less predictable than when the inoculum is given as nasal drops

On several days following the administration of virus, nasal washings are performed from which virus isolation is later attempted. A total of 10 ml of Hank's saline is used. It is dropped into the nose

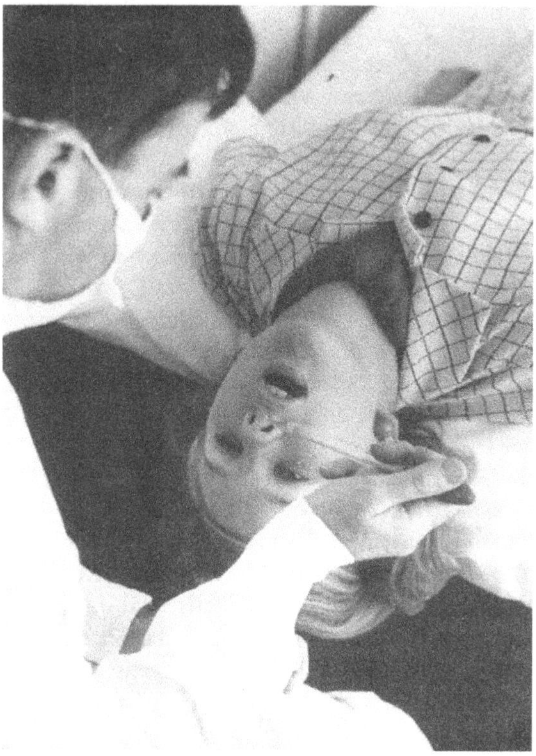

FIGURE 5. Method of administration of virus inoculum in the form of nasal drops.

in approximately 1.0-ml amounts and expelled into a petri dish. It is then taken up with a pipette and discharged into a universal container which is kept on melting ice. On return to the laboratory, the washings are mixed with an equal volume of sterile nutrient broth and stored in triplicate at -70°C in 2.0-ml vials until required for testing.

Protocols are, of course, varied as necessary to suit individual experiments, but it is desirable to keep to the general routine as much as possible so as to achieve the highest reproducibility in results. Numbers of volunteers required for investigations also vary. Where two viruses of sharply contrasting virulence are being compared, or when a drug of high antiviral activity is being studied, relatively few people may be sufficient to provide a definite answer. However, when one is attempting to demonstrate marginal differences only, much larger numbers may be required.

E. Safety Testing of the Viral Inocula

Knight[5] lists the tests necessary to exclude the possibility of contaminating agents in the material given to volunteers. In Salisbury, where it is sometimes necessary to use viruses with a minimum of formality, exclusion of infection in the first instance is stressed. Viruses used for volunteer experiments are handled in a special vaccine suite which is sterilized by ultraviolet light. Unauthorized personnel are excluded from this area. Material is harvested and stored at -70°C in sealed ampoules. When large pools of rhinoviruses are prepared, standard checks for sterility are performed and a sample of the material is neutralized by a potent antiserum before inoculation into a variety of tissue cultures. Large pools of influenza viruses are not prepared, and it is usually only possible to check the material for sterility and to confirm its identity with a hyperimmune serum.

F. Volunteer Experiments in Historical Perspective

Knight[5] describes how volunteers have come to be used in medical research and experiments ranging from those of Jenner with smallpox and Finlay with yellow fever to the more benign procedures we and our colleagues pursued in the United States. The American method often consists of inviting prison inmates to submit to medical experiments subject to very strict safeguards. A criticism often leveled against this procedure is that prisoners can never be "genuine" volunteers, either because pressure can be brought to bear upon them or because their general conditions are unpleasant enough to make them volunteer without giving proper consideration to attendant risks. There is little substance in this objection, but it should be noted that it may be difficult to convince the layman that medical ethics are constantly being observed in dealings with a prison population. Jackson et al.[6] have used students at the University of Chicago with great success, and a similar approach has been adopted by other groups, including Sir Charles Stuart-Harris and colleagues[7] at the University of Sheffield in Britain. If the volunteers are not in isolation, increasing the number of subjects can compensate for any inexactitude in measuring clinical responses. The high intelligence of the student population and their easy accessibility has enabled investigations to be carried on for considerable periods and with remarkable success.

III. DESIGN OF ANTIVIRAL TRIALS

A. Design of the Experiment

Double-blind, placebo-controlled experiments

are used, with approximately equal numbers of subjects receiving active compound or placebo with virus challenge. Preliminary screening tests can be used to select subjects who are serologically susceptible; in this case, the maximum amount of information can be obtained from the minimum number of volunteers. Random coded compound and placebo can then be used, with individual courses of treatment being identified only by a number; in these circumstances no attempt is made to match or balance the groups of subjects receiving active compound or placebo. This procedure is suitable for larger numbers of volunteers. Alternatively, when there are fewer volunteers, drug-treated and placebo groups may be matched for age, sex, weight, and antibody status.[8] If subjects are not all antibody-free, matching for prechallenge antibody level is particularly important. If the treated and placebo groups are unequal in antibody status and, therefore, in their susceptibility to challenge, interpretation of the results of the experiment becomes more difficult. However, imbalance of groups generally evens out if the numbers are large enough.

In addition to the two main groups of subjects who receive virus challenge and either the test compound or placebo, third and fourth groups are included; these receive saline "challenge" and either the test compound or placebo. This serves two purposes: the double-blind is preserved in respect of whether or not a subject has received any virus challenge at all, and a check can be made on any possible symptoms related to drug administration, unobscured by the symptoms of virus infection. This may be particularly important if the medication is given intranasally because a mildly irritant formulation can easily confuse the assessment of coryza induced by the virus.

The numbers of subjects used will depend on many factors, but if rather slight antiviral effects are to be detected, the numbers cannot be small. Groups of 10 to 15 volunteers of equal antibody status, receiving active compound and virus or placebo and virus, will provide a reasonable chance of detecting moderate antiviral effects; the saline-"challenged" groups can generally be smaller than this.

Trials are usually designed to cover both prophylactic and therapeutic possibilities in the initial experiments, i.e., the medication is given both before and after the virus challenge, and this design provides the maximum possible sensitivity in a single experiment. If a positive result is obtained, optimum dosage schedules can be defined later. However, the timing of medication will basically depend on the characteristics of individual antiviral compounds.

If the volunteers' cooperation is to be assured, they must be kept fully informed of the daily schedule of the experiment and the timing of doses of medication, and collection of samples. Every volunteer receives written information and instructions, as well as a full oral explanation of what is required. Preferably, all medication is administered under the direct supervision of the staff; however, when dosing is frequent, highly cooperative volunteers may be allowed to medicate themselves provided that frequent checks are made to ensure that this is done satisfactorily. The importance of an indistinguishable placebo cannot be overemphasized, particularly when volunteers are of an intelligent and inquiring disposition and may not be able to resist the temptation to try to break the code. A breach of the code may have a distinct influence on the more subjective aspects of clinical assessment.[9,10]

B. Routes and Methods of Drug Administration

The choice of oral or local medication will depend on the pharmacological properties and, to some extent, on the toxicity of the compound under study. An orally administered compound must be adequately absorbed and must reach the mucous membrane of the upper respiratory tract in adequate concentration. If there is doubt as to whether an orally administered compound does in fact do this, or if toxicological considerations suggest that the dose should be kept low, local administration may achieve adequate concentrations on the nasal mucous membrane even though the total dose is relatively small. However, mucociliary clearance mechanisms must be taken into account, since their efficiency may rapidly reduce the local concentration of a nasally administered drug. Figure 6 shows radiographically the distribution and clearance of nasally administered drops of a solution of human albumin labeled with $Tc^{99}m$. Aoki and Crawley[106] showed that 0.1- to 0.75-ml drops were better and more consistently distributed through the nasal cavity than the same solution delivered as droplets from a coarse nasal spray. They also showed that the half-life of the radioactivity initially present in

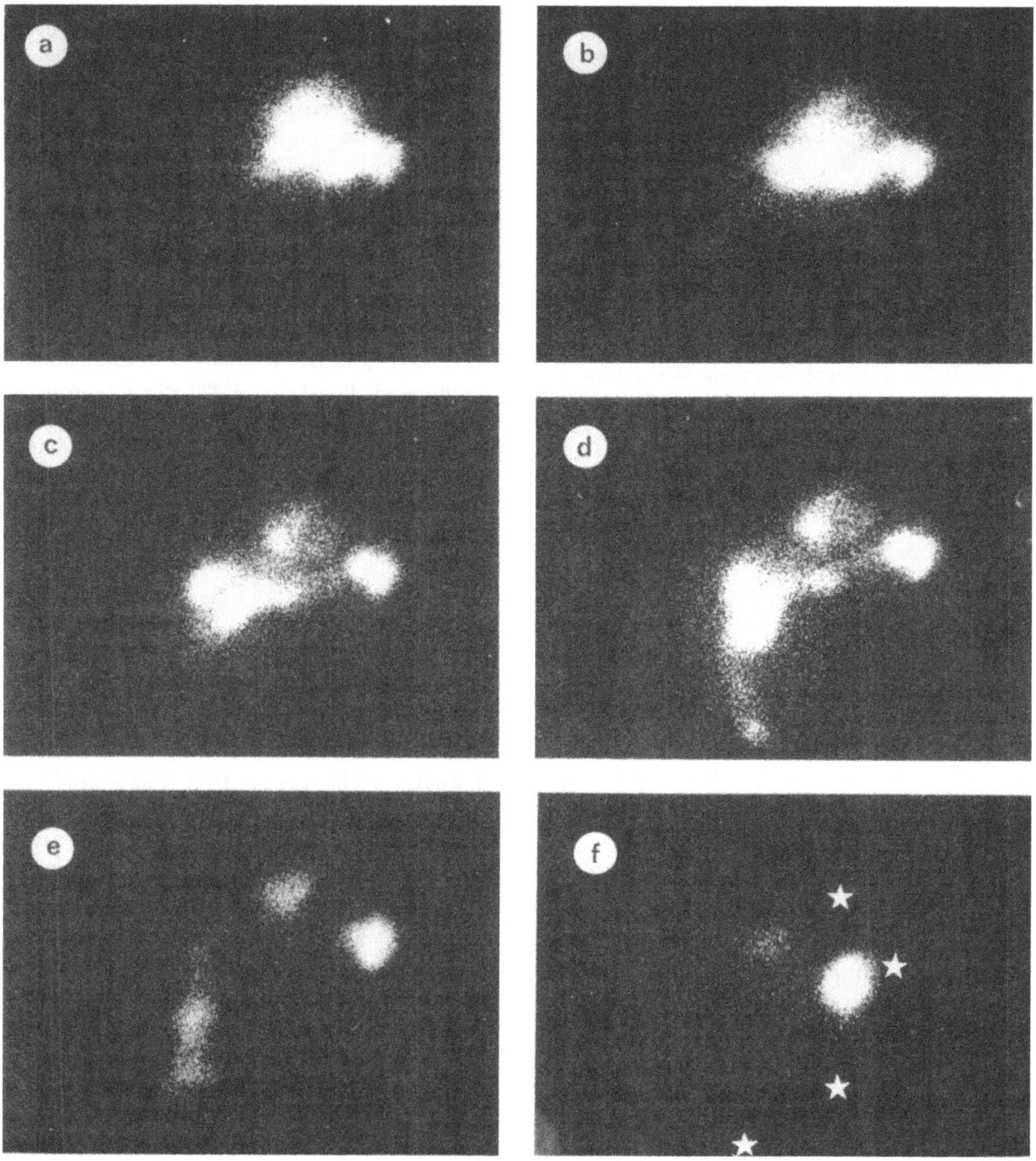

FIGURE 6. Scintigrams showing clearance of technetium 99m-labeled human serum albumin from the nose. The labeled albumin (30% solution in isotonic saline) was given as 0.25-ml nasal drops, and pictures were taken with a gamma camera at the following times thereafter: (a) 2 min; (b) 4 min; (c) 9 min; (d) 12 min; (e) 18 min; (f) 23 min. Reference markers were in the same position for each picture but are shown only in f and g, a sagittal section of the nasal cavity, for comparison. (Data of F. Y. Aoki and J. C. Crawley.)

the nasal cavity was approximately 15 min although some activity persisted much longer. These findings suggest that to be successful, nasally administered antiviral therapy may need to be given at frequent intervals, although if the substance under test has a prolonged antiviral effect after its initial contact with the cell (like interferon and its inducers), the apparent efficiency of the clearance mechanisms of the nose will not present such a problem. Some success has indeed been achieved using nasally administered interferon[11] and its inducers.[12,13] The results of

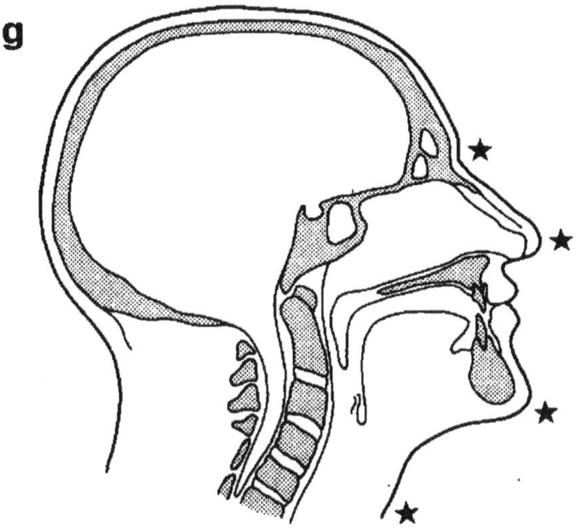

FIGURE 6G.

Aoki and Crawley suggest that drops administered with the subject supine may give a better intranasal distribution of drug than some types of sprays which produce large-particle aerosols. However, sprays have the advantage of convenience when medication is frequent.

C. Methods of Assessment of Results

Clinical – The overall number and severity of clinical responses (assessed by methods described in Section II) in drug-treated and placebo groups may be compared. However, smaller differences between the experimental groups may be detected by evaluation of clinical scores (Section II) on a daily basis, with the mean daily score for each group being calculated (Figure 7a). The incubation periods, time of maximal symptoms, duration of illness, and incidence of specific symptoms can then be compared. For example, Togo et al.[14] found that a compound that did not reduce the overall frequency or severity of illness did reduce nasal symptoms at the peak of illness. Merigan et al.[11] found that prophylaxis with interferon delayed but did not otherwise diminish infection with influenza B.

Virus shedding – The frequency and amount of virus shedding are also compared in treated and placebo groups. The total number of virus isolates obtained on the days following virus infection are compared; daily titration of virus in nasal washings gives a more accurate comparison (Figure 7b). As a control, virus isolations should be carried out on the washings from subjects who have received saline challenge instead of virus.

If medication is given intranasally, the collection of nasal washings for virus isolation must be timed so that contamination of the specimen with residual medication is minimal. Optimally, specimens for virus isolation are collected before each day's treatment begins.

Antibody response – The levels of serum antibody against the challenge virus before and after the experiment are compared by whatever methods are appropriate for the virus. The number of volunteers showing significant (fourfold or greater) rises in antibody titer is noted, and the mean rise in titer can be calculated. In some studies nasal antibody[12] and even sputum antibody[15] have been measured and will generally give results similar to those obtained from serum. The appropriate serological tests are also, of course, done on the saline-"challenged" controls.

Interferon response – In experiments involving interferon or its inducers, it may be desirable to measure the interferon content of nasal washings and, perhaps, serum. The challenge virus will itself induce interferon in a proportion of infected volunteers.[16-18] This complicates the interpretation of the significance of interferon in nasal washings collected from volunteers who receive both virus challenge and medication; the inherent interferon-inducing capacity of the medication can only be properly assessed in the absence of a virus challenge.

IV. METHODS FOR ANTIVIRAL TRIALS WITH ORTHOMYXOVIRUSES

A. Plan of Trials

Influenza viruses are now often used in human experiments. In general, it would seem that most influenza A strains produce an influenzal illness, but that influenza B viruses are more variable and induce reactions ranging from the influenzal to the imperceptible.[1] The free use of virulent wild viruses is not, however, recommended. Attenuated

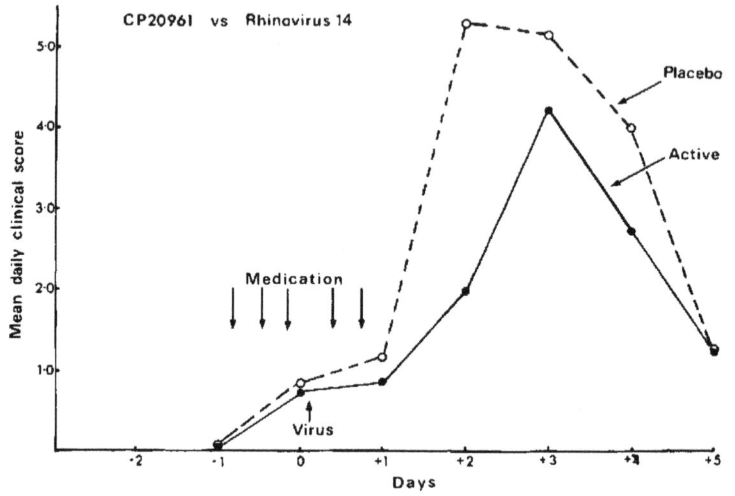

A.

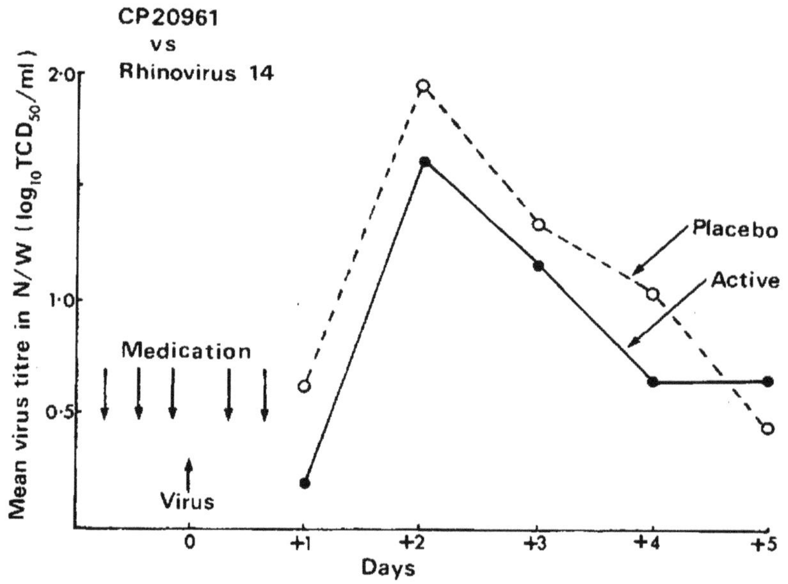

B.

FIGURE 7. Results obtained in a trial of an interferon inducer in volunteers. Volunteers were medicated with compound CP20961 and challenged with rhinovirus type 14. (a) Mean daily clinical scores in groups receiving CP20961 or placebo. (b) Mean daily titer of rhinovirus in nasal washings of groups of volunteers receiving CP20961 or placebo. (From Reed, S. E., Matthews, T. H. J., and Tyrrell, D. A. J., unpublished data.)

variants are frequently available and may be as fully susceptible to the action of a drug as the epidemic strains from which they were derived. Nevertheless, the effect of the drug on symptoms produced by an attenuated virus may not be identical to the effect it would have on symptoms produced by the wild parent, and final evaluation will depend on a large-scale field trial.

While the general outline of the antimyxovirus trial resembles that of other antiviral trials, certain differences are constantly encountered. The epidemic situation is, for instance, never static. When a new influenza A virus appears with a marked change in the hemagglutinin antigen, a high proportion of volunteers will be found to be susceptible to infection. As the virus circulates in nature, however, it becomes progressively more difficult – and finally impossible – to use this virus in volunteer trials. It may then be necessary to wait for another antigenic change before the trials can be recommenced. Furthermore, each new influenza virus infection produces antibodies against itself as well as many of its quite distantly related predecessors;[19] therefore, it is seldom possible to go backwards and use the earlier versions of a current serotype. It may, however, be noted that influenza B viruses often circulate slower than influenza A viruses and that it is frequently possible to perform influenza B trials when influenza A trials are impracticable.

The plan of the trials is outlined in Section III. For trials involving influenza virus, volunteers' sera are screened for hemagglutination-inhibiting (HI) antibodies against the virus to be used, and this is an adequate measure of susceptibility to infection.[20] Ideally, all volunteers should be antibody free; if not, good results can be obtained by balancing individuals of comparable susceptibility to infection in the drug-treated and placebo-treated groups.[20,21]

Ideally, the selection of volunteers should consider neuraminidase as well as hemagglutinin antibodies since they too exercise an influence on resistance to infection.[22,23] However, this ideal is very difficult to attain. In nature the neuraminidase antigen changes less often than the hemagglutinin antigen, and the N2 serotype has been circulating since 1957. Children are almost the only members of the population free of N2 antibody. Neuraminidase antibodies are much less important in protection against influenza than hemagglutinin antibodies, and they generally do not inhibit the replication of the virus even though they modify the clinical effects once infection has taken place. Hobson et al.[20] seem to place the significance of HI antibody in proper perspective and show that it affords an adequate criterion of protection subject to the acceptance of occasional anomalies.

Surveillance consists of isolation and sometimes also titration of virus from nasal washings collected 2 to 5 days following attempted infection of the volunteers, and of HI antibody titrations in sera collected before and 2 to 3 weeks after the trials. With recent strains of influenza virus, it has been customary to perform virus isolation and titrations in embryonated hen's eggs or chorio-allantoic membrane bit cultures, but it is sometimes necessary to use monkey kidney tissue cultures when the growth capacity of the virus in eggs is low.[3]

B. Choice of Virus and Virus Dose

Assuming that an attenuated virus is used, it should be as virulent as is ethically permissible and should have been shown to be as sensitive to the action of the drug in vitro as the wild strain from which it was made. It should have an acceptable passage history and been passed only in approved tissue cultures. Theoretically, a small dose should be given so as to increase the sensitivity of the test. However, it is not always easy to arrive at this dose since titrations must be performed in laboratory systems and not in antibody-free people. It is possible that $10^{5.0}$ 50% egg-infecting doses (EID_{50}) represent about 10 human-infecting doses of a virus which grows well in eggs. Attenuated viruses are, however, less infective than wild strains, and it is doubtful if one can ever use fewer than $10^{6.0}$ EID_{50} to obtain a high rate of infection in the controls. We would strongly recommend that characterization of the virus in volunteers should as far as possible be carried out before the antiviral experiment is started because of the highly deceptive results which are sometimes observed. As the virus dose is decreased, there is no sharp cutoff in the incidence of virus infection, although clinical symptoms are entirely abolished. In our experience, a dose as small as 10 EID_{50} may produce occasional abortive infections, but increasing the dose beyond what is strictly necessary does not aggravate the severity of reactions.[3,23,24] Therefore, when administering live influenza viruses for immunization purposes, it is customary to err on the side of excess. For an antiviral trial, however, it is worth going to some trouble to define the minimum necessary to produce a desired effect in the control groups, and this is probably the equivalent of five to ten 50% human-infecting doses.

C. Therapeutic Trials against Influenza

A therapeutic effect in an established influenza virus infection seems to be much more difficult to achieve than a prophylactic effect. In Section VI we shall review the work which has already been done in this field. This section is confined to describing an appropriate method of carrying out such a trial in isolated volunteers. In Salisbury this type of trial raises problems because of the relatively short stay of the volunteers in the Unit. The virus used must retain at least some virulence since an abolition or modification of symptoms may be the only effect which is observed. Therapy usually begins with drug or placebo 2 days after the administration of virus, which is commonly the time of onset of symptoms and of the commencement of virus excretion. Therapy continues for at least 4 days, serial nasal washings are collected daily and titrated for virus content, and clinical symptoms are analyzed from the day on which the drug was first taken (see Figure 8). A clinical grading or overall clinical score is now of little value, and it is necessary to assess the criteria of virus infection on single days and to compare, among other things, the duration of symptoms in the drug- and placebo-treated groups.

V. METHODS FOR ANTIVIRAL TRIALS WITH PICORNAVIRUSES

A. Availability and Selection of Susceptible Subjects

Usually one third or more of any group of people possess enough antibody against any specific rhinovirus serotype to give some protection against challenge with that type. This proportion tends to increase with increasing age;[25] therefore, younger subjects are more likely to be useful as volunteers. The frequency of antibody against any serotype varies in different communities at different times, depending on whether that particular virus has recently been current in the area. Coxsackievirus A21 can also be used as a challenge virus that produces a mild upper respiratory tract illness, and this virus has the advantage that 90 to 95% of most populations are susceptible to it. The symptoms it produces are very similar to those of rhinovirus-induced colds.

Measurement of serum neutralizing antibody is usually the method by which susceptible subjects are selected, and this generally proves satisfactory; many different observers have found that serum antibody levels correlate with degree of protection against rhinovirus infection. Perkins et al.[26] suggest that nasal antibody is more protective than serum antibody; since nasal and serum levels are themselves usually correlated, however, serum antibody assays usually suffice for selection of volunteers. A neutralization test is usually used. Measurement of type-specific rhinovirus hemagglutination-inhibiting antibody can also be used to assess resistance to challenge.[27] This method is quicker than a tissue culture neutralization test, but is only applicable to certain hemagglutinating serotypes.[28]

It is preferable not to use volunteers who have recently had an upper respiratory infection, since some infections may be followed by a period of "nonspecific" or heterologous resistance.[16]

B. Choice of Challenge Virus

The virus preferred for challenge will be one to which a high proportion of the potential volunteers are susceptible. There are also other important requirements. The strain must have an acceptable passage history and should have successfully undergone the necessary safety tests (Section II.E). It should also be known to induce an acceptable level of symptoms severe enough to enable any reduction in severity produced by treatment to be measured, yet not so severe as to be unacceptable. The various rhinovirus serotypes all produce very similar symptoms. Some types may produce slightly more severe clinical effects than others,[29] but to our knowledge no strain has been considered unsuitable for volunteers on grounds of excessive severity of symptoms. Coxsackievirus A21 in high doses has sometimes produced more severe symptoms, but in lower doses has proved a satisfactory challenge virus in many studies in volunteers.[30,31,87] It is now generally agreed that the dose of virus should be kept low in order to mimic a natural infection as closely as possible, although in natural colds the infecting dose presumably varies considerably.

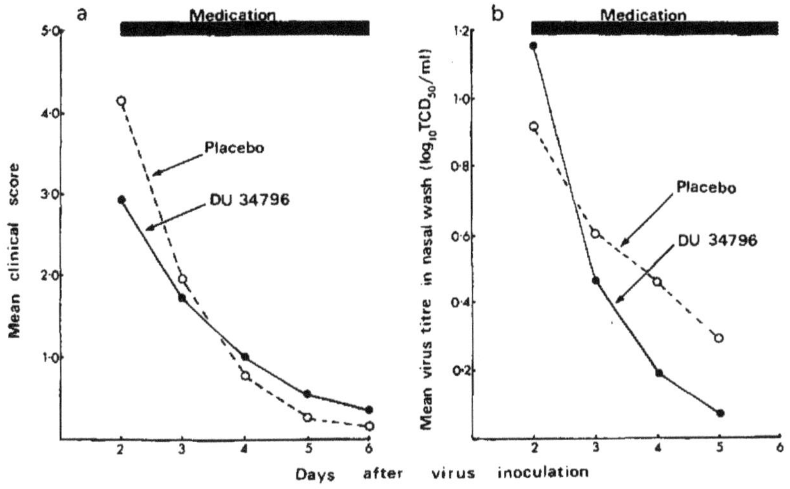

FIGURE 8. Therapeutic trial of spiroamantadine in volunteers challenged with influenza A/England/42/72, showing decline of (a) mean clinical score and (b) mean virus titer in nasal washings during treatment with spiroamantadine (DU 34796) or placebo.

There is some difficulty in assessing the 50% human infectious dose (HID_{50}) of a rhinovirus inoculum. Some attempt has been made to do so using large numbers of volunteers,[29] although the great heterogeneity of human populations makes this no simple matter. These authors assessed the HID_{50} of RV14 as 5.7 $TCID_{50}$ for WI-38 cells and of RV39 as 0.4 $TCID_{50}$, in each case using antibody-free volunteers. Of course, it must be remembered that estimates of rhinovirus infectivity for tissue culture may vary markedly, depending on the cell line used for titration[32] and the passage history of the virus.

The laboratory passage history of the rhinovirus used for challenge should be very short, so that the virus is as near as possible to its "natural" state; infectivity for man declines rapidly when an isolate is passed in the laboratory, as happened in early studies on RV1A[33] and has also been demonstrated more recently.[34-36] Preparations of nasal washings from infected subjects, preferably volunteers inoculated with defined serotypes and housed in isolation, have been used as inoculum for further volunteer studies[4] since the earliest days of the Salisbury Unit. However, virus passaged once or twice in cells acceptable by present day standards of safety may also be used, although further passages may produce a marked drop in infectivity for man.[35] It is a great advantage if a large pool of challenge virus suitable for use on repeated occasions can be prepared, so that the optimum challenge dose and the effects it produces can be well defined.

C. Laboratory Methods for Assessing Rhinovirus Infection

Nasal washings collected and stored by methods described in Section II.D are tested and titrated in rhinovirus-sensitive cell cultures. Diploid human embryo lung fibroblasts, especially WI-38, are frequently used, but sensitive HeLa cells may give improved rates of isolation and higher virus titers.[32] A titration assessing $TCID_{50}$ in tubes is more suitable than a plaque assay for virus unadapted to tissue culture, since the plaques produced by unadapted virus tend to be small.

Antibody in serum or nasal secretion is generally measured by a neutralization method in tube cultures of WI-38 cells or HeLa cells or by micro-methods.[37,38] In the case of hemagglutinating serotypes, antibody responses can also be measured by the hemagglutination-inhibition method.[27]

Coxsackievirus A21 suitable for use in man has been grown in primary cultures of human cells or human embryo fibroblasts. Reisolation of virus can be done in human embryo fibroblasts, monkey kidney cells, or continuous cells such as HeLa or HEp-2, and antibody can be assayed by neutralization or hemagglutination-inhibition methods.

VI. REVIEW OF ANTIVIRAL TRIALS USING AN INFLUENZA VIRUS CHALLENGE

No really successful anti-influenza drug has yet been synthesized, although a number of substances have been evaluated in man against a deliberate virus challenge. One of the earliest experiments was that of Jackson and colleagues[39] in which the substance ABOB was assessed against a strain of Asian influenza. The result was considered negative on the basis of clinical illness, virus excretions, viral fluorescence in exfoliated cells, and rises of serum antibody. In the following paragraphs we shall consider some of the more notable compounds that have reached the stage of clinical trial in recent years.

A. Amantadine and Other Adamantane Compounds

Amantadine is probably the only antiviral drug that has managed to achieve a place in the prevention and therapy of human type A influenza (Chapter 3). Jackson et al.[40] showed protection against an Asian virus challenge in university students who had been treated with the drug beforehand. The criterion of infection was mainly rises of serum antibody, and these provided evidence of a clear-cut effect. Tyrrell et al.,[41] however, found no evidence of protection in Salisbury volunteers challenged with an Asian virus which had previously been shown to be sensitive to the drug in vitro. Togo and colleagues[42] reported a therapeutic effect by the drug against naturally occurring Asian influenza in 1967 to 1968, manifested mainly by faster recovery and alleviation of clinical symptoms. A prophylactic effect against artificially induced Asian influenza in Russian medical students was shown about the same time by Smorodintsev et al.,[43] and a prophylactic effect against naturally occurring Hong Kong influenza in Finnish medical students was shown by Oker-Blom and co-workers a short while later.[44] Galbraith et al.[45] confirmed Togo's observation of a therapeutic effect in natural influenza, this time against the Hong Kong serotype. Their main positive finding was a reduction in the average duration of fever from 75 to 47 hr. Wingfield et al.[46] found a therapeutic effect against naturally occurring Asian influenza in inmates of Virginia State Penitentiary by both amantadine HCl and a related compound (rimantadine HCl), and this also showed itself principally in accelerated resolution of symptoms.

It is interesting to note that the antiviral effect of amantadine in vivo is not confined to human influenza. Bryans et al.[47] found evidence of protection in horses artificially challenged with A/equine 2 virus, a virus antigenically related to the human Hong Kong strain; Lang et al.[48] found a protective effect in turkeys exposed to the virus A/turkey/Ontario/7732/66. Nevertheless, it would seem that amantadine is not a highly active compound and that in man it tends to have only a marginal influence even against viruses highly sensitive to its action in vitro. Attempts to synthesize related substances with greater activity have not been very promising. Beare et al.[49] found that an adamantane spiro compound reduced the incidence of virus excretions, clinical reactions, and serum antibody rises in antibody-free volunteers challenged with A/Hong Kong/68 virus. Arroyo et al.[50] saw only a slight therapeutic effect with the same drug in volunteers infected with the A/England/42/72 serotype. There was no influence on symptoms, but virus shedding decreased more rapidly in the treated group than in the controls (Figures 8a and 8b). A probable final judgment on this group of substances, as shown in volunteer experiments, is that they produce a definite but restricted effect on some influenza A subtypes.

B. Isoquinolines

Isoquinoline compounds have been tested in man on a limited scale. They were the first compounds to show activity against influenza B, and Beare et al.[21] found a prophylactic effect in Salisbury volunteers against challenges with a 1962 type B strain and a 1965 Russian live influenza A vaccine virus. Two compounds, U.K. 2054 and U.K. 2371, were tested, and the limited degree of activity shown by both was confirmed by Williamson and Jackson,[51] Meenan and Hillary,[52,53] and Reed et al.[54] (see also Section VII).

C. Other Anti-influenza Studies in Volunteers

The prophylactic effect of the compound cyclooctylamine, given intranasally, was shown by Togo et al.[55] in a standard volunteer trial with an artificial virus challenge. Again, the main effect was on clinical symptoms, and little change was observed in objective criteria. Fruchtman et al.[56]

reported a similar study with commercial gamma globulin which they considered promising even though the results were not statistically significant. Merigan et al.[11] delayed the onset of symptoms in Salisbury volunteers inoculated with influenza B virus by local prophylactic application of leucocyte interferon, and Longley et al.[57] observed only a slight reduction in virus shedding in people treated with isoprinosine and then challenged with live A/Hong Kong/68 virus. More recently, Togo et al.[58] have given the drug ribavirin (virazole) to volunteers whom they later challenged with an influenza B virus. This compound effectively inhibits influenza A and B in tissue cultures and in mice. However, the only effect seen in volunteers was a reduction in the severity of clinical symptoms.

VII. REVIEW OF ANTIVIRAL TRIALS USING A PICORNAVIRUS CHALLENGE

There have mainly been two kinds of antiviral experiments using rhinovirus-infected volunteers. The volunteers may be used to assess compounds which have specific antirhinovirus activity or, alternatively, the human rhinovirus infection may act as a convenient model for testing substances with much wider therapeutic potential, such as interferon and its inducers. As has already been pointed out, the likelihood of obtaining a useful result from experiments in volunteers is high.

Perhaps a dozen different rhinovirus types have been used as challenge viruses, and little difference has been found among them. Most authors now seem aware of the need to use a virus that is not adapted to tissue culture, and to use it in low dosage, so that the effects of a weakly active drug will not be swamped.

In most experiments the compound under test has been given both before and after the virus challenge, not necessarily in imitation of field conditions, but to give the maximum chance of obtaining a positive result in the first experiment. A number of studies have been done with orally administered medication, but many antirhinovirus compounds are of extremely low solubility or poorly absorbed, and for these and other reasons local treatment has often been used.

Some of the earliest experiments on antiviral therapy were those of Walker et al.[9] using ascorbic acid prophylaxis. Evidence was not clear as to whether ascorbic acid could prevent colds, and no antirhinovirus effect could be shown in vitro. These investigators used rhinoviruses, coronavirus B814, and an influenza B virus to test the prophylactic effect of ascorbic acid (3 g daily) against placebo. Results were assessed clinically and in the case of the rhinovirus and influenza virus also by virus isolation and serological responses. No significant effect was found. Some years later a similar experiment using rhinovirus 44 was reported by Schwartz et al.,[59] and despite more prolonged dosage than in Walker's experiments, there was no prophylactic effect. Attempts have been made using field trials to demonstrate effects not shown by virus challenge studies. The trials of Anderson et al.[60,61] and Coulehan et al.[62] monitored a wide variety of natural infections in open communities and showed minor positive effects which might not have been detected using individual virus infections in smaller numbers of subjects. Another large-scale field trial[10] showed no effect.

Coxsackievirus A21 was used by Acornley et al.,[63] to test the effect of the steroid-related antibiotic sodium fusidate which was active in vitro against this virus and against rhinoviruses. However, neither oral nor local therapy effectively suppressed the infection despite the fact that adequate levels of the compound were found in serum after oral dosage and in nasal secretion after intranasal dosage. Challenge with Coxsackie A21 was also used by Buthala et al.[64] who tested the effect of intranasal gamma globulin given before and after challenge. There was some beneficial effect on symptoms, especially the amount of nasal secretion.

In clinical practice any antiviral compounds having a wide spectrum of activity would have an enormous advantage, and there has been great interest in compounds that seem to be active against more than one group of viruses. A number of experiments were done in volunteers using the isoquinoline compounds UK 2054 and UK 2371. These compounds showed some activity against influenza viruses in man[21,51-54] (see Section VI.B). Since UK 2054 was also slightly active against rhinoviruses in vitro, it was tested in volunteers given a rhinovirus challenge, but with negative results.[65] Another isoquinoline, 3,4-dihydroisoquinoline acetamide (DIQA), which was active against a range of viruses in mice but not in tissue culture,[66,67] was tested in volunteers challenged with rhinovirus 24.[14] The symptoms and the virus excretion were apparently marginally reduced in the treated group compared with the controls, but additional trials have not been reported although the authors suggest that these were warranted.

Isoprinosine is another compound that has been thought to have wide-spectrum potential as an antiviral drug, although evidence of activity in vitro and in animals is slight.[68] This compound, the *p*-acetamidobenzoic acid salt of inosine-dimethylaminoisopropanol (1:3 *M* ratio), given both before and after inoculation of volunteers with rhinovirus, was tested by Soto et al.[69] In this trial RV9 and RV31 were used together as a double challenge, and there was no beneficial effect on symptoms or on laboratory evidence of infection. Trials of isoprinosine reported by Pachuta et al.[70] were similar except that rhinovirus 44 or rhinovirus 32 were used in separate experiments. Symptoms produced by each of the viruses were slightly but not significantly less in the treated groups as compared with the placebo controls, and virus shedding was not affected. In the experiments of both Soto et al. and Pachuta et al. antibody responses in the treated groups were sometimes greater than in the controls, but the differences were not significant.

Another compound that has activity against influenza and rhinovirus, at least in vitro, is the adamantane spiro compound DU 34796 (spiroamantadine). However, the antirhinovirus effect was less than the anti-influenzal effect; although volunteer experiments using challenge with influenza virus gave a positive result,[49] those using rhinovirus proved negative.[71]

A number of compounds whose chief activity is against rhinovirus have come to the stage of trials in man. Of these, the substituted triazinoindole series, described by Gladych et al.[72] and Gwaltney,[73] has been among the most active in vitro. However, these too have been disappointingly negative in man, despite an apparently encouraging result in gibbons.[74] One of these derivatives, SKF 21687, was given intranasally to volunteers challenged with rhinovirus 44 without antiviral effect.[15] Tyrrell and Reed, after earlier negative studies using small numbers of volunteers and the derivatives SKF 21687 and SKF 30097 given orally, later obtained no encouragement using a solution of SKF 40491 as a nasal spray.[75] In man problems imposed by the relative insolubility of the compounds, and their potential toxicity, have limited the dosage that can safely be given. However, in the same report[75] covering studies of several antirhinovirus compounds, Reed and Tyrrell found that the imidazo-thiazole compound RP 19326 significantly reduced volunteers' shedding of rhinovirus 9, the symptoms also being somewhat reduced in the treated group.

Only one study using human interferon against challenge with respiratory viruses has so far been reported. Merigan et al.[11] used two different dosage regimes for their studies with human leucocyte interferon and influenza B or rhinovirus 4. In the former case the medication (800,000 units of interferon or placebo) was given before the virus challenge, and only a marginal effect on symptoms was obtained. In the rhinovirus trial the total dose of interferon was much higher (14×10^6 units), and the medication was given both before and after challenge. This produced a significant improvement in symptoms and a highly significant reduction in virus shedding. Further trials to confirm and extend these observations await adequate supplies of human interferon.

The theoretical advantages of interferon inducers for prophylaxis or therapy of human viral infections need no emphasis. One of the best-studied inducers, at least in animals, is the synthetic double-stranded RNA polyinosinic-polycytidylic acid (poly I:poly C). Initially there was some hesitancy over the possible toxicity of this compound, but judicious testing allayed these doubts at least in part, and the compound has been used in volunteers challenged with influenza virus or rhinovirus 13.[12] Medication was given as nasal drops on the day before virus challenge and the following 6 days. Surprisingly, little nasal interferon was induced. The symptoms of viral infection were reduced by the treatment, but not very strikingly, and virus shedding was not consistently reduced. No further major trials of this compound in human respiratory infections have been reported, but stabilized forms of poly I:poly C resistant to enzymic breakdown are being developed.[76]

More recently, some low molecular weight interferon inducers which are not absorbed and show little risk of toxicity have also been tested in volunteers. The compound CP 20,961 (N,N-dioctadecyl-N′-bis-(hydroxyethyl)-propanediamine) was given on the day before and the day of challenge with rhinovirus 21.[77] Evidence of improvement of symptoms or laboratory evidence of infection in comparison with the control group was very slight, although nasal interferon was apparently induced. Further reports on this compound have been somewhat conflicting. Panusarn et al.[13] found that a slightly different formulation of CP 20,961, again used against rhinovirus 21, reduced both symptoms and virus shedding, whereas Douglas and Betts,[78] using rhinovirus 13, and Matthews et al.,[107] using rhinovirus 14, found only very limited improvement. Douglas et al.,[79] used the related compound CP 28,888 against influenza challenge with similarly restricted success. Overall, studies of interferon inducers in volunteers seem to have shown less effect than animal experiments would suggest; the reasons for this are, unfortunately, far from clear.

VIII. MISCELLANEOUS RESPIRATORY VIRUSES USED IN VOLUNTEERS

We will now consider a number of viruses which, although seldom or never tested in man against an antiviral compound, are capable of being used if the indication should arise.

A. Adenoviruses

This group of viruses causes outbreaks of acute respiratory illness in man, generally in closed communities[80-83] (see Chapter 1, Volume I). Service recruits have frequently been infected in nature with types 3, 4, 7, 14, and 21 and infants and children with types 1, 2, 3, 5, and 7. In 1960 Hitchcock et al. inoculated Salisbury volunteers intranasally with adenoviruses 4 and 7 grown in tissue cultures of pig kidney.[84] Little clinical effect was observed but viruses were recovered both from throat swabs and feces. Kasel and colleagues[85] inoculated adult volunteers by the nasal and conjunctival routes with adenovirus types 26 and 27 that had been recovered in tissue cultures of human embryo kidney from children's fecal specimens. The virus dose (assayed in Hep-II cells) was $10^{3.2}$ to $10^{4.0}$ TCD_{50} in the eye and $10^{4.2}$ to $10^{4.7}$ TCD_{50} in the nose. Both methods of administration were followed by prolonged virus excretion in the feces. Intraocular inoculation produced some conjunctivitis. Couch and co-workers[86,87] showed that adenovirus type 4 was more infectious when given in a small-particle aerosol, which deposited mainly in the lower respiratory tract, than when given as nose drops. In another experiment Kasel et al.[88] vaccinated volunteers with type-specific and group-specific antigens of adenovirus type 1 and then challenged them with live virus. The challenge was performed by moistening a cotton applicator with a 1:10 dilution of a virus suspension containing $10^{8.5}$ TCD_{50}/ml and applying this to the lower conjunctival sac. Surveillance consisted of clinical examinations of the eye and attempted virus isolation from throat swabs and feces. Chanock and colleagues[89] incorporated live adenovirus type 4 from a diploid cell culture in enteric-coated capsules and gave these orally to volunteers as a potential vaccine. A later study by Smith et al.[90] demonstrated that routes of vaccine administration were all important, and that resistance to infection by nasal challenge could be induced by intranasal but not by intraoral vaccination. Couch et al.[91] produced antibody responses in volunteers with crystalline hexon and fiber subunits of adenovirus 5 administered parenterally and found that the latter gave better protection against a homologous intranasal live virus challenge. A paper by Dudding et al.[92] described enteric immunization with live adenovirus 21 and established the effective immunizing dose as $10^{4.6}$ TCD_{50}, which was lower than that of adenovirus types 4 or 7. Finally, Schwartz and colleagues[93] used monovalent live oral enteric-coated preparations of adenovirus types 1, 2, and 5 to induce protection in adult prisoners.

Although adenoviruses have been widely used in volunteer experiments without apparent mishap, the procedure as a whole remains questionable since a number of serotypes are undoubtedly oncogenic in animals. Adenoviruses have not yet been associated with human tumors; nevertheless, their freedom from any potential oncogenicity remains unproven. Probably the approach of Couch et al.,[91] using purified proteins without genetic material, is the best hope for the adenovirus vaccine of the future. The use of live adenoviruses for human challenge should be practiced with great restraint.

B. Paramyxoviruses

Paramyxoviruses are mainly responsible for disease in children and infants (Chapter 1, Volume I). Parainfluenza type 1 was given to adult volunteers by Smith et al.[94] in an investigation of the protective influence of local antibody, and the antiviral effect of amantadine was later tested in other volunteers.[95] No protection was demonstrated. Taylor-Robinson and Bynoe[96] gave volunteers 2 X

10^4 to 2×10^6 TCD_{50} of parainfluenza type 2 and induced coryza, sore throat, and nasal congestion. Tyrrell and others[4,97] performed similar studies with parainfluenza types 2, 3, and 4 with much the same effect. Parainfluenza 2 has also been used successfully as a challenge virus in antiviral studies in volunteers using the isoquinoline compound UK 2371.[54]

C. Influenza C

Joosting et al.[98] gave a new isolate of influenza C, C/Johannesburg/1/66, to volunteers in doses ranging from $10^{4.25}$ to $10^{5.7}$ EID_{50} and produced a number of afebrile colds. The virus was easily recovered from nasal washings. Its infrequent isolation in nature is difficult to understand in view of the widespread presence of antibodies in man, but this may be related at least partly to the lability of the virus.

D. Coronaviruses

The members of this group are cultivated with difficulty (Chapter 5), and the inoculation of volunteers was initially extensively used for detecting their presence.[99] Strain 229E, which was propagated in WI-38 cells by Hamre and Procknow,[100] was inoculated into man by Bradburne et al.[101] and shown to induce colds with a slightly longer incubation period than that of rhinoviruses. Mathur et al.[71] found that 229E was sensitive to the action of a spiroadamantane substance in vitro, but Reed[108] detected no prophylactic effect in a volunteer trial in which results were assessed by clinical responses, virus isolation, and rises in serum-neutralizing antibody levels. Coronavirus B814 was also used in volunteers in studies of ascorbic acid,[9] the results being assessed by clinical responses only.

E. Respiratory Syncytial Virus

This virus is primarily a pathogen of babies (Chapter 1, Volume I). Johnson et al.[102] performed early volunteer experiments in adults, and Friedewald et al.[103] compared wild virus with virus adapted to growth at low temperatures. The former produced mild upper respiratory illness only. Other studies to develop a system for the characterization of live vaccines in adults are described by Mills et al.[104] and Wright et al.[105] To our knowledge, no antiviral experiments involving challenge of volunteers with this virus have yet been done.

IX. CONCLUSIONS

Volunteer studies should not be undertaken without careful initial preparation and planning. Maximum effort should be made to exploit human capacity to cooperate and to provide data which could not be obtained from animal experiments. These considerations, coupled with the fact that man is the definitive experimental species, will amply compensate for small numbers.

Among factors to be taken into account in the design of a trial are mechanisms of virus infection, likely incidence of protective antibodies in different sections of the population, pathogenesis of the disease in question, and clinical signs and symptoms that may be abolished or modified by a prophylactic or therapeutic drug. In the interest of the investigator and the volunteers, it is desirable that experimental procedures be no more complex than can be efficiently managed and that laboratory testing not exceed the bounds of efficient and reliable performance. As knowledge and experience increase, it will prove possible to design more efficient trials; nevertheless, each human trial will remain a unique event requiring its own special evaluation.

APPENDIX A*

Diseases of the respiratory system are the commonest ailments of man, and the commonest minor respiratory ailment is the common cold. A team of scientific workers under the Medical Research Council, in collaboration with Department of Health & Social Security, is making an attack on the difficult problem of colds and other respiratory diseases, especially influenza. One of the main obstacles to progress in research in colds has been that guinea pigs and other small laboratory animals cannot be infected with the common cold in the laboratory; only man and the chimpanzee have so far proved susceptible. As explained overleaf our present investigations are still at the stage where we need to use human volunteers for many parts of our work. For this purpose a Research Unit has been established in pleasant surroundings at Harvard Hospital, Salisbury, where research is carried out not only on colds but also on influenza, with special reference to the production of vaccines against colds and against influenza and on the efficacy of medicinal drugs against these illnesses.

We are appealing for volunteers to come and live here for periods of 10 days. Up to 30 volunteers at a time are needed. During the experiment they are isolated from each other and from the outside world. As solitary confinement is uncongenial to most people, the volunteers are normally isolated in pairs (married couples or pairs of one sex) or in threes, in centrally heated flats, each flat having a living room, kitchenette, bathroom, and either two or three bedrooms, so that each person has his or her own bedroom. Many volunteers plan to come with a friend, but it is usually possible to arrange a suitable companion for those who come on their own. The age limits are 18 to 50 years. Children cannot be accommodated, and no pets are allowed. Board and lodgings are free and 35 p/day pocket money is given to each volunteer.

The expenses of traveling by the most direct route, by train or coach, to and from Salisbury are paid, from anywhere within Great Britain. Volunteers who wish may travel by car or motorcycle and may claim the cost of a second-class return rail fare from their home or nearest railway station. Only one fare may be claimed, irrespective of the number of volunteers traveling in the car.

Over 12,000 volunteers have already passed through the Unit. Nearly all of them have enjoyed their stay and have thought it worth the risk of catching a cold or a mild attack of influenza, and many have made repeated visits to the Unit. By no means all volunteers develop symptoms, as many of the materials used prove to be inactive or the volunteer proves resistant to the inoculum given. The physician in charge and the matron visit each volunteer daily.

There is opportunity for country walks but volunteers must keep 30 ft away from anyone whom they may meet, so they may not go into Salisbury City or into any residential area. Amenities include indoor and outdoor games involving not more than two or three persons, a radio in each living room and television if required, and Post Office and internal telephone for use during office hours. After office hours there is a coin box telephone which volunteers may use for outgoing calls only, provided they carry out the simple instructions concerning prevention of cross infection. Urgent incoming messages for volunteers will, however, be passed on to them. Library books and newspapers are provided. There is ample opportunity for study, and a visit to the Unit is especially popular with students working for examinations. There is no restriction on the books which they may bring with them. Cooked meals are brought three times daily to the isolation quarters, and there are also facilities for making toast, tea, coffee, etc. in the quarters. The maintenance of isolation precludes access by the domestic staff to the volunteers' quarters, and volunteers are therefore requested to keep the flats clean and tidy and to make their own beds. The success of each trial depends on the willingness of volunteers to conform to the rules of isolation, which are by no means arduous, and to maintain strict discipline in essential matters affecting the experiment. Those unwilling to do so should not apply, and those who break isolation will be sent home, and their pocket money will be withheld.

*From Medical Research Council and Department of Health & Social Security, Clinical Research Centre, MRC Common Cold Unit, Harvard Hospital, Salisbury, Wilts.

APPENDIX B*
RESEARCH ON COLDS AND OTHER RESPIRATORY DISEASES: WHAT IT IS ALL ABOUT

Volunteers may like to have some idea why their help is needed. We cannot study colds in laboratory animals, and chimpanzees are difficult to work with, expensive, and scarce. We can test for the presence of cold viruses by dropping material up a volunteer's nose and seeing whether he gets a cold. From the time the Unit was started in 1946 until 1959 that was the only test for cold viruses that we had, and our first objective was to find a laboratory test which would serve instead. In 1959 and 1960 workers at Salisbury discovered how to grow "cold" viruses in cultures of living tissues and to measure their activity by the changes produced in those cultures. Thus our first objective was attained. We still need your help however for several reasons. We still cannot grow viruses from all the colds we investigate and we need volunteers, just as we did before, to help us to discover how to grow viruses from these colds. We need you for studies on immunity to colds, to find out why some people get colds and others do not, and for work aiming at developing a vaccine against colds. We need you also for testing the efficacy of medicinal drugs in the prevention and treatment of colds. All these studies usually necessitate the taking of two small blood samples (5 to 10 ml) from the volunteers at suitable intervals.

Influenza is also caused by viruses, but they are different from those causing colds. In 1933 workers at the National Institute for Medical Research discovered that if garglings from a "flu" patient were put into a ferret's nose, that ferret, 2 days later, would develop a fever, would sneeze and, in fact, would catch "flu." That finding provided a laboratory method which has led to the disentangling of a lot of confusion about influenza, to the accumulation of much knowledge about it, and to the development of protective vaccines that are now in general use. The work of the World Influenza Centre at the National Institute for Medical Research has been made possible as a result of these laboratory studies. We are now working on the possibility of producing an influenza vaccine, containing live influenza virus, which can be given in the form of drops up the nose or by nasal spray. Such a vaccine would give better protection against influenza than the present one. Our aim is to produce a mild infection in volunteers without producing symptoms of influenza, but which will lead, nevertheless, to the production of antibody with protection against subsequent infection. However, since the vaccines contain live virus, complete absence of all symptoms cannot be guaranteed. In past trials about one in three of those given the vaccine have developed symptoms of a cold, and in about one in ten of these the symptoms have been bad enough to be graded as mild influenza. In future trials we plan to use viruses which we expect to be milder than those used previously, but at present this is the best guide we can give you as to what participating in such a trial might involve. The effect of the vaccine will be judged from two blood tests taken at suitable intervals.

In recent years several new synthetic drugs have been developed which prevent the multiplication of cold and influenza viruses in laboratory cultures, and it will be necessary from time to time to test whether they can prevent colds and influenza in human volunteers or are beneficial in the treatment of these illnesses. In these experiments we would also use "mild" viruses which would be expected to cause only mild symptoms. In addition, it is also intended to study the effect of human interferon on colds and influenza.

These then are the procedures for which we need the help of volunteers. They may be required to take part in any of the investigations described above, be it a common cold trial, an influenza trial, or a medicinal drug trial, and anyone not willing to do so will not be accepted as a volunteer and should not apply. It is emphasized however that no vaccine or medicinal drug will be given to a volunteer until we are satisified that it is completely safe to do so.

It is possible to catch colds and influenza from apparently normal people; so that we may be sure that any symptoms that you may get while at Salisbury come from the material dropped up your nose and from nowhere else, you have to be kept away from everybody (except the doctors and

*From Medical Research Council and Department of Health & Social Security, Clinical Research Centre, MRC Common Cold Unit, Harvard Hospital, Salisbury, Wilts.

matron who are gowned and masked). You may also become infected by breathing air in a room in which other people have recently been. Since you could have become infected by a fellow passenger on your way to Salisbury, and would not show signs of this until a day or two later, you do not have test material dropped up your nose for 2 or 3 days after arrival. This again helps to ensure that any infection you get at the Unit comes from the material used and not from anywhere else. The success of our experiments and the accurate interpretation of our results depend upon the volunteers' strict observance of the isolation rules.

Some of you will not be given virus but will receive instead what is called "control" material, that is, material known *not* to contain virus. Neither you, nor the doctor nor others who see you will know who gets infective material and who noninfective material: only at the end of the trial will all the information be put together and only then will you be told. There are several reasons for this: it acts as a check on the efficiency of our system for keeping out all respiratory viruses other than those in the test materials. It also ensures that we interpret correctly any symptoms you develop and do not imagine that we see signs and symptoms of an infection that is not really there. So don't imagine that you are bound to get an illness after the test is made; the chances are that you may not.

We hope that this explanation will make the trial more interesting for you and will show how, by obeying the isolation rules throughout your stay, you can help to make it a success.

APPENDIX C*

VOLUNTEER'S APPLICATION FORM

Name (in full, surname in block capitals)
Mr./Mrs./Miss ______________________________

Home address
and telephone no. ______________________________

Occupation ______________________________

University, College (in case of students), and faculty
(give degrees if any) ______________________________

Date of birth (Age limits 18–50 years) ______________________________

Dates of previous visits (if any) to this Unit ______________________________

Name of partner (in full – see sheet C.1.)
(if none, state interests for purpose of pairing) ______________________________

Railway Station from which you will be coming† ______________________________

Railway Station to which you will be returning* ______________________________

If you have a history of asthma, hay fever, pulmonary tuberculosis, repeated attacks of sinusitis or of middle ear disease, or chronic nasal catarrh you would not be suitable for the trial and should not volunteer. Please do not volunteer if you are pregnant.

Have you suffered from any serious illness? ______________________________
If so, what and when? ______________________________
____

Have you ever been in hospital? ______________________________
If so, give brief particulars. ______________________________
____

Are you taking any special tablets? ______________________________
If so, give details. ______________________________

I (Name) ______________________________ of (Address) ______________________________
have read Form C.1 dated July 1973 and understand the nature of the investigations proposed which are described therein. I offer my services as a volunteer and confirm my willingness to take part in any of the trials described, be it a common cold trial, an influenza vaccine trial, or a medicinal drug trial.

Signed ______________________________
Date ______________________________

I am available for 10 days between the following dates:
(Give as many dates as possible from the list attached) ______________________________
____

Name and address of next-of-kin and telephone no: ______________________________
____

*From Medical Research Council and Department of Health & Social Security, Clinical Research Centre, MRC Common Cold Unit, Harvard Hospital, Salisbury, Wilts.
†To enable travel warrants to be prepared. Volunteers traveling by coach may claim a refund after arrival at the Unit.

APPENDIX D

Summary of Harvard Clinical Scoring System (1973)

Each day, for 3 days before virus inoculation and 6 days after inoculation:

1. Count handkerchiefs.
2. Record each sign or symptoms as 0, ±, +, or ++.
3. Record oral temperature (twice daily).

At the end of the trial, score these records numerically.

1. For daily *handkerchief count,* score as follows.

Handkerchiefs per day	Score
< 4	0
4	1
8	2
12	3
16	4
20	5
25	6
30	7
35	8
40	9
50	10
60	11
70	12

2. The following signs and symptoms should be scored at full rate.

0 = 0
± = 1
+ = 2
++ = 3

a. Mucopurulent nasal discharge
b. Nasal obstruction
c. Postnasal discharge
d. Sinus pain
e. Fauces (noticeable abnormality compared with the preinoculation state)
f. Cervical adenitis
g. Hoarseness
h. Cough
i. Sputum
j. Headache
k. Malaise
l. Myalgia
m. Chill

Symptoms j to m can be scored at half value (see below) if the observer doubts the real existence of these subjective but significant findings.

3. Symptoms (mainly subjective) which should be scored at half rate:

0 = 0
± = ½
+ = 1
++ = 1½

a. Sneezing
b. Watering eyes
c. Nasal stuffiness
d. Sore throat (subjective)

4. Pyrexia – score one point for each 0.5°C above normal. Pyrexia is scored twice daily (unlike other symptoms and signs).

5. Volunteer retired to bed. Score either one or two points per hour at the observer's discretion based on knowledge of the individual volunteer's personality.

6. Other signs and symptoms which should be recorded and scored include:

a. Earache
b. Suffusion of conjunctiva
c. Dizziness
d. Shivering
e. Sweating
f. Stiff neck
g. Nausea and vomiting

7. The total score for the 9-day period is calculated.

REFERENCES

1. **Beare, A. S.**, Live viruses for immunization against influenza, *Prog. Med. Virol.*, 20, 49, 1975.
2. **Andrewes, C. H.**, *In Pursuit of the Common Cold*, William Heinemann Medical Books Limited, London, 1973.
3. **Beare, A. S., Bynoe, M. L., and Tyrrell, D. A. J.**, Investigation into the attenuation of influenza viruses by serial passage, *Br. Med. J.*, 4, 482, 1968.
4. **Tyrrell, D. A. J.**, *Common Colds and Related Diseases*, Edward Arnold Ltd., London, 1965.
5. **Knight, V.**, The use of volunteers in medical virology, *Prog. Med. Virol.*, 6, 1, 1964.
6. **Jackson, G. G., Muldoon, R. L., Johnson, G. C., and Dowling, H. F.**, Contributions of volunteers to studies on the common cold. Conference on newer respiratory disease viruses, *Am. Rev. Respir. Dis.*, 88, (Suppl.), 120, 1963.
7. **Potter, C. W., Jennings, R., McLaren, C., and Clarke, A.**, Immunity following intranasal administration of an inactivated, freeze-dried A/England/42/72 vaccine, *Arch. Virol.*, 48, 307, 1975.
8. **Bloomfield, S. S., Gaffney, T. E., and Schiff, G. M.**, A design for evaluation of antiviral drugs in influenza, *Am. J. Epidemiol.*, 91, 568, 1970.
9. **Walker, G. H., Bynoe, M. L., and Tyrrell, D. A. J.**, Trial of ascorbic acid in prevention of colds, *Br. Med. J.*, 1, 603, 1967.
10. **Karlowski, T. R., Chalmers, T. C., Frenkel, L. D., Kapikian, A. Z., Lewis, T. L., and Lynch, J. M.**, Ascorbic acid for the common cold, *JAMA*, 231, 1038, 1975.
11. **Merigan, T. C., Reed, S. E., Hall, T. S., and Tyrrell, D. A. J.**, Inhibition of respiratory virus infection by locally applied interferon, *Lancet*, 1, 563, 1973.
12. **Hill, D. A., Baron, S., Perkins, J. C., Worthington, M., Van Kirk, J. E., Mills, J., Kapikian, A. Z., and Chanock, R. M.**, Evaluation of an interferon inducer in viral respiratory disease, *JAMA*, 219, 1179, 1972.
13. **Panusarn, C., Stanley, E. D., Dirda, V., Rubenis, M., and Jackson, G. G.**, Prevention of illness from rhinovirus infection by a topical interferon inducer, *N. Engl. J. Med.*, 291, 57, 1974.
14. **Togo, Y., Schwartz, A. R., and Hornick, R. B.**, Antiviral effect of 3,4-dihydro-1-isoquinoline acetamide hydrochloride in experimental human rhinovirus infection, *Antimicrob. Agents Chemother.*, 4, 612, 1973.
15. **Togo, Y., Schwartz, A. R., and Hornick, R. B.**, Failure of a 3-substituted triazinoindole in the prevention of experimental human rhinovirus infection, *Chemotherapy*, 18, 17, 1973.
16. **Cate, T. R., Douglas, R. G., and Couch, R. B.**, Interferon and resistance to upper respiratory infection, *Proc. Soc. Exp. Biol. Med.*, 131, 631, 1969.
17. **Jao, R. L., Wheelock, E. F., and Jackson, G. G.**, Production of interferon in volunteers infected with Asian influenza, *J. Infect. Dis.*, 121, 419, 1970.
18. **Smorodintsev, A. A., Beare, A. S., Bynoe, M. L., Head, B., and Tyrrell, D. A. J.**, The formation of interferon during acute respiratory virus infection of volunteers, *Arch. Gesamte Virusforsch.*, 33, 9, 1971.
19. **Morita, M., Suto, T., and Ishida, N.**, Antigenic memory in man in response to sequential infections with influenza A viruses, *J. Infect Dis.*, 126, 61, 1972.
20. **Hobson, D., Curry, R. L., Beare, A. S., and Gardner, A. W.**, The role of serum haemagglutination-inhibiting antibody in protection against challenge infection with influenza A2 and B viruses, *J. Hyg.*, 70, 767, 1972.
21. **Beare, A. S., Bynoe, M. L., and Tyrrell, D. A. J.**, Prophylaxis of influenza with a synthetic isoquinoline, *Lancet*, 1, 843, 1968.
22. **Murphy, B. R., Kasel, J. A., and Chanock, R. M.**, Association of serum antineuraminidase antibody with resistance to influenza in man, *N. Engl. J. Med.*, 286, 1329, 1972.
23. **Couch, R. B., Kasel, J. A., Gerin, J. L., Schulman, J. L., and Kilbourne, E. D.**, Induction of partial immunity to influenza by a neuraminidase-specific vaccine, *J. Infect. Dis.*, 129, 411, 1974.
24. **Beare, A. S., Maassab, H. F., Tyrrell, D. A. J., Slepushkin, A. N., and Hall, T. S.**, A comparative study of attenuated influenza viruses, *Bull. WHO*, 44, 593, 1971.
25. **Hamre, D.**, Rhinoviruses, *Monogr. Virol.*, 1, 44, 1968.
26. **Perkins, J. C., Tucker, D. N., Knopf, H. L. S., Wenzel, R. P., Kapikian, A. Z., and Chanock, R. M.**, Comparison of protective effect of neutralizing antibody in serum and nasal secretions in experimental rhinovirus type 13 illness, *Am. J. Epidemiol.*, 90, 519, 1969.
27. **Reed, S. E. and Hall, T. S.**, Hemagglutination-inhibition test in rhinovirus infection of volunteers, *Infect. Immun.*, 8, 1, 1973.
28. **Stott, E. J. and Killington, R. A.**, Haemagglutination by rhinoviruses, *Lancet*, 1, 1369, 1972.
29. **Hendley, J. O., Edmondson, W. P., and Gwaltney, J. M.**, Relation between naturally acquired immunity and infectivity of two rhinoviruses in volunteers, *J. Infect. Dis.*, 125, 243, 1972.
30. **Buckland, F. E., Bynoe, M. L., and Tyrrell, D. A. J.**, Experiments on the spread of colds. II. Studies in volunteers with Coxsackievirus A21, *J. Hyg.*, 63, 327, 1965.
31. **Couch, R. B., Douglas, R. G., Lindgren, K. M., Gerone, P. J., and Knight, V.**, Airborne transmission of respiratory infection with Coxsackievirus A type 21, *Am. J. Epidemiol.*, 91, 78, 1970.
32. **Strizova, V., Brown, P. K., Head, B., and Reed, S. E.**, The advantages of HeLa cells for isolation of rhinoviruses, *J. Med. Microbiol.*, 7, 433, 1974.
33. **Tyrrell, D. A. J.**, The use of volunteers, *Am. Rev. Respir. Dis.*, 88, 128, 1963.

34. **Douglas, R. G. and Couch, R. B.**, Attenuation of rhinovirus type 15 for humans, *Nature*, 223, 213, 1969.
35. **Douglas, R. G., Couch, R. B., Baxter, B. D., and Gough, M.**, Attenuation of rhinovirus type 15: relation of illness to plaque size, *Infect. Immun.*, 9, 519, 1974.
36. **Holmes, M. J., Reed, S. E., Stott, E. J., and Tyrrell, D. A. J.**, Studies of experimental rhinovirus type 2 infections in polar isolation and in England, *J. Hyg.*, 76, 379, 1976.
37. **Stott, E. J. and Tyrrell, D. A. J.**, Some improved techniques for the study of rhinoviruses using HeLa cells, *Arch. Gesamte Virusforsch.*, 23, 236, 1968.
38. **Kriel, R. L., Wulff, H., and Chin, T. D. Y.**, Microneutralization test for determination of rhinovirus and Coxsackievirus A antibody in human diploid cells, *Appl. Microbiol.*, 17, 611, 1969.
39. **Jackson, G. G., Muldoon, R. L., Akers, L. W., Liu, O., Johnson, G. C., and Engel, C.**, Effect of N',N'-anhydrobis-(β-hydroxyethyl) biguanidehydrochloride on Asian influenza virus in volunteers, *Antimicrob. Agents Chemother.*, p. 883, 1961.
40. **Jackson, G. G., Muldoon, R. L., and Akers, L. W.**, Serological evidence for prevention of influenzal infection in volunteers by an anti-influenzal drug adamantanamine hydrochloride, *Antimicrob. Agents Chemother.*, p. 703, 1963.
41. **Tyrrell, D. A. J., Bynoe, M. L., and Hoorn, B.**, Studies on the antiviral activity of 1-adamantanamine, *Br. J. Exp. Pathol.*, 46, 370, 1965.
42. **Togo, Y., Hornick, R. B., Felitti, V. J., Kaufman, M. L., Dawkins, A. T., Jr., Kelpe, V. E., and Claghorn, J. L.**, Evaluation of therapeutic efficacy of amantadine in patients with naturally occurring A2 influenza, *JAMA*, 211, 1149, 1970.
43. **Smorodintsev, A. A., Zlydnikov, D. M., Kiseleva, A. M., Romanov, J. A., Kazantsev, A. P., and Rumovsky, V. I.**, Evaluation of amantadine in artificially induced A2 and B influenza, *JAMA*, 213, 1448, 1970.
44. **Oker-Blom, N., Hovi, T., Leinikki, P., Palosuo, T., Pettersson, R., and Suni, J.**, Protection of man from natural infection with influenza A2 Hong Kong virus by amantadine: a controlled field trial, *Br. Med. J.*, 3, 676, 1970.
45. **Galbraith, A. W., Oxford, J. S., Schild, G. C., Potter, C. W., and Watson, G. I.**, Therapeutic effect of 1-adamantanamine hydrochloride in naturally occurring influenza A2/Hong Kong infection, *Lancet*, 2, 113, 1971.
46. **Wingfield, W. L., Pollack, D., and Grunert, R. R.**, Therapeutic effect of amantadine HCl and rimantadine HCl in naturally occurring influenza A2 respiratory illness in man, *N. Engl. J. Med.*, 281, 579, 1969.
47. **Bryans, J. T., Zent, W. W., Grunert, R. R., and Boughton, D. C.**, 1-Adamantanamine hydrochloride prophylaxis for experimentally induced A/equine 2 influenza virus infection, *Nature*, 212, 1542, 1966.
48. **Lang, G., Narayan, O., and Rouse, B. T.**, Prevention of malignant avian influenza by 1-adamantanamine hydrochloride, *Arch. Gesamte Virusforsch.*, 32, 171, 1970.
49. **Beare, A. S., Hall, T. S., and Tyrrell, D. A. J.**, Protection of volunteers against challenge with A/Hong Kong/68 influenza virus by a new adamantane compound, *Lancet*, 1, 1039, 1972.
50. **Arroyo, M., Beare, A. S., Reed, S. E., and Craig, J. W.**, A therapeutic study of an adamantane spiro compound in experimental influenza A infection in man, *J. Antimicrob. Chemother.*, 1 (Suppl.), 87, 1975.
51. **Williamson, G. M. and Jackson, D.**, The antiviral activity of the isoquinolines, famotine and memotine, in respiratory infections in man, *Bull. WHO*, 41, 665, 1969.
52. **Meenan, P. N. and Hillary, I. B.**, Prophylaxis of influenza with U.K. 2054, *Lancet*, 2, 614, 1969.
53. **Meenan, P. N. and Hillary, I. B.**, Field study of U.K. 2054 against natural influenza with A2/Hong Kong/68 influenza virus, *Lancet*, 2, 641, 1969.
54. **Reed, S. E., Beare, A. S., Bynoe, M. L., and Tyrrell, D. A. J.**, Studies of isoquinoline derivatives (U.K. 2371 and U.K. 2054) in respiratory infections of volunteers, using influenza viruses, a parainfluenza virus and a rhinovirus, *Ann. N.Y. Acad. Sci.*, 173, 760, 1970.
55. **Togo, Y., Schwartz, A. R., Tominaga, S., and Hornick, R. B.**, Cyclooctylamine in the prevention of experimental human influenza, *JAMA*, 220, 837, 1972.
56. **Fruchtman, M. H., Mauceri, A. A., Wigley, F. M., and Waldman, R. H.**, Aerosol administration of human gamma globulin as prophylaxis against influenza virus challenge, *Clin. Med.*, 79, 17, 1972.
57. **Longley, S., Dunning, R. L., and Waldman, R. H.**, Effect of isoprinosine against challenge with A/(H3 N2)/Hong Kong influenza virus in volunteers, *Antimicrob. Agents Chemother.*, 3, 506, 1973.
58. **Togo, Y. and McCracken, E. A.**, Double blind clinical assessment of ribavirin (virazole) in the prevention of induced with type B influenza virus, *J. Infect. Dis.*, Suppl. 133, A109, 1976.
59. **Schwartz, A. R., Togo, Y., Hornick, R. B., Tominaga, S., and Gleckman, R. A.**, Evluation of the efficacy of ascorbic acid in prophylaxis of induced rhinovirus 44 infection in man, *J. Infect. Dis.*, 128, 500, 1973.
60. **Anderson, T. W., Reid, D. B. W., and Beaton, G. H.**, Vitamin C and the common cold: a double-blind trial, *Can. Med. Assoc. J.*, 107, 503, 1972.
61. **Anderson, T. W., Suranyi, G., and Beaton, G. H.**, The effect on winter illness of large doses of vitamin C, *Can. Med. Assoc. J.*, 111, 31, 1974.
62. **Coulehan, J. L., Reisinger, K. S., Rogers, K. D., and Bradley, D. W.**, Vitamin C prophylaxis in a boarding school, *N. Engl. J. Med.*, 290, 6, 1974.
63. **Acornley, J. E., Bessell, C. J., Bynoe, M. L., Godtfredsen, W. O., and Knoyle, J. M.**, Antiviral activity of sodium fusidate and related compounds, *Br. J. Pharmacol. Chemother.*, 31, 210, 1967.

64. **Buthala, D. A., Damiano, R., and Eidson, E. E.**, Studies on Coxsackie A-21 (Coe) virus-infected volunteers: effect of local therapy with gamma globulin, *Ann. N.Y. Acad. Sci.*, 173, 794, 1970.
65. **Reed, S. E. and Bynoe, M. L.**, The antiviral activity of isoquinoline drugs for rhinoviruses *in vitro* and *in vivo*, *J. Med. Microbiol.*, 3, 346, 1970.
66. **Grunberg, E. and Prince, H. N.**, The antiviral activity of 3,4-dihydro-1-isoquinolineacetamide hydrochloride *in vitro, in ovo,* and in small laboratory animals, *Proc. Soc. Exp. Biol. Med.*, 129, 422, 1968.
67. **Murphy, B. R. and Glasgow, L. A.**, The antiviral activity of 3,4-dihydro-1-isoquinolineacetamide hydrochloride on Columbia SK virus infection in mice and in tissue culture, *Ann. N.Y. Acad. Sci.*, 178, 255, 1970.
68. **Glasgow, L. A. and Galasso, G. J.**, Isoprinosine: lack of antiviral activity in experimental model infections, *J. Infect. Dis.*, 126, 162, 1972.
69. **Soto, A. J., Hall, T. S., and Reed, S. E.**, Trial of the antiviral action of isoprinosine against rhinovirus infection of volunteers, *Antimicrob. Agents Chemother.*, 3, 332, 1973.
70. **Pachuta, D. M., Togo, Y., Hornick, R. B., Schwartz, A. R., and Tominaga, S.**, Evaluation of isoprinosine in experimental human rhinovirus infection, *Antimicrob. Agents Chemother.*, 5, 403, 1974.
71. **Mathur, A., Beare, A. S., and Reed, S. E.**, In vitro antiviral activity and preliminary clinical trials of a new adamantane compound, *Antimicrob. Agents Chemother.*, 4, 421, 1973.
72. **Gladych, J. M. Z., Hunt, J. H., Jack, D., Haff, R. F., Boyle, J. J., Stewart, R. C., and Ferlauto, R. J.**, Inhibition of rhinovirus by isatin thiosemicarbazone analogues, *Nature*, 221, 286, 1969.
73. **Gwaltney, J. M.**, Rhinovirus inhibition by 3-substituted triazinoindoles, *Proc. Soc. Exp. Biol. Med.*, 133, 1148, 1970.
74. **Pinto, C. A., Bahnsen, H. P., Ravin, L. J., Haff, R. F., and Pagano, J. F.**, The antiviral effect of a triazinoindole (SK & F 40491) in rhinovirus infected gibbons, *Proc. Soc. Exp. Biol. Med.*, 141, 467, 1972.
75. **Reed, S. E., Craig, J. W., and Tyrrell, D. A. J.**, Four anti-rhinovirus compounds studied *in vitro* and in volunteers, *J. Infect. Dis. Suppl.*, June, 1976.
76. **Levy, H. B., Baer, G., Baron, S., Buckler, C. E., Gibbs, C. J., Iadarola, M. J., London, W. T., and Rice, J.**, A modified polyriboinosinic-polyribocytidylic acid complex that induces interferon in primates, *J. Infect. Dis.*, 132, 434, 1975.
77. **Gatmaitan, B. G., Stanley, E. D., and Jackson, G. G.**, The limited effect of nasal interferon induced by rhinovirus and a topical chemical inducer on the course of infection, *J. Infect. Dis.*, 127, 401, 1973.
78. **Douglas, R. G. and Betts, R. F.**, Effect of induced interferon in experimental rhinovirus infections in volunteers, *Infect. Immun.*, 9, 506, 1974.
79. **Douglas, R. G., Betts, R. F., Little, J. W., Hogan, P. W., and Roth, F. K.**, Evaluation of a low molecular weight interferon inducer in experimental rhinovirus type 13 infection in normal volunteers, in Antivirals with Clinical Potential, Proc. Symp. at Stanford University, Stanford, California, August 1975, to be published.
80. **Bell, J. A., Rowe, W. P., Engler, J. I., Parrott, R. H., and Huebner, R. J.**, Pharyngoconjunctival fever. Epidemiological studies of a recently recognized disease entity, *JAMA*, 157, 1083, 1955.
81. **Bell, J. A., Huebner, R. J., Rosen, L., Rowe, W. P., Cole, R. M., Mastrota, F. M., Floyd, T. M., Chanock, R. M., and Shvedoff, R. A.**, Illness and microbial experiences of nursery children at Junior Village, *Am. J. Hyg.*, 74, 267, 1967.
82. **Bell, T. M., Turner, G., Macdonald, A., and Hamilton, D. A.**, Type-3 adenovirus infection, *Lancet*, 2, 1327, 1960.
83. **Van Der Veen, J. and Dijkman, J. H.**, Association of type 21 adenovirus with acute respiratory illness in military recruits, *Am. J. Hyg.*, 76, 149, 1962.
84. **Hitchcock, G., Tyrrell, D. A. J., and Bynoe, M. L.**, Vaccination of man with attenuated live adenovirus, *J. Hyg.*, 58, 277, 1960.
85. **Kasel, J. A., Evans, H. E., Spickard, A., and Knight, V.**, Conjunctivitis and enteric infection with adenovirus types 26 and 27: responses to primary, secondary and reciprocal cross-challenges, *Am. J. Hyg.*, 77, 265, 1963.
86. **Couch, R. B., Cate, T. R., Douglas, R. G., Gerone, P. J., and Knight, V.**, Effect of route of inoculation on experimental respiratory viral disease in volunteers and evidence for airborne transmission, *Bacteriol Rev.*, 30, 517, 1966.
87. **Hamory, B. H., Couch, R. B., Douglas, R. G., Black, S. H., and Knight, V.**, Characterization of the infectious unit for man of two respiratory viruses, *Proc. Soc. Exp. Biol. Med.*, 139, 890, 1972.
88. **Kasel, J. A., Huber, M., Loda, F., Banks, P. A., and Knight, V.**, Immunization of volunteers with soluble antigens of adenovirus type 1, *Proc. Soc. Exp. Biol. Med.*, 117, 186, 1964.
89. **Chanock, R. M., Ludwig, W., Huebner, R. J., Cate, T. R., and Chu, L.-W.**, Immunization by selective infection with type 4 adenovirus grown in human diploid tissue culture, *JAMA*, 195, 445, 1966.
90. **Smith, T. J. Buescher, E. L., Top, F. H., Jr.,. Altemeier, W. A., and McCown, J. M.**, Experimental respiratory infection with type 4 adenovirus vaccine in volunteers: clinical and immunological responses, *J. Infect. Dis.*, 122, 239, 1970.
91. **Couch, R. B., Kasel, J. A., Pereira, H. G., Haase, A. T., and Knight, V.**, Induction of immunity in man by crystalline adenovirus type 5 capsid antigens, *Proc. Soc. Exp. Biol. Med.*, 143, 905, 1973.
92. **Dudding, B. A., Bartelloni, P. J., Scott, R. M., Top, F. H., Jr., Russell, P. K., and Buescher, E. L.**, Enteric immunization with live adenovirus type 21 vaccine. I. Tests for safety, infectivity, immunogenicity and potency in volunteers, *Infect. Immun.*, 5, 295, 1972.
93. **Schwartz, A. R., Togo, Y., and Hornick, R. B.**, Clinical evaluation of live, oral types 1, 2 and 5 adenovirus vaccines, *Am. Rev. Respir. Dis.*, 109, 233, 1974.

94. **Smith, C. B., Purcell, R. H., Bellanti, J. A., and Chanock, R. M.,** Protective effect of antibody to parainfluenza type 1, *N. Engl. J. Med.*, 275, 1145, 1966.
95. **Smith, C. B., Purcell, R. H., and Chanock, R. M.,** Effect of amantadine hydrochloride on parainfluenza type 1 virus infections in adult volunteers, *Am. Rev. Respir. Dis.*, 95, 689, 1967.
96. **Taylor-Robinson, D. and Bynoe, M. L.,** Parainfluenza 2 virus infections in adult volunteers, *J. Hyg.*, 61, 407, 1963.
97. **Tyrrell, D. A. J. and Bynoe, M. L.,** Studies on parainfluenza type 2 and 4 viruses obtained from patients with common colds, *Br. Med. J.*, 1, 471, 1969.
98. **Joosting, A. C. C., Head, B., Bynoe, M. L., and Tyrrell, D. A. J.,** Production of common colds in human volunteers by influenza C virus, *Br. Med. J.*, 2, 153, 1968.
99. **Bradburne, A. F. and Tyrrell, D. A. J.,** Coronaviruses of man, *Prog. Med. Virol.*, 13, 373, 1971.
100. **Hamre, D. and Procknow, J. J.,** A new virus isolated from the human respiratory tract. *Proc. Soc. Exp. Biol. Med.*, 121, 190, 1966.
101. **Bradburne, A. F., Bynoe, M. L., and Tyrrell, D. A. J.,** Effects of a "new" human respiratory virus in volunteers, *Br. Med. J.*, 3, 767, 1967.
102. **Johnson, K. M., Chanock, R. M., Rifkind, D., Kravetz, H. M., and Knight, V.,** Respiratory syncytial virus. IV. Correlation of virus shedding, serologic response, and illness in adult volunteers, *JAMA*, 176, 663, 1961.
103. **Friedewald, W. T., Forsyth, B. R., Smith, C. B., Gharpure, M. A., and Chanock, R. M.,** Low-temperature grown RS virus in adult volunteers, *JAMA*, 204, 690, 1968.
104. **Mills, J., Van Kirk, J. E., Wright, P. F., and Chanock, R. M.,** Experimental respiratory syncytial virus infection of adults, *J. Immunol.*, 107, 123, 1971.
105. **Wright, P. F., Mills, J., and Chanock, R. M.,** Evaluation of a temperature-sensitive mutant of respiratory syncytial virus in adults, *J. Infect. Dis.*, 124, 505, 1971.
106. **Cohen, A., Togo, Y., Khakoo, R., Waldmann, R., and Sigel, M.,** Comparative clinical and laboratory evaluation of the prophylactic capacity of ribavirin, amantadine hydrochloride and placebo in induced human influenza type A, *J. Infect. Dis.*, Suppl. 133, 1976
107. **Matthews, T. H. J., Reed, S. E., and Tyrrell, D. A. J.,** unpublished observations.
108. **Reed, S. E.,** unpublished.

Chapter 3

CLINICAL TESTING OF ANTIVIRAL COMPOUNDS ON A LARGE SCALE IN GENERAL PRACTICE

A. W. Galbraith and G. I. Watson

TABLE OF CONTENTS

I. INTRODUCTION

The great majority of patients with viral infections are not treated in hospitals, but from start to finish by general practitioners. Therefore, any drugs that are developed for prophylactic or therapeutic use against human viral diseases will have to be studied either in general practice or under comparable conditions, for example, in schools, factories, or the armed services.

Many formerly serious viral diseases of man (notably poliomyelitis, measles, and yellow fever) have been or are now being controlled by vaccination. Two main groups of viral infections (namely common colds and epidemic influenzas), each of great economic importance, remain uncontrolled. In the case of common colds, various complications cause an increase in morbidity and mortality at either end of life. Influenza is a highly infectious disease, affecting all ages and crippling industry in a widespread epidemic. Why is it, then, that more pharmaceutical companies do not seek the financial advantages of producing commercially successful antiviral agents? Although there may be many reasons, three principally come to mind.

The first reason concerns the problems of research, including the expense of virological testing and the clinical evaluation of any compound (for further discussion of these aspects, see Chapter 6). The second reason involves the marketing of a drug which, in the case of influenza, may only be prescribed for a limited period in any one year. The third reason concerns the attitudes of doctors and the public to influenza itself. While a broad-spectrum antiviral agent for the treatment of rhinovirus or other respiratory infections would be an attractive proposition, the public appears to accept "flu," like frozen pipes or disrupted communications, as one of several irritating hazards of winter. Those particularly susceptible receive immunization, but one's general impression is that the man in the street remains unconcerned about influenza until an epidemic appears. Inevitably, therefore, unless the public is made aware of the true medical and economic consequences of epidemic influenza, the significance of the disease is lost, which does little to encourage participation in field studies.

A further consideration at present is the absence of sufficient quantities of inactivated influenza vaccines for attempts at mass immunization. Therefore, it is not possible to control influenza by vaccination, and this situation is not likely to change in the near future. The advent of live influenza vaccines will hardly influence the epidemiological situation because it is unlikely that enough vaccine of the correct subtype will be available. This is particularly pertinent in the possible emergence of swine influenza virus A/New Jersey/76 ($H_{swine\ 1}N1$) in the U.S. at Fort Dix.

Although large quantities of inactivated vaccine will be produced for the U.S., this will probably not be possible for most European countries. Since the virus has been shown to be inhibited by amantadine (see Chapters 5, Volume I and 1, Volume II), chemoprophylaxis may have an important place in the control of this virus if it becomes epidemic or pandemic.

A doctor's attitude towards influenza prophylaxis is influenced by his area of clinical responsibility. The general practitioner (who sees the majority of patients suffering from the infection), the school doctor (whose charges may have important examination schedules to complete), or the industrial medical officer (who is answerable to a manager intent on keeping his labor force at work) all may prescribe an anti-influenza agent on clinical trial or on the market, but each has constraints placed on him. His decision to evaluate a new compound or regularly prescribe a recent drug during an epidemic will depend on the time involved, the cost of the medication, and the extra effort needed when his work load is already increased. The lower the incidence of the disease, the greater the need for collective inquiries.

Postulating that resistance is that state of the body which prevents infection, while immunity includes a variety of reactions that set in only after infection has occurred, the lines of future research might include attempts to

1. Stimulate human resistance to a virus.
2. Reduce the infectiousness of a virus within the body.
3. Alter human reactions to infection while immunity is building up within the body.
4. Increase immunity without inducing clinical reactions.

Work on interferon is an example of the first objective. Amantadine and ABOB come under the second heading. The use of steroids in glandular fever or infections of the central nervous system are attempts of the third objective. Vaccination is the usual method under group 4, but prophylactic amantadine has also shown some effect.

A number of antiviral agents have been produced by various research departments for the prophylaxis or treatment of respiratory infections.[1-3] N′, N′-Anhydrobis (α-hydroxyethyl-diaguanide HCl), abbreviated ABOB, was previously marketed in the United Kingdom as virugon. Initial field trials demonstrated some mild efficacy as an influenza prophylactic, but ABOB was subsequently withdrawn from the market following negative studies[7,8] after achieving relatively little commercial success. To date, the only compound that has any claim to scientific or commercial success is amantadine hydrochloride, marketed as Symmetre®.[9] Davies and his colleagues[10] found this compound to possess antiviral properties, principally against influenza A (see Chapter 5, Volume I). Safety testing in animals was satisfactory (see Chapter 1), and small-scale human volunteer studies demonstrated no adverse effects[11] (see Chapter 2).

The next stage in the evaluation of the drug is similar in principle for any compound, differing only in matters of detail. The various regulatory bodies (for example, the Food and Drug Administration in the United States or the Committee on Safety of Medicines in Great Britain) demand clearance by their secretariat before further clinical evaluations may proceed. In Britain, a Clinical Trial Certificate (CTC) must be obtained and a detailed protocol of each trial is required before the CTC is granted (details are presented in Chapter 6).

After the initial clinical testing, amantadine was investigated on a larger scale in America among volunteer students who were challenged with an attenuated strain of influenza A virus. Stanley and colleagues[12] demonstrated the prophylactic efficacy of amantadine on the attenuated strain they were using, thus confirming the direct advantages of the challenge method over the naturally occurring epidemic.

Further investigations with amantadine (against naturally occurring influenza) included challenge studies among prison volunteers[13,17] and the evaluation of patients in the armed services,[14] hospitals,[15] and boarding schools.[16] The obvious advantage of studying such groups lies in their environmental convenience. However, challenge studies have their own limitations; in the naturally occurring infection, the groups selected for study may not succumb during the epidemic. Again, the very fact that the groups have been selected renders them atypical of the general population.

II. CLINICAL EVALUATION OF AMANTADINE IN THE UNITED KINGDOM

When amantadine became available in the United Kingdom, a small challenge study was performed at the Common Cold Research Centre.[18] The attenuated strain of influenza A was sensitive to amantadine in in vitro laboratory studies, but clinical and serologic trials did not demonstrate a prophylactic action of the drug. Nevertheless, further clinical evaluation of amantadine was planned; the following venues were involved, although it could not be known that there would be very little epidemic influenza in the next 2 years.

A. Hospital for the Mentally Subnormal

This hospital was chosen for the evaluation of amantadine as an influenza prophylactic for two reasons. First, the hospital was situated close to the office of the medical adviser supervising the study, which underlines the advantage of accessibility when planning such a clinical investigation. Second, the patient population was a long-stay, relatively closed community of all ages and both sexes. The permission of the parents or relatives of these mentally subnormal patients was obtained before involving them in the study. Signed permission was sought at visiting times or by letter. Approximately 200 patients were included in the investigation. A serum specimen was obtained from each and was tested for resting antibody to the most recent influenza A virus. Sampling took place during October 1966; all wards were provided with code-numbered bottles holding either amantadine capsules or syrup or identical capsules or syrup containing inert material. In addition, record cards were prepared for recording the temperature and any respiratory symptoms, should the patient develop an infection. The medical and nursing staff then waited for the appearance of a naturally occurring influenza epidemic.

In the middle of December 1966, one patient was reported to be suffering from a possible influenzal illness. Nose and throat swabs were taken, and influenza A was confirmed by the Public Health Laboratory. Meanwhile, two other patients developed symptoms of influenza; therefore, instructions were given for the capsules and syrup to be distributed to the patients. As the sex, age, and resting antibody titer of each patient in the study was known, allocation of active and placebo material was performed by the statistician in charge, so that the groups were balanced for these characteristics. The amantadine capsules contained 100 mg of active drug and the syrup contained 50 mg of amantadine in 5 ml. The dose was one capsule every 12 hr for adults, one capsule daily for children 10 to 15 years old, and a proportional dose of syrup for children between 2 and 10 years old. A record card was completed for each patient; the study continued for 2 weeks. During that time no further spread of influenza occurred; consequently, no drug effect could be demonstrated. Tolerance of both capsules and syrup was good, however, and the relatives agreed to another study to take place during the following winter.

Another blood sample was taken from all patients for the estimation of resting antibody in the second study during October 1967. Once again, freshly coded capsules and syrup (allocated on the basis of sex, age, and antibody status so that active and placebo treated groups matched) were distributed to the wards. The expected influenza epidemic was awaited patiently, but no patient contracted influenza.

Although the convenient locality and the enthusiasm of the medical and nursing staff of the hospital were advantageous, this type of study illustrated all too clearly the disadvantages of studying an epidemic disease in a closed community.

Studies in the U.S.[19] have demonstrated the prophylactic effectiveness of amantadine in preventing hospital-acquired infection (Tables 1 and 2). Amantadine prophylaxis resulted in a 79.7% protective effect against A (H3N2) virus, and the authors stated that serious consideration should be given to its use in hospitalized patients during epidemics.

B. Psychiatric Hospital

During the winter of 1966 to 1967, resident patients in a large psychiatric hospital were included in a similar study. The prior permission of all relatives was sought during visiting hours or by circular letter with prepaid return postage. Capsules of amantadine or placebo, code labeled and randomly distributed, together with appropriate record cards were left with the ward sisters. In the event of patients contracting influenza, venepuncture was to be performed by the medical staff at the beginning of the illness and 10 days later. Antibody estimations to influenza A were to be performed by a central virus laboratory. However, influenza did not occur in the hospital during the winter of 1966 to 1967 or the following winter. The disadvantage of selecting a closed community for such a study is again demonstrated. In a natural or open community there are more chances for the introduction of the epidemic agent by those entering or leaving for shopping, work, or entertainment.

C. The Armed Services

Cooperation with the armed services in evaluating an influenza prophylactic has several advantages, such as a high ratio of medical observers to test subjects and regular documentation. However, old-style barrack life included some of the disadvantages encountered in hospitals. Now, however, the modern peacetime serviceman reports for duty in the morning, returns to his married quarters or small barrack block in the evening, and is out of camp when off duty or during the weekend. His chances of encountering a scanty epidemic are thereby increased. He may however report ill elsewhere than in camp, thus depriving the study of some positive records.

TABLE 1

Comparison of Amantadine-treated and Untreated Patients Considered in the Analysis of Hospital- acquired Influenza

	No. of patients	
	Treated	Not treated
No. of patients	50	61
Previous vaccination	1	2
Recent respiratory illness	17	18
Age (years)[a]	58.1 ± 13.0	56.2 ± 16.8
Duration of hospitalization (days)[b]	13.9 ± 8.2[b]	16.1 ± 8.4[b]

[a]Mean ± S.D.
[b]1.393 (Stadent's test); not significant.

From O'Donoghue, J. M., Ray, C. G., Terry, D. W., and Beaty, H. N., *Am. J. Epidemiol.*, 97, 276, 1973. With permission.

TABLE 2

Hospital-acquired Influenza A (H3N2) Infection Among Amantadine-treated and Untreated Patients

Hospital-acquired influenza A	Amantadine treated	Not treated
Clinical	0/50[a]	7/61[a]
Subclinical	2/50	5/61
Total	2/50[b]	12/61[b]

[a]$p < 0.02$ (χ^2 analysis).
[b]$p < 0.05$ (χ^2 analysis).

From O'Donoghue, J. M., Ray, C. G., Terry, D. W., and Beaty, H. N., *Am. J. Epidemiol.*, 97, 276, 1973. With permission.

D. Prisoners

Although volunteer prisoners in America contributed much useful information on the prophylactic effect of amantadine during a naturally occurring influenza epidemic, special permission from the Home Office is required before such clinical investigation may be conducted in British prisons. Under certain circumstances permission may be granted for the study of a new drug in prison volunteers. However, access to the prisons is strictly controlled, and it was decided to study amantadine elsewhere.

E. Boarding School

The medical officer at a large boarding school for boys had been interested in influenza prophylaxis for over 20 years. When amantadine was first being clinically evaluated in Britain, he agreed to perform a double-blind trial if an outbreak of influenza occurred. The headmaster, who had only recently been appointed to the school, felt that at that stage he could not give his permission for such a study. Subsequently during the evaluation of amantadine as a therapeutic agent in influenza A, agreement was reached, written permission was obtained from parents of the boys, and a successful investigation was conducted.[20]

During the Easter term over 90 boys with respiratory infection were admitted to the school sanatorium, the greater proportion suffering from influenza. Upon admission, throat swabs were taken together with acute and subsequently convalescent sera. A positive culture or a fourfold or greater rise in hemagglutination-inhibiting or complement-fixing antibody was taken as proof of influenza infection.

The spread of infection among children in a boarding school is generally accepted to be rapid. The children in this instance came from all over the country, were visited regularly by friends and relatives, and mixed freely with people outside the school.

The school records revealed that influenza occurred more often during the Easter term. The response to infection of the age group 13 to 18 years may not be typical of the rest of the population, but an added advantage in this instance was the experience of the medical officer and the nursing staff, resulting in the availability of complete clinical records and serology.

F. Local Authority Accommodation

In the major conurbations in Great Britain, the local authorities have provided hostels for retired persons requiring such accommodation. Each resident has a separate bedroom, with communal living rooms. The numbers in such a home vary: in the five studied during the amantadine influenza investigation, the average was 38 persons. The consultant geriatrician, who was responsible for supervising the health care of the residents, was approached, and his permission was obtained to proceed with the trial. The general practitioners of the residents were circularized, informing them that such a study was to take place and requesting their agreement to include their patients. Serum was collected from all persons during October 1967, and the HAI antibody titer to influenza A was measured. Code-labeled amantadine and placebo capsules were allocated to the volunteers by random number. Capsules and record cards were delivered to each home prior to the expected epidemic. Influenza did not appear in any of the five homes during that winter.

The trial was repeated in 1968, with specimens of blood being again taken during October and the allocation of capsules as before. Although this population was elderly, friends and relatives frequently visited the residents, many of whom went out during the day. Therefore, the chances of spread of influenza were good, and it was unfortunate for the purposes of the study that these particular communities were not affected. Future studies would be of particular interest in this old age group, at special risk in an influenza outbreak. An investigation of any effect of amantadine in preventing virus pneumonia and death could be evaluated.

G. Nurses

It would appear to be wise to protect key personnel by influenza prophylaxis. No consistent policy for the protection of nurses is short-sighted.

Following a winter influenza epidemic, during which one large general hospital suffered an acute shortage of nurses, the medical officer in charge of the nursing staff carried out a prophylactic trial with amantadine. The response for volunteers was disappointing, record keeping was indifferent, and the nurses were hesitant to come forward for blood sampling. Such a group, potentially well motivated to become involved in an influenza study and trained to record medical data, proved difficult to supervise and failed to produce worthwhile results during a small but localized epidemic.

H. General Practice

If a person contracts influenza, he is most likely to see his general practitioner. Relatively few cases are admitted to a hospital and then only if complications arise. For many years the Royal College of General Practitioners has had an Epidemic Observation Unit consisting of family doctors with a particular interest in epidemiology. While exploring the traditional venues for the investigation of amantadine as a prophylactic during an influenza epidemic, it became apparent that instead of earmarking certain groups of the population for study and then awaiting an influenza outbreak to strike the group, one should study the infection in places where it occurs. The Director of the Unit agreed to invite the cooperation of members in an amantadine influenza prophylactic study during the winter of 1967 to 1968.

I. Rationale and Design of the First Prophylactic Study

The design of this study was unusual. Provided that 100 mg of amantadine was ingested 4 hr prior to the exposure of the person to influenza A virus, it was known that a prophylactic effect might be observed.

By the time a patient suffering from influenza contacts his family doctor, infection has already occurred and the patient is presumed to be infectious to others. In this study the first such patient treated in each separate household would be considered the "index case." He would be treated only with placebo capsules (or syrup) so that his infectivity would not be reduced. All other members of the household would then be included in the study. This is illustrated in Figure 1.

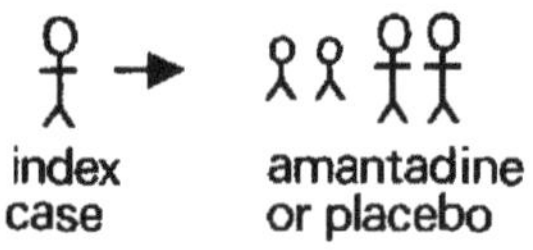

FIGURE 1. In this design of trial the index case can be left untreated, but the immediate family is given active drug or placebo. The effectiveness of the drug in preventing spread of virus from the index case to the rest of the family can then be estimated.

Prior to the epidemic, the cooperation of volunteer families known to the doctor was sought and obtained. No one included in the trial had been vaccinated against influenza in the preceding year. Each doctor was provided with code-labeled, family packs, containing all he needed for the trial, with an equal proportion of active and placebo packs to be distributed in a centrally arranged random order.

Again, so as not to reduce the spread of infection within control households and to render maximum prophylaxis in the treated households, all members of the same household apart from the index case would receive either active or placebo medication. Adults would take capsules containing 100 mg of either amantadine or placebo every 12 hr; children under 2 years were not included in the study, while those between 2 and 10 years received a proportional dose of syrup (50 mg of amantadine in 5 ml).

An addition to the design, which was in fact difficult to execute, related to the provision of material for a "contact family." It was assumed that in some instances an "index case family" might be visited frequently by a friend or relative who might spread the infection to another (volunteer) household. In the event, the number of contact families included was insufficient to demonstrate spread of infection by this route.

Provision was made to establish the diagnosis in each index case both serologically and, if possible, by virus isolation. Each doctor was provided with a wide-mouthed thermos flask, swabs, and transport medium. The Director of the Public Health Laboratory nearest to each doctor had consented to cooperate in attempting virus isolation, but in practice, difficulties such as distance from the laboratory greatly reduced the number of swabs submitted.

When the index case was first seen, a 5-ml sample of venous blood was to be taken from him or her and from each member of the household, as young as practicable. Two or three weeks later a second sample was to be collected from those already bled. Special syringes and needles, packing material, and prepaid addressed labels were provided in the family packs. The Virus Research Laboratory at Sheffield very kindly agreed to perform all antibody examinations. This centralized and reliable standardization was a most successful feature of the study. The time involved for taking paired sera from each member of a study household was estimated by one doctor to be about 1 hr.

A record card was provided for each household to describe symptoms in anyone who became ill as well as to record temperatures.

1. Comments on the Design

During the winter of 1966 to 1967, an attempt was made to run the study, but insufficient time was available for preparation and execution. The following winter the study went ahead as planned and was extraordinarily successful in spite of some misgivings.[21] A general practitioner is extremely busy during an influenza epidemic. In spite of this, 22 doctors included 52 families in the study, comprising 208 contacts divided among the treated and placebo groups. They were well matched in age and sex (Table 3). The novelty and complexity of the investigation appealed to the doctors, and the provision of every item of equipment necessary to include large families in the study (Figure 2) contributed to this appeal.

Two disadvantages in the material arrangements became apparent. The bulk of the four family boxes and four contact family boxes sent to each doctor was excessive (Figure 3). The special syringes (Figure 4) which formed blood containers, when nozzles were capped and barrels were removed, were more difficult to handle than standard syringes and plastic containers. The consensus of opinion among the family doctors who participated in the prophylactic study was that although it was time-consuming to include and follow up a family, it was possible to study four or more such families, even at the height of an influenza epidemic. The second specimen of blood was taken 14 to 21 days after the first and required perserverence and good record keeping. To include young children in the investigation was often difficult because of their dislike of venepuncture.

2. Analysis of the Study

While many degrees of sophistication can be introduced into the analysis of results obtained in clinical trials, two basic plans were available. In one, the so-called within-patient study, each patient acts as his own control. In the other, the between-patient trial, the performances of two differently treated groups are compared. The former is patently unsuitable as a design for a study of acute influenza.

The between-patient or group-comparative trial requires that some patients receive the treatment under investigation while others act as controls. In the amantadine study previously described, one group was to receive the drug while the control group took medication which resembled it, but contained no active compound. Both patients and doctors would be unaware of the composition of the medication, the trial thus being double blind.

To compare the performance of one group with the other, the two groups should be comparable in certain essential characteristics. These have to be anticipated, and arrangements must be made for the inclusion into each group of patients possessing these characteristics in approximately equal numbers. In the case of influenza it was felt that age, the severity of the influenzal infection, and the amount of preformed antibody could all be important factors in determining the response of any one patient to an influenzal infection. Among individuals living in a closed community and earmarked for study should influenza strike (for example, the long-stay patients in hospital, residents of local authority homes and school children), matching could take place in a leisurely fashion after estimations of resting antibody, and the allocation of drug or placebo could be made prior to any epidemic. In the general practice study, however, it was of course impossible to predict which patients would succumb to the infection, and it became necessary to rely on chance. It was assumed that all the family doctors taking part in the study would include the same type of patient in the trial; this was subsequently proved correct.

3. Results

Observations were made by 22 doctors distributed generally in England (15), Wales (4), and Scotland (3). Of the serum pairs from the 52 index cases, 35 (67%) showed serological evidence of influenza A infection (Table 4). For analysis of the results of the study, the 208 contacts of index cases were divided into two groups according to the presence or absence of serological evidence of influenza in the index case of the same household.[21] During the study, influenza A virus immunologically related to A/England/10/67 (H2N2) was isolated from seven of the index cases, all of whom also showed serological evidence of influenza infection. These isolations were made in several different areas in Great Britain. In 35

TABLE 3

Ages and Sex of Contacts in Aminoadamantane and Placebo Groups

Group	No. of contacts whose ages were known	No. of contacts in each age group (year)									Mean age (year)	Proportion of males and females in each group	
		2–10	11–20	21–30	31–40	41–50	51–60	61–70	71–80	81+		Males	Females
Aminoadamantane	102	16	29	13	11	10	18	2	2	1	33.0	50.9%	49.1%
Placebo	100	7	30	13	9	19	9	9	3	1	28.9	42.6%	57.4%

From Galbraith, A. W., Oxford, J. S., Schild, G. C., and Watson, G. I., *Lancet*, 2, 1026, 1969. With permission.

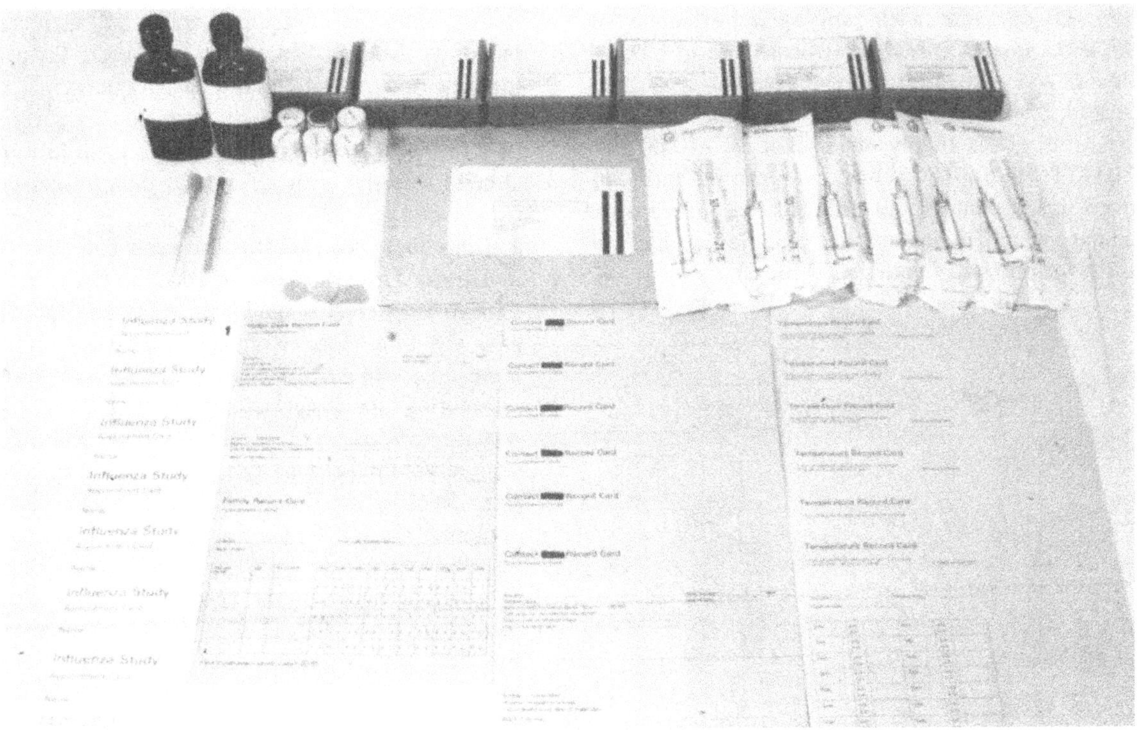

FIGURE 2. Items of equipment in family boxes for amantadine trial in general practice.

FIGURE 3. Family box and insulated flask for amantadine trial in general practice.

households in which there was serological evidence of influenza A2 infection in the index case, two of 55 (3.6%) contacts in the amantadine-treated group developed clinical symptoms of influenza (cough and temperature of 100°F or more), while 12 of 85 (14.1%) contacts in the corresponding placebo group developed a similar illness within 10 days of the index case (Table 4). The difference in incidence of illness in these two groups was of marginal significance (P = 0.07). However, in the two individuals of the amantadine-treated group who developed symptoms of influenza, there was no serological evidence of infection with influenza A2 virus. By contrast, in the placebo group 10 of the 69 contacts from whom paired sera were available developed antibody rises to influenza A2 (Table 5). Thus, the difference in incidence of serologically confirmed influenzal illness in the amantadine and control groups was statistically significant (P = 0.05 to 0.01). When the proportion of contacts with serological evidence of influenza type A infection (irrespective of clinical illness) was compared for the placebo- and drug-treated groups (Figure 5), the difference between the groups was highly significant (P = 0.001 to 0.01) (Table 5). The geometric mean HI antibody levels in the control and drug-treated groups were similar, which indicated that amantadine had no immunosuppressive activity (Table 6).

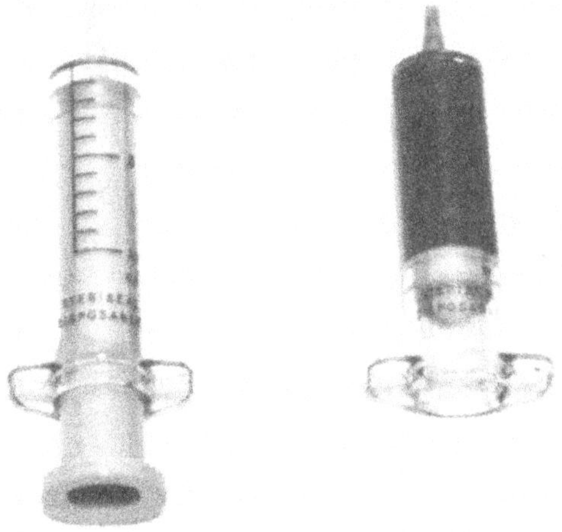

FIGURE 4. Syringes sent to general practitioners could be capped and then used as blood containers suitable for transmission through the mail to the serology laboratory.

Although the numbers in this study were small in comparison with previous large-scale influenza investigations, by eliminating all the index patients who did not demonstrate a fourfold or greater rise in CF of HI antibody, there was no dilution of clinical material with cases other than influenza A.

J. Subsequent Amantadine Studies

During January to March 1969, in Great Britain an influenza type A epidemic occurred due to the new strain A/Hong Kong/68 (H3N2). A double-blind placebo controlled investigation closely resembling the design of the 1967 to 1968 study was performed.[22] A total of 72 general practitioners volunteered to take part in the investigation. Because of the patchy distribution of the epidemic, only 29 doctors actually contributed results to the study, but they were well distributed throughout the country (Figure 6) and included one practice in Belfast. The design differed from the first trial, in one important respect. All the members of one family, including the index case, received the same treatment of drug or placebo. The dose of amantadine was 100 mg twice daily, and children received the syrup as before. Paired blood samples were taken; records of fever and symptoms were kept, with virus isolation performed where possible. The A2 virus strain isolated during the epidemic was sensitive to amantadine as measured by quantitative hemadsorption (QH). The amantadine- and placebo-treated groups were compared. The age distribution of the individuals was closely similar, and there was a good

TABLE 4

Effect of Aminoadamantane on Incidence of Clinical Influenza in Contacts of Index Cases

Laboratory evidence of influenza A infection in index case	Treatment	No. of families	Contacts who developed clinical influenza within 10 days of entering the study: All cases No.	%	P	Confirmed serologically[a] No.	%	P
Present	Aminoadamantane	13	2/55	3.6		0/48	0	
	Placebo	22	12/85	14.1	0.07	10/69	14.5	0.05–0.01
Absent	Aminoadamantane	11	1/45	2.2		0/43	0	
	Placebo	6	3/23	13.0	0.3–0.2	2/21	9.5	0.10

[a]A fourfold or greater rise in antibody titer in either or both hemagglutination-inhibition tests with A2/England/10/67 virus or in complement-fixation tests with influenza A "S" antigen was taken as a serological evidence of influenza A2 infection.

From Galbraith, A. W., Oxford, J. S., Schild, G. C., and Watson, G. I., *Lancet*, 2, 1026, 1969. With permission.

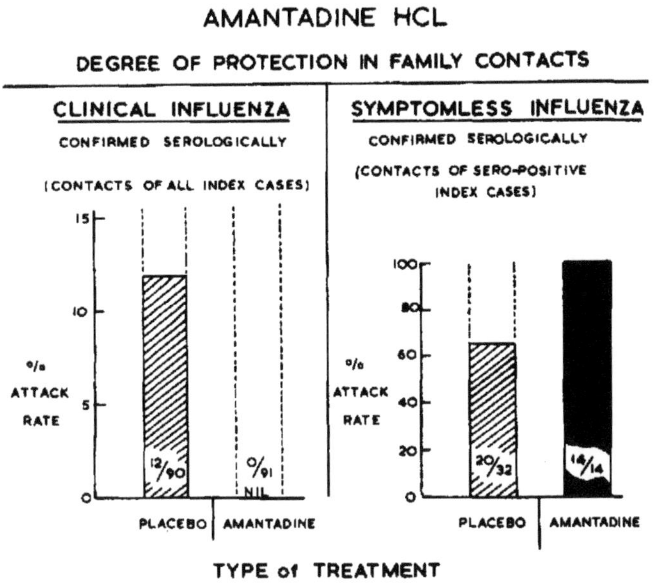

FIGURE 5. Prophylaxis against influenza type A.

TABLE 5

Effect of Aminoadamantane on the Incidence of Clinical and Subclinical Influenza A Infections

Laboratory evidence of influenza A infection in index case	Treatment	Contacts with serological evidence of influenza infection: Clinical and subclinical infections No.	%	P	Subclinical infections only No.	%	P
Present	Aminoadamantane	7/48	14.6		7/48	14.6	
	Placebo	27/69	39.1	0.001–0.01	17/69	24.6	0.2
Absent	Aminoadamantane	7/43	16.3		7/43	16.3	
	Placebo	5/21	23.8	0.7	3/21	14.3	0.8

From Galbraith, A. W., Oxford, J. S., Schild, G. C., and Watson, G. I., *Lancet*, 2, 1026, 1969. With permission.

TABLE 6

Geometric Mean HI Antibody Levels in Aminoadamantane-treated and Placebo Groups

Treatment	Geometric mean antibody titers in contacts[a] with laboratory evidence of infection: No.	Acute sera	Convalescent sera	Geometric mean antibody titers in contacts with no laboratory evidence of infection: No.	Acute sera	Convalescent sera
Aminoadamantane	7	4.92	35.67	41	13.81	12.07
Placebo	27	9.12	54.57	42	16.65	17.21

[a]Contacts of index cases with laboratory evidence of influenza A infection.

From Galbraith, A. W., Oxford, J. S., Schild, G. C., and Watson, G. I., *Lancet*, 2, 1026, 1969. With permission.

FIGURE 6. Distribution of doctors throughout the United Kingdom for amantadine trials in general practice.

correspondence in the mean ages of individuals in each group. Treated and control groups contained closely similar proportions of males and females. In addition, the antibody titer to the virus in the initial serum samples was similar in the two groups.

1. Results

Observations were made by 29 doctors on 58 families comprising 176 individuals. Of the 58 index cases, 36 (62%) showed serological evidence of influenza A infection (Table 7). When the groups were compared it was apparent that in this instance amantadine had conferred no protective effect on those individuals compared with the contacts receiving placebo.

2. Discussion

A possible explanation of the negative result, in view of the sensitivity of the virus to amantadine, lies in the low transmission rate of infection from index case to the other members of the family, associated with an absence of coughing.

Another difference between the first study and this one concerns the low initial antibody levels to the virus. In the 1967 to 1968 study, 40% of the contacts studied possessed initial HI titers below 1:12, whereas in the present study the figure was 90% (Table 8). It was suggested that there is an enhancement of the inhibitory effect of amantadine on influenza A2 infection in individuals with preexisting antihemagglutinin antibody for A2 virus. In view of this, possible advantages might

TABLE 7

Infection Groups

	Amantadine	Controls	Total
Influenza A	15	13	28
Influenza B	16	14	30
Unproven	17	29	46
Totals	48	56	104

From Galbraith, A. W., in *Chemotherapy and Control of Influenza*, Oxford, J. S. and Williams, J. D., Eds., Academic Press, London, 1975. With permission.

TABLE 8

Mean Duration of Fever (hr)

Influenza A	Amantadine	Control	Difference	Significance
1972 to 1973	45.0 (21)	62.5 (18)	17.5	$0.1 > P > 0.05$
1973 to 1974	60.5 (15)	92.1 (12)	31.6	$0.1 > P > 0.05$
Combined	51.4 (36)	74.3 (30)	22.9	$0.2 > P > 0.01$
Influenza B	80.9 (15)	82.4 (14)	1.5	$P > 0.9$

From Galbraith, A. W., in *Chemotherapy and Control of Influenza*, Oxford, J. S. and Williams, J. D., Eds., Academic Press, London, 1975. With permission.

be gained by combining amantadine treatment with immunization of maximum prophylaxis.

III. THERAPEUTIC STUDIES

If amantadine acts by inhibiting viral penetration or uncoating of the cell[23] one might be justified in assuming that once clinical influenza is already apparent, the drug would prove ineffective. Alternatively, it could be argued that if the patient is treated soon after symptoms appear, further cellular involvement might be prevented. Be that as it may, evidence was presented demonstrating therapeutic activity in laboratory studies[24] during experimental infections in mice[25] (see also Chapter 1) and in volunteers[26] with experimentally induced influenza A2 infection.

In a therapeutic study in the U.S. with naturally occurring influenza A (H3N2) both amantadine and rimantadine were found to be effective in increasing clinical improvement, the rate of defervescence, and the rate of disappearance of signs and symptoms. There was a statistically significant increase in the number of rapidly improving patients in each drug group (Figure 7). The effect of the compounds on fever was also significant and is illustrated in Figure 8. By probit analysis, the median time for the temperatures of the amantadine patients to fall to less than 100°F was 23 hr as compared to 45 hr for the placebo group. The effect of the compounds on signs and symptoms was also significant, and relevant data are presented in Figures 9 and 10. By probit analysis, the median time for the amantadine and rimantadine patients to show improvement was 35 and 31 hr, respectively, as compared to 69 hr in the placebo group. The signs and symptoms (Figure 10) included dry cough, swelling of the nasal mucosa, sore throat, nasal congestion, eye pain, myalagia, malaise, fatigue, headache, chills, and feverishness. Although significantly more of the amantadine patients had all these signs at the start of medication, treatment with the drug resulted in a significant decrease over the placebo patients.

Similar results were obtained in a natural outbreak of influenza A (H2N2) in a state penitentiary.[28] Nine patients treated with rimantadine had a shorter course of influenza disease than a placebo group of 9 patients when judged by analysis of duration of fever (Figure 11) and overall clinical effects, including the duration and severity of the signs and symptoms.

A modification of the design used for investi-

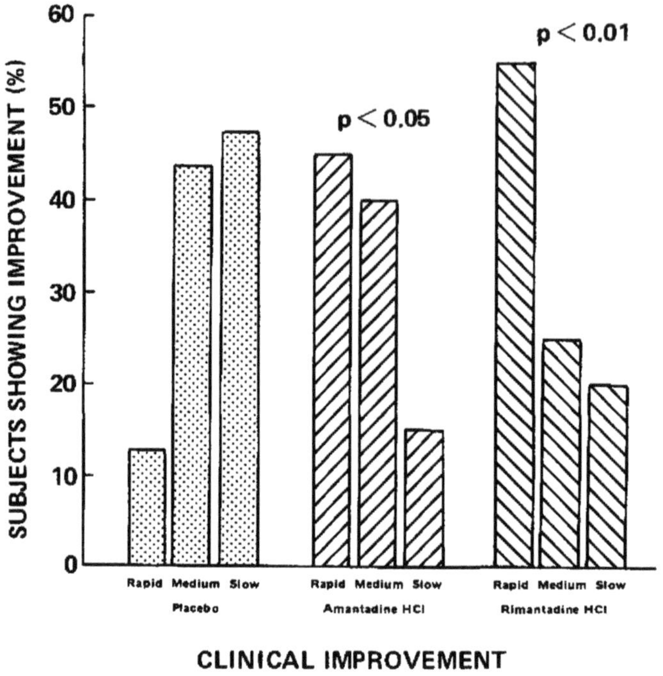

FIGURE 7. Effect of amantadine HCl and rimantadine HCl on the percentage of patients showing rapid, medium, and slow rates of clinical improvement. Probability by chi-square analysis. (From Wingfield, W. L., Pollack, D., and Grunert, R. R., *N. Engl. J. Med.*, 281(11), 579, 1969. With permission.)

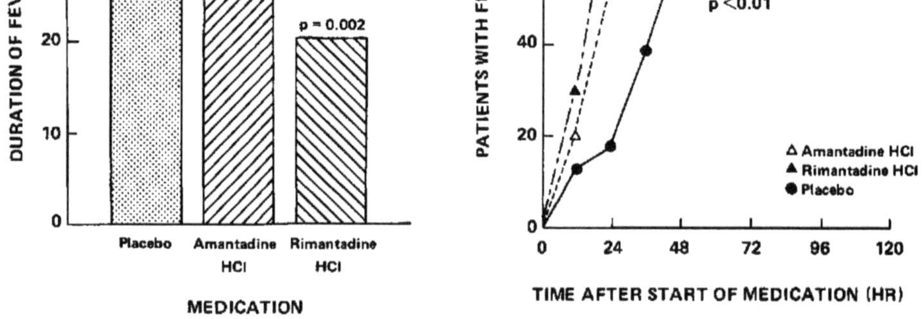

FIGURE 8. Effect of amantadine HCl and rimantadine HCl on the duration of fever (temperature over 100°F) and on the percentage of patients with fever (temperature under 100°F) after the start of medication. Probability by rank and probit analyses, respectively. (From Wingfield, W. L., Pollack, D., and Grunert, R. R., *N. Engl. J. Med.*, 281(11), 579, 1969. With permission.)

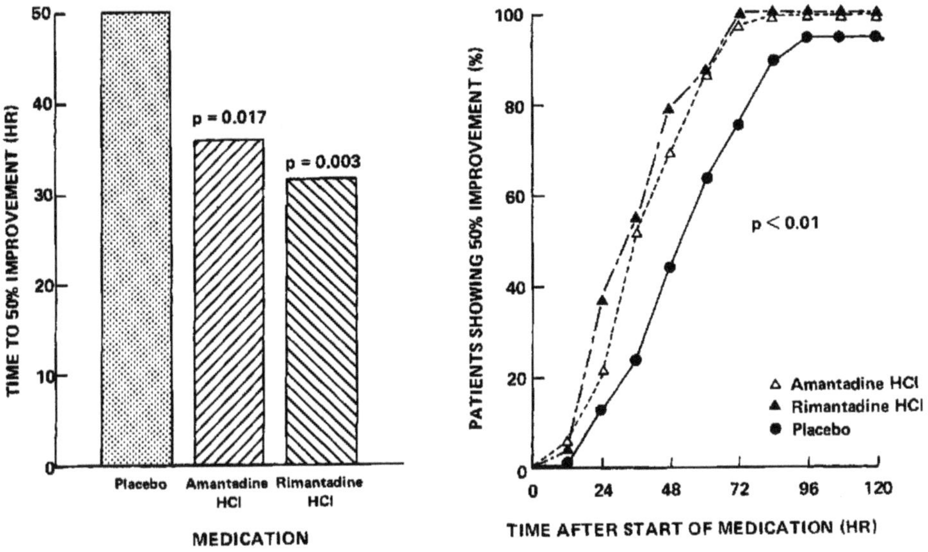

FIGURE 9. Effect of amantadine HCl and rimantadine HCl on the time required for a 50% improvement in composite signs and symptoms of illness and on the percentage of patients showing 50% improvement after the start of medication. Probability by rank and probit analyses, respectively. (From Wingfield, W. L., Pollack, D., and Grunert, R. R., *N. Engl. J. Med.*, 281(11), 579, 1969. With permission.)

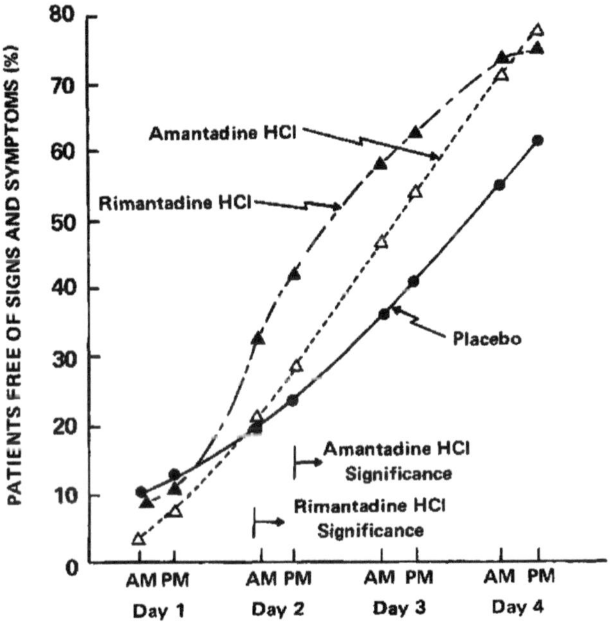

FIGURE 10. Effect of amantadine HCl and rimantadine HCl on the percentage of patients free of 11 signs and symptoms consistent with influenzal illness during the first 4 days of medication. Significance by principal-component analysis. (From Wingfield, W. L., Pollack, D., and Grunert, R. R., *N. Engl. J. Med.*, 281(11), 579, 1969. With permission.)

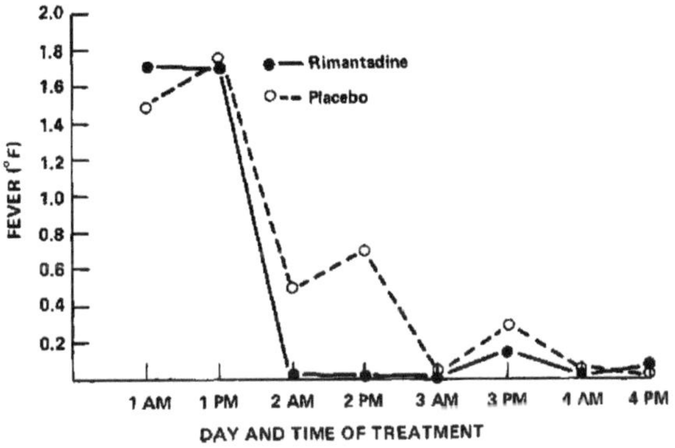

FIGURE 11. Comparison of temperatures in patients with influenza receiving rimantadine or placebo.

gating the prophylactic action of amantadine in influenza A2 seemed to lend itself easily to allow study of the drug's possible therapeutic activity in general practice. It should also be noted that the Royal College of General Practitioners is well suited to carry out therapeutic drug studies, and a list of volunteer members is held in a "therapeutic trials register."

The protocol was simpler because only the "index" case was involved. A double-blind placebo-controlled trial was envisaged involving the practitioners who had previously volunteered with such good heart together with new recruits to the group.

Each doctor was asked to include in the study any patients suffering from clinical influenza of not more than 48-hr duration. He was provided with sufficient equipment to treat five patients, but, as before, could obtain additional material if he wished to treat more cases. Also, two or more patients from one household could be included.

At the first visit, blood was taken and sent in the prepared box to the Virus Research Laboratory in Sheffield. Code-labeled drug or placebo-containing capsules were taken, the drug dose being 100 mg twice a day from the time of the doctor's visit. The treatment kit for each patient was much neater (Figure 12), a suitable standard box being available in quantity (the example shown was designed for cosmetic samples); record cards were then made to fit the box.

Temperature readings were taken, with the thermometer provided, three times a day before meals or mealtime drinks, and selected symptoms were recorded daily as present or absent. In the subsequent analysis, the duration of fever up to the time at which the temperature fell below 99°F (37.2°C) was measured in hours. A convalescent serum sample was obtained 2 to 3 weeks after onset.

A. Results

Observations were made on 203 patients with clinically diagnosed influenza. Fifty-seven doctors participated in the study,[29] 46 in England, 4 in Scotland, 6 in Wales, and 1 in Belfast. Of the 203 patients, 30 did not meet the serological criteria of infection with influenza A2 virus, namely a fourfold or greater rise in HI or CF titer in the paired serum specimens. In addition, 20 records were incomplete, leaving 153 records for analysis.

B. Fever

The duration of fever after starting treatment

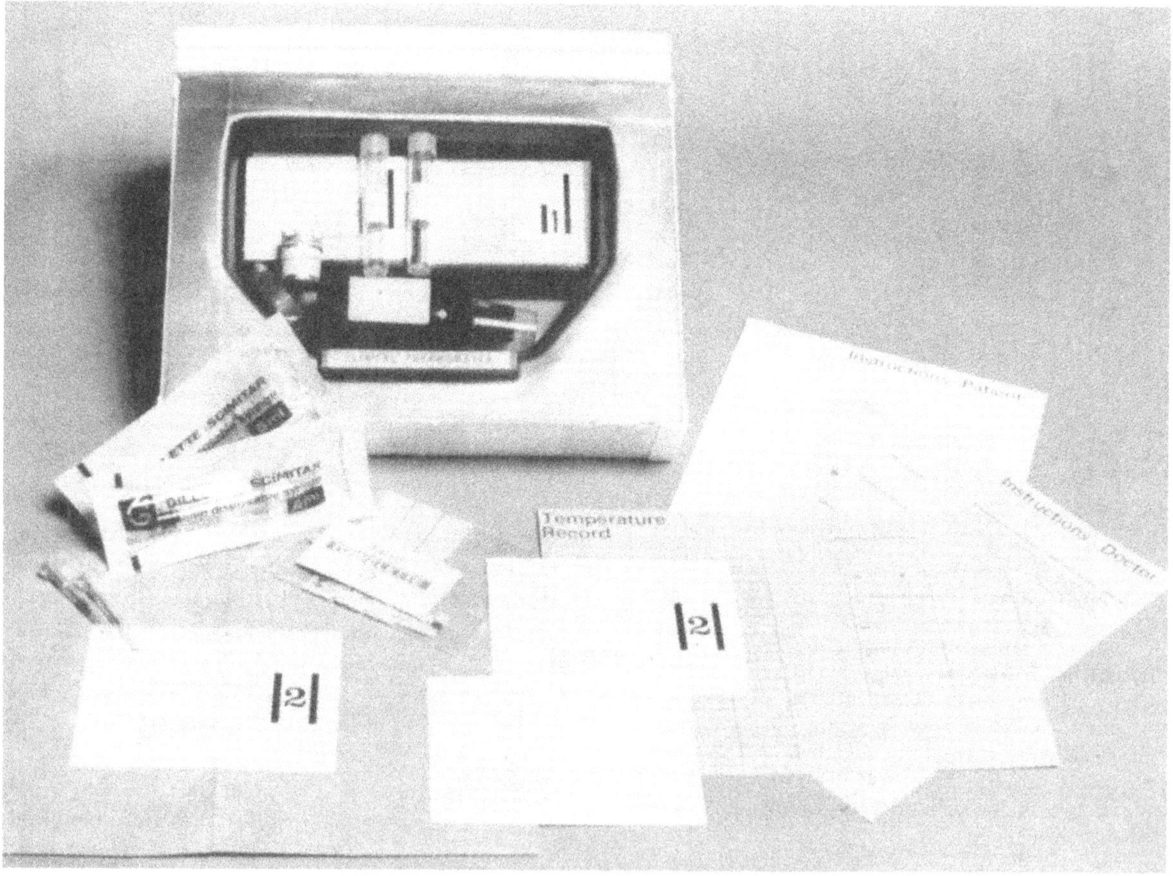

FIGURE 12. Revised kit sent to general practitioners in the amantadine therapeutic trial.

was significantly shorter in the drug-treated group than among those receiving placebo ($P < 0.01$).

C. Symptoms

There was no significant difference in the rate of symptom clearance between the control and drug-treated groups.

D. Sensitivity of the Trial Design

In a more recent study,[30] some patients suffering from influenza B were included. The doctors could not distinguish these cases clinically from those with influenza A infection, and the diagnosis was revealed when the sera were tested. In the analysis (Tables 7, 8, and 9) it was interesting to note that those patients suffering from influenza A2 had a shorter illness when treated with amantadine, while those with influenza B behaved, as would be expected from the insensitivity of the virus, like the controls.

TABLE 9

Days in Bed During Illness (Mean)

Influenza A	Amantadine	Control	Significance
1972 to 1973	2.13	2.95	$0.1 > P > 0.05$
1973 to 1974	3.27	4.15	$0.2 > P > 0.1$
Combined	2.58	3.44	$P > 0.01$
Influenza B	3.45	3.67	$0.9 > P > 0.8$

From Galbraith, A. W., in *Chemotherapy and Control of Influenza,* Oxford, J. S. and Williams, J. D., Eds., Academic Press, London, 1975. With permission.

E. Discussion

In the clinical trials described in general practice, certain features are worthy of comment.

1. Family doctors with an interest in the epidemiology of respiratory infection are particularly well placed to study the effect of prophylaxis and treatment of the disease.
2. Through the trust that the family has in their general practitioner, involvement in a study which includes each member, even though many individuals are not ill, can take place quite voluntarily.
3. Good record keeping and accurate blood sampling present no problems to the family doctor, provided that he is presented with a kit complete to the smallest detail.
4. As far as the majority of practices are concerned, virus isolation is difficult to perform. Transporting the specimens in a viable condition to the nearest laboratory capable of doing the work may have to be by hire car or individual messenger. These are costly and are often impossible to obtain in the more remote practices.
5. Paired serum specimens provide definite evidence of infection. The convalescent sample may sometimes be difficult to obtain if the first venepuncture has been painful and the patient is reluctant to subject himself to a further assault; or it may be inconvenient for the doctor and patient to meet at the appropriate time. However, with the great majority of patients these difficulties were overcome.
6. While this type of study may appeal to many of the practitioners, payment of expenses associated with the extra work involved may well be necessary to ensure a good response. The level of payment is a matter for negotiation but note should be taken of the considerable time involved, particularly in a prophylactic investigation.
7. The participation of a laboratory capable of handling the serum specimens is vital to the success of a study involving virus infection. The prior agreement and cooperation of the virologist is essential, for without consistant antibody measurement the accuracy of the investigation is lost.
8. Fever is reliably measured in practice. With the provision of standardized thermometers to patients, regular objective assessment of the response to therapy is obtained.
9. Symptoms are easily recorded, but individual response is extremely variable. When does cough, a common symptom of influenza, disappear? A compromise was reached by one team of investigators who calculated the time taken for patients to experience a 50% reduction in their "symptom score" and compared this time between the drug- and placebo-treated groups.[27] Another team of investigators considered the progress of patients classified as "slow, medium, or rapid resolvers." The responses of drug- and placebo-treated patients within these groups were compared.[31] The authors concluded that the method produced a more realistic approach to the analysis.
10. In any multicenter general practitioner trial in influenza, it is impossible to predict which patients will be included in the study prior to the epidemic. It is necessary, therefore, to provide

each doctor with approximately equal quantities of the substance under investigation and the comparative preparation, together with instructions to admit only patients defined for inclusion by the protocol. It has been our experience that by this method the active and placebo-treated groups were well matched for age, sex, and initial antibody status. A comparison of their response could therefore legitimately be made.

In conclusion, we suggest that further research in the field of antiviral chemotherapy be performed among patients directly under the care of the family doctor. Every age and social class, urban dwellers, and remote homesteaders can be represented if the appropriate practitioners are included. The general practitioner is particularly well placed to involve individuals who might otherwise not wish to become involved in a study, if they were to be approached by anyone other than their own family doctor.

With the recent emphasis on community health care, general practice should become an increasingly important field for the study of preventive and therapeutic advances.

REFERENCES

1. **Anon.**, Specific drugs against influenza, *Br. Med. J.*, 3, 428, 1969.
2. **Schofield, K. P., Potter, C. W., Edey, D., Jennings, R., and Oxford, J. S.**, in *Antiviral Activity of Ribavirin on Influenza Infection in Ferrets, in Chemotherapy and Control of Influenza,* Oxford, J. S. and Williams, J. D., Eds., Academic Press, London, 1975, 63.
3. **Lundahl, K., Schut, J., Schlatmann, J. L. M. A., Paerels, G. B., and Peters, A.**, Synthesis and antiviral activities of adamantane spiro compounds. I. Adamantane and analogous spiro-3′-pyrrolidines, *J. Med. Chem.*, 15, 129, 1972.
4. **Sjöberg, B.**, *Conference on ABOB and "Flumidin,"* Stockholm, October 28, 1960.
5. **Weinstein, L.**, Advances in the treatment of virus diseases, *Practitioner,* 185, 500, 1960.
6. General Practitioner Clinical Trials, Trial of an antiviral agent, *Practitioner,* 185, 795, 1960.
7. **Meiklejohn, G., McCann, W. P., and Shkolnik, S.**, Field trial with ABOB (SK 8898-A) as a prophylactic agent against influenza A_2 and adeno-virus infection, *Mil. Med.*, 128, 890, 1963.
8. **Watson, G. I.**, "Virugon" ("Flumidin"): No benefit in two respiratory diseases, *Br. Med. J.*, 2, 1785, 1960.
9. **Wink, C. A. S., Ed.**, *Symmetrel in Virology,* Geigy Pharmaceuticals, Macclesfield, Cheshire, 1975.
10. **Davies, W. L., Grunert, R. R., Haff, R. F., McGahen, J. W., Neumayer, E. M., Paulshock, M., Watts, J. C., Wood, T. R., Hermann, E. C., and Hoffman, C. E.**, Antiviral activity of 1-adamantanamine (amantadine), *Science,* 144, 862, 1964.
11. **Jackson, G. G., Muldoon, R. L., and Akers, L. W.**, Serological evidence for prevention of influenzal infection in volunteers by an anti-influenzal drug, adamantanamine hydrochloride, *Antimicrob. Agents Chemother.*, 703, 1963.
12. **Stanley, E. D., Muldoon, R. L., Akers, L. W., and Jackson, G. G.**, Evaluation of antiviral drugs: the effect of amantadine on influenza in volunteers, *Ann. N.Y. Acad. Sci.*, 130, 44, 1965.
13. **Togo, Y., Hornick, R. B., and Dawkins, A. T.**, Studies on induced influenza in man. I. Double blind studies designed to assess prophylactic efficacy of amantadine hydrochloride against A_2/Rockville/1/65 strain, *JAMA,* 203, 1089, 1968.
14. **Peckinpaugh, R. O., Askin, F. B., Pierce, W. E., Edwards, E. A., Johnson, D. P., and Jackson, G. G.**, Field studies with amantadine: acceptability and protection, *Ann. N.Y. Acad. Sci.*, 173, 1, 62, 1970.
15. **Nafta, I., Turcanu, A. G., Braun, I., Companetz, W., Simionescu, A., Birt, E., and Florea, V.**, Administration of amantadine for the prevention of Hong Kong influenza, *Bull. WHO,* 42, 423, 1970.
16. **Finklea, J. F., Hennessy, A. V., and Davenport, F. M.**, A field trial of amantadine prophylaxis in naturally-occurring acute respiratory illness, *Am. J. Epidemiol.*, 85(3), 403, 1967.
17. **Wendel, H. A., Snyder, M. T., and Pell, S.**, Trial of amantadine in epidemic influenza, *Clin. Pharmacol. Ther.*, 7(1), 38, 1966.
18. **Tyrrell, D. A. J., Bynoe, M. L., and Hoorn, B.**, Studies on the antiviral activity of 1-adamantanamine, *Br. J. Exp. Pathol.*, 46, 370, 1965.
19. **O'Donoghue, J. M., Ray, C. G., Terry, D. W., and Beaty, H. N.**, Prevention of nosocomial influenza infection with amantadine, *Am. J. Epidemiol.*, 97, 276, 1973.
20. **Sparks, J. P.**, Symmetrel and influenza outbreaks in a large residential school, in *Symmetrel in Virology,* Wink, C. A. S., Ed., Geigy Pharmaceuticals, Macclesfield, Cheshire, 1975, 40.

21. **Galbraith, A. W., Oxford, J. S., Schild, G. C., and Watson, G. I.,** Protective effect of 1-adamantanamine hydrochloride on influenza A_2 infections in the family environment, *Lancet,* 2, 1026, 1969.

22. **Galbraith, A. W., Oxford, J. S., Schild, G. C., and Watson, G. I.,** Study of 1-adamantanamine hydrochloride used prophylactically during the Hong Kong influenza epidemic in the family environment, *Bull. WHO,* 41, 677, 1969.

23. **Kato, N. and Eggers, H. J.,** Inhibition of uncoating of fowl plague virus by 1-adamantanamine hydrochloride, *Virology,* 37, 632, 1969.

24. **Grunert, R. R., McGahen, J. W., and Davies, W. L.,** The in vivo antiviral activity of 1-adamantanamine (amantadine). I. Prophylactic and therapeutic activity against influenza viruses, *Virology,* 26, 262, 1965.

25. **McGahen, J. W. and Hoffmann, C. E.,** Influenza infection of mice. I. Curative activity of amantadine HCl, *Proc. Soc. Exp. Biol.,* 129, 678, 1968.

26. **Bloomfield, S. S., Gaffney, T. E., and Schiff, G. M.,** A design for the evaluation of anti-viral drugs in human influenza, *Am. J. Epidemiol.,* 91, 568, 1970.

27. **Wingfield, W. L., Pollack, D., and Grunert, R. R.,** Therapeutic efficacy of amantadine HCl and rimantadine HCl in naturally occurring influenza A_2 respiratory illness in man, *N. Engl. J. Med.,* 281(11), 579, 1969.

28. **Rabinovich, S., Baldini, J. T., and Bannister, R.,** Treatment of influenza. The therapeutic effect of rimantadine HCl in a naturally occurring influenza A_2 outbreak, *Am. J. Med. Sci.,* 257, 328, 1969.

29. **Galbraith, A. W., Oxford, J. S., Schild, G. C., Potter, C. W., and Watson, G. I.,** Therapeutic effect of 1-adamantanamine hydrochloride in naturally occurring influenza A_2/Hong Kong infection, *Lancet,* 2, 7716, 113, 1971.

30. **Galbraith, A. W.,** Therapeutic trials of amantadine (Symmetrel) in general practice, in *Chemotherapy and Control of Influenza,* Oxford, J. S., and Williams, J. D., Eds., Academic Press, London, 1975.

31. **Togo, Y., Hornick, R. B., Felitti, V. J., Kaufman, M. L., Dawkins, A. T., Kilpe, V. E., and Claghorn, J. L.,** Evaluation of therapeutic efficacy of amantadine in patients with naturally occurring A_2 influenza, *JAMA,* 211(7), 1149, 1970.

Chapter 4

THE STRATEGY OF THE PICORNAVIRUS GENOME: SOME IMPLICATIONS FOR CHEMOTHERAPY

Peter D. Cooper

TABLE OF CONTENTS

I. INTRODUCTION

A great deal of information is now available on the molecular mechanisms of picornavirus replication. Unfortunately, we do not know much about the details of many of the individual steps, particularly the nature and actions of the replicases, and we understand others (such as penetration, regulation, and assembly) rather imperfectly. Nevertheless, the general outline of their overall scheme of gene function and replication ("genome strategy") can be almost completely discerned at the present time. The concept of strategy can most directly be categorized and described in terms of the total information capacity of the genome, i.e., of the genetic map.

Accordingly, this chapter outlines the strategy of the picornavirus genome in genetic terms and seeks to emphasize those aspects that should be susceptible to selective chemotherapeutic attack.

II. PROPERTIES AND GROUPING OF PICORNAVIRUSES

Picornavirus particles comprise a single strand of RNA (vRNA) of molecular weight 2.6×10^6 daltons, enclosed in an icosahedral capsid made up of 60 fundamental structural units. Each particle is 20 to 40 nm in diameter and is nonenveloped and ether resistant. Each structural unit contains several different polypeptides whose size can vary even between closely related strains,[1] but whose total molecular weight is about 100,000 daltons. Viral growth processes occur on membranes in the cytoplasm, and the translation products are not internally punctuated but are cleaved after translation by proteolytic enzymes that are probably specified by the host cell. There is genetic recombination between RNA strands, and the only gene functions identified so far are (from the 5′ end) for structural protein and for one or two replicases or replicase factors. The gene products often have more than one function. Virus particles are ultimately released in a burst by rupture of the cell membrane.

A case has been made[2] for including within a particular family only those viruses having substantially identical genome strategies. The strategies of the adequately studied members of the Family Picornaviridae seem very similar indeed. Member groups are *Enterovirus* (polio, coxsackie, echoviruses), *Cardiovirus* (EMC, Columbia, SK group), *Rhinovirus* (common cold group), and *Aphthovirus* (foot-and-mouth disease group). These are sufficiently well delineated to be regarded as separate genera, as the main criterion separating them is various combinations of properties of the capsid (Table 1). However, the Study Group on Picornaviridae[3] recommends for the time being that the aphthoviruses be included in the genus *Rhinovirus* and cardioviruses in the genus *Enterovirus*. The Caliciviruses are suggested as a genus for possible inclusion only, as they have larger particles with characteristic projections, only one major capsid protein, and no evidence of posttranslational cleavage.[84] Thus they may in fact comprise a separate family.[4]

III. THE SCHIZON

This chapter will consider the molecular biology of picornaviruses largely in terms of a catalogue of gene functions. However, before proceeding further, it is desirable to clarify a point of ambiguity regarding the term "gene." In the past two decades the gene has become equated by molecular biologists with the cistron,[5] that functional genetic unit defined by the cis-trans complementation test and specifying a particular discrete polypeptide. However, "cistron" is ambiguous for picornaviruses. Their large RNA was originally regarded as a polycistronic message, which was subsequently shown to have its translation product subdivided after translation by a system of proteolytic enzymes (see below), rather than by the initiation and termination ("punctuation") characteristic of classic cistrons. This implies that the RNA is monocistronic (although as developed below, there is now evidence that it may in fact comprise two cistrons). Thus, a new concept and a novel type of genetic element is involved, for which a new term is desirable.

For this new genetic element we have proposed[6] the term "schizon" (Greek: *schizein*, to cleave). Schizon means the element of genetic material coding for a "schizomer," which is a particular polypeptide created by posttranslational cleavage. In what follows, the terms gene, gene function, and gene product will be used in a general sense, but schizon and cistron will be distinguished where appropriate. The schizons in a translation sequence may comprise separate, adjacent, or nonadjoining nucleotide sequences, or the nucleotide sequences of distinct schizons may overlap or be entirely contained in one another. Picornavirus vRNAs present themselves as poly-

Distinguishing Features of Picornavirus Genera

Property	Enterovirus	Cardiovirus	Rhinovirus	Aphthovirus	Calicivirus (possible genus)	Probably others
Main members	Polio-, echo-, and coxsackievirus	Columbia SK group	Common cold	Foot-and-mouth disease	Vesicular exanthema (pigs), "picorna" (cats)	Equine "rhinovirus"?, others?
Serotypes	Many	One or few	>90	Several	Several	
Main hosts	Human, vertebrate	Human, rodent, vertebrate	Human, vertebrate	Rodents, ungulates	Pig, cat, sealion	Horses, insects, others?
Host range and tissue tropisms of individual strains	Narrow	Wide	Narrow	Wide	Wide (pig), narrow (cat)	
Main habitat	Gut	CNS, heart	Upper respiratory tract	Generalized	Generalized (pig), respiratory (cat)	
pH 3	Stable	Stable	Labile	Labile	Labile	
pH 5–6	Stable	Labile in 0.1 *M* halide	Stable	Becoming labile	Variable	?
Ceiling temperature	High	High	Usually low	High	High	
EM, surface diameter (nm)	Featureless 24–28	Featureless 24–30	Featureless 20–30	Featureless 23–25	Large "cups" 30–40	?
RNA						
C	20–24	24–26	19	28	25	?
A	27–30	25–27	35	25	29	
G	23–28	23–24	19	24	21	
U	23–25	24–27	27	22	25	
Virion %	29	31	30	31.5	22–24	
CsCl density	1.33–1.35	1.34	1.38–1.43	1.43	1.37–1.38	?

schizonic mono- (or possibly di-) cistronic messages.

IV. THE GENETIC MAP OF PICORNAVIRUSES

The strategy of the picornavirus genome may be summarized in simple terms by means of its total of gene functions as revealed by the genetic map. In theory, the genome itself provides the upper limit of the virus' functional capability.[7] However, subsequent experience has shown that very small virus genomes are capable of considerable subtlety in effecting an economical organization of their genetic material, and a simple linear picture of genes may be only half the story (see below).

Fortunately, picornaviruses undergo reproducible genetic recombination, which seems small (1% per 1250 bases, so that < 2% of the total progeny is recombinant) but which in fact lies between the recombination frequencies for phage T4 and *Escherichia coli* (1% per 250 and 1750 nucleotide pairs respectively); therefore, it is substantial when regarded in terms of recombination per codon.[6] The properties of the picornavirus recombination process argue strongly in favor of a precisely ordered molecular exchange or crossing over,[6,8-12] similar in end result to the DNA systems. This allows a linear, additive genetic recombination map to be constructed which seems likely to closely reflect the genome of picornaviruses. Such a map has been worked out extensively only for poliovirus, but aphthovirus is similar.[13,14]

Figure 1 shows the map for poliovirus. About 40 *ts* mutants have been mapped; only the main ones are shown, together with five loci of resistance to certain inhibitors of poliovirus replication.

V. THE PRIMARY GENE FUNCTIONS OF POLIOVIRUS

The characterization of the defects of those mutants with a single *ts* lesion, in respect to a substantial number of properties, revealed a strict correlation between the mutants' positions in the map and their physiological defect. Those mutants to the extreme left of the map were defective in their synthesis of progeny "plus" or viral RNA (vRNA) and double-stranded (ds) RNA,[15] which itself is a measure of the "minus" or complementary RNA (cRNA). Those on the right of the map all had a defect in structural protein, expressed as an altered thermostability,[16] or some defect in assembly.[17] A small group in the middle had a defect in synthesis of vRNA but not cRNA, showing that these RNA strands were made by genetically separable activities; this provided genetic evidence that they are functionally distinct enzymic sites. The two RNA synthetic activities are contained in two physically separable replication complexes.[18,19]

The structural protein and RNA synthesis functions are concerned with production of virion components and thus may be termed "primary"

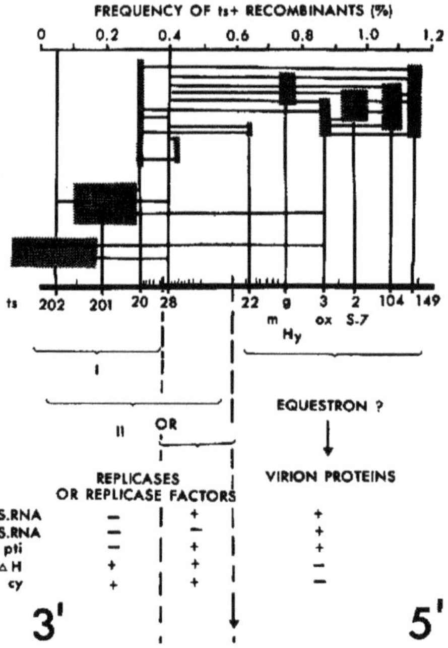

FIGURE 1. Genetic recombination map of poliovirus temperature-sensitive mutants. The horizontal bars (top half) show the mean standardized recombination frequencies between the main mutants (*ts*-202, etc.), and the hatched boxes show the maximum discrepancies in additivity. The loci for resistance to four inhibitors of poliovirus are also shown. (m) Dextran sulfate; (Hy) 5-methyl-5-3,4-dichlorophenylhydantoin; (ox) 2-(3-chloro-*p*-tolyl)-5-ethyl-1,3,4-oxadiazole; (S-7) ethyl-2-methylthio-4-methyl-5-pyrimidine carboxylate. The lower half summarizes the main tests on which the identification of map position with gene function is based, namely synthesis of double-stranded (DS) and single-stranded (SS) RNA, prevention of host DNA-synthesis (pti), thermal lability (ΔH), and cystine dependence (cy); (+) and (–) indicate wild-type or mutant response, respectively, at restrictive temperature. The 5′ end of the map is to the right. (From Cooper, P.D., in *Comprehensive Virology*, Fraenkel-Conrat, H. and Wagner, R. R., Eds., Plenum Press, New York, in press. With permission.)

gene functions. They are generally accepted as being carried out by polypeptides specified by the genome, the evidence for this resting on their absence from uninfected cells, their induction (even in cells blocked in host syntheses) on adding purified virus or its RNA, and their exhibition of specific mutations after mutagenic treatment of purified virus or its RNA.

VI. THE SCALE OF THE *ts* MAP

Surprisingly, the recombination map appears to be completely filled with these three primary gene functions, despite the fact that poliovirus shows many other degenerative and synthetic activities within infected cells. The suspicion was accordingly raised that it may represent only a small portion of the genome.

This possibility was tested in two ways. The first was a chemical estimation of the scale of the *ts* map. The size of the structural protein schizon was measured by a count of tryptic peptides from virions labeled with the rare amino acids methionine or histidine, in comparison with a redetermination of the contents of these amino acids in virions.[20] These showed, respectively, a total of 21 and 25 uniquely placed residues of histidine and methionine and 21 and 27 residues of histidine and methionine recovered per 1000 amino acids. Thus, the structural protein schizon is equivalent to about 1000 amino acids. An identical result was obtained by a semiquantitative estimate of the N-terminal amino acids in the virion[21] using ^{125}I-labeled *p*-iodophenylisothiocyanate.[22] A redetermination of poliovirus RNA[23,24] gave as the molecular weight 2.6×10^6 daltons, with a maximum coding capacity of 2500 amino acids. Thus, the structural protein schizon comprises about 40% of the genome. Subsequent work[25] showed that the structural protein primary schizomer (NCVPIA or p110) comprises about 50% of the primary polyprotein (NCVPOO or p210). As the structural protein region of the map comprises 48% of the total map,[12] it is clear that the primary functions occupy virtually all the genome.

The second test was a fresh *ts* mutant isolation series, giving emphasis to extending the boundaries of the map.[12] This almost doubled the number of mutants in the map, but extended its boundaries only by 25%, indicating that the map was almost totally saturated with *ts* mutants and that it did indeed represent the major portion of the genome.

VII. THE SECONDARY GENE FUNCTIONS OF POLIOVIRUS

Although the genetic map is fully occupied by primary gene functions, poliovirus has many other well-documented effects on cells, which are accordingly not accounted for by this map. The following paragraphs summarize recent evidence showing that most, if not all, of these are secondary effects arising from additional activities of two of the primary gene functions.

A. Early Metabolic Repression by Poliovirus

Early in the infectious cycle of most picornaviruses, the synthesis of host protein is repressed. This effect may be very rapid and can lead to the total disappearance of label entering host protein in 2 to 3 hr.[25] It appears to depend on synthesis of a viral protein that prevents entry of host mRNA into polysomes.[26,27] Previous searches for poliovirus mutants defective in this function, which might reveal the nature of the operator protein, had failed;[9] fortunately, the implication of a product of the structural protein schizon in possible regulation mechanisms of poliovirus (see below) allowed the prediction that mutants defective in repression of host protein synthesis might also be defective in structural protein. This proved to be the case, and when the leak rate (small expression of gene function under restrictive conditions) was minimized by the presence of guanidine, low multiplicities of infection, and the absence of cystine, *ts* mutants defective in repressing host protein synthesis were found.[27] These mutants were all *ts* in structural protein, and those mutants not defective in repression of host protein synthesis were not defective in structural protein. The association was confirmed by a study of revertants in both characters.

Although picornavirus double-stranded (ds) RNA strongly inhibits translation in vivo and in vitro,[28,29] dsRNA does not appear to be operative in vivo because the translation of both cellular and viral RNA is affected, dsRNA does not appear early in the cycle, and mutants defective in dsRNA synthesis were not defective in repression of host protein synthesis. There are several studies showing that repression of host protein synthesis is a gene function distinct from that causing the late CPE,[30] and there is strong evidence that the late CPE is caused by dsRNA.[31] Thus, dsRNA cannot be regarded as the actual in vivo effector of repression of host protein synthesis in picornavirus-infected cells.

The repression of host RNA synthesis by poliovirus follows shortly after that of protein.[32,33] Since puromycin stops synthesis of RNA soon after that of protein,[34] the two processes are interdependent. It is likely that repressions of host protein and RNA synthesis are the result of the same picornavirus gene function, but this has not been directly demonstrated.

B. Late Cytopathic Effects

This group of effects of picornavirus is comprised of several very characteristic cytological changes, including rounding of the cell, changes in chromatin and eventual rupture of the cell membrane. As mentioned, they appear to depend on accumulation of some viral gene product distinct from that effecting the early changes.[30]

Recently, Cordell-Stewart and Taylor[29,35,36] showed that purified picornavirus dsRNA is cytotoxic and rapidly reproduces the membrane damage typical of late picornavirus growth. The kinetics and properties are almost precisely those expected if these changes were caused by dsRNA. Garwes and colleagues[31] report the inverse correlation, namely that *ts* defects in membrane damage are restricted to poliovirus mutants specifically *ts* in making dsRNA. Thus, the CPE and dsRNA defects are strictly covariant. These results appear to provide the functional test to confirm that dsRNA is the in vivo effector of the late CPE, which accordingly appears to be a secondary effect of the replicase I schizon.

C. Repression of Host DNA Synthesis

This effect occurs after the repression of host protein synthesis but before the final rupture of the cell membrane. Its cause is currently unexplained, but since all mutants defective in repression of host DNA synthesis are defective in replicase I and vice versa (Figure 1), this effect also appears to be a secondary activity of the replicase I schizon and may possibly be related to a virus involvement with host enzymes concerned with DNA synthesis.[6]

D. Other Possible Gene Functions

A few possibly distinct gene functions of poliovirus are suggested by its effects on the infected cell, but these are now seen to be relatively nonspecific. A possible maturation factor[37,38] probably reflects a facilitation of assembly by added membranes already carrying structural protein;[39,40] much of the picornavirus growth process occurs on, and is undoubtedly facilitated by, intracellular membrane surfaces. The slight derangement of cellular lipid synthesis manifested by small increases in choline and glycerol uptake[41–44] may be due to stimulation of phospholipid synthesis by lysolipids.[45] Virulence does not appear to be associated with any specific gene function.[6] In summary, these effects cannot be ascribed to any hitherto undetected gene function, so that the catalogue of known poliovirus gene activities seems to be fully accounted for by the primary and secondary functions of the genetic map.

VIII. REGULATION MECHANISMS

A. The Need for a Regulator

The preceding sections show that the primary schizons specifying the components of picornavirus particles fully occupy the genetic map, and the functions resulting in various cellular changes and losses of activity appear to cover it almost fully a second time. However, there is another aspect of picornavirus replication that must be considered. Close examination of the growth kinetics of picornaviruses reveals a surprising number of anomalies. These can be summarized as follows.

1. The vRNA replication is at first exponential (Figures 2 and 3), but then changes to a constant rate during which the bulk of the replicative events occur.
2. At first the rate is very fast (doubling time 1 to 2 min), but later slows while still exponential (doubling time 15 min).
3. The vRNA synthesis ceases while the cell is still intact, at a constant level that is independent of the input multiplicity.
4. The vRNA and cRNA contents can be deduced to be equal initially, but later the proportion of vRNA increases to a five- to tenfold excess.
5. The bulk of the genetic recombination events occur early in the cycle,[8] which is the reverse of expectations from topographical factors.
6. The rate of ribosome attachment to vRNA can be deduced to be rather slow initially (1/11 min), a situation to be predicted from the fact that poliovirus RNA is not capped (see below). In

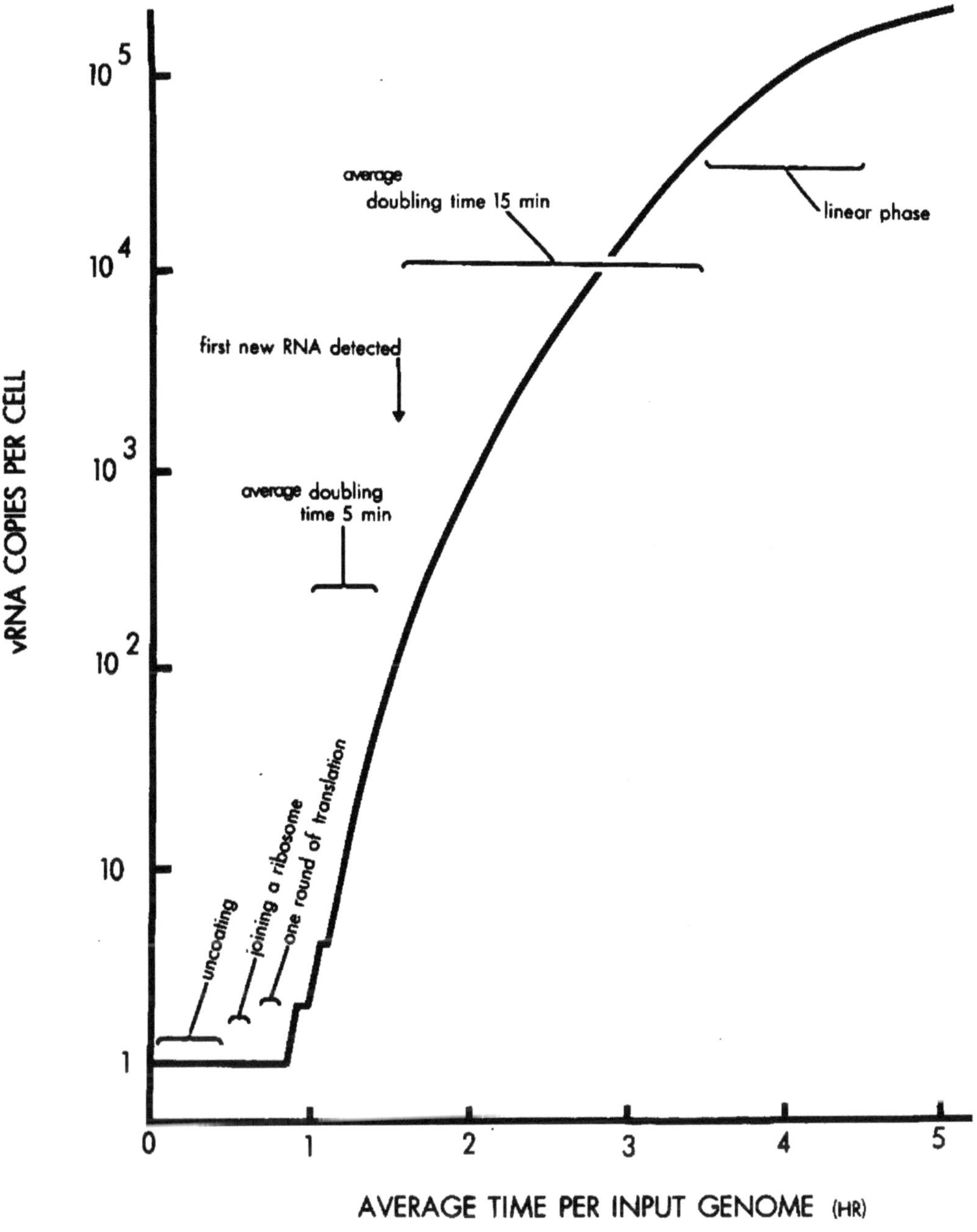

FIGURE 2. Kinetics of vRNA increase from a single input poliovirus particle. The lag period comprises the average time for uncoating (30 min) plus average time predicted for first joining a ribosome (10 min) plus average time to complete one round of translation and transcription (12 min). (From Cooper, P. D., in *Comprehensive Virology*, Fraenkel-Conrat, H. and Wagner, R. R., Eds., Plenum Press, New York, in press. With permission.)

midcycle, ribosomes are observed to attach to vRNA much more rapidly (38/11 min), a rate comparable to that of cell message despite the lack of a cap on vRNA.

7. In contrast, the rate of ribosomal interaction with host message markedly decreases during the cycle.

8. At the beginning of infection, vRNA is made as fast as or faster than viral protein, but this ratio is changed by midcycle, when vRNA and protein are made in the correct ratio (1:60) for assembly.

9. The increase in the pool of soluble antigen

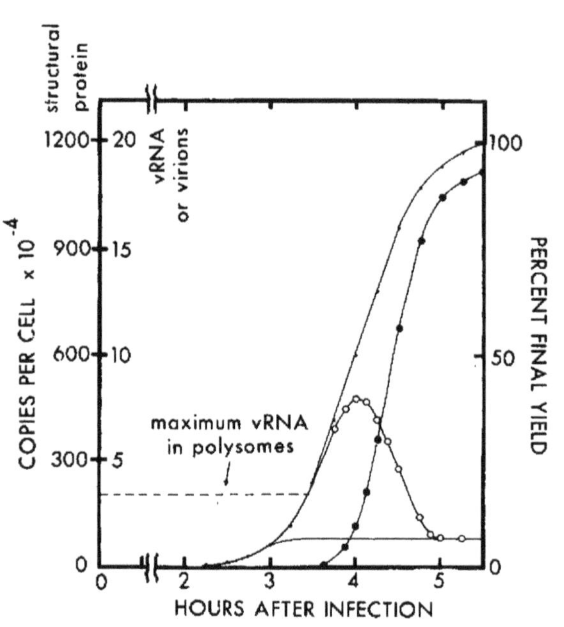

FIGURE 3. Idealized single-multiplicity one-step growth cycle of poliovirus. (——) vRNA or total structural protein; (•——•) total equestrons; (——) equestrons bound to 45S ribosomal subunits; (•——•) virions or virion structural units. The broken line indicates the vRNA capacity of ribosomes. (From Cooper, P. D., in *Comprehensive Virology*, Fraenkel-Conrat, H. and Wagner, R. R., Eds., Plenum Press, New York, in press. With permission.)

is suddenly reversed,[46] although overall viral protein synthesis is not affected.

10. The bulk of the vRNA is progressively found in the replication complex, then the translation complex, and finally in mature virions. This is the correct sequence for efficient replication. Some RNA enters dsRNA late in the cycle.

11. The growth process is well synchronized and economical.

A most important point is that many relationships *change* during the growth process, so that specific affinities alone (e.g., of replicase for vRNA) are not adequate to explain these paradoxes. It is hard to avoid the conclusion that a very effective viral regulation system is involved. All this is achieved without differential translation of genes, as the vmRNA is polyschizonic, and in almost all picornavirus systems all gene products increase in equivalent molar ratio, at the same rate at different times of the cycle.[25,47] (Some exceptions to this[48,49] are quantitatively small and are not general and therefore seem unlikely to represent a regulation mechanism common to all picornaviruses.) However, virtually the whole genome is occupied by the primary gene functions of poliovirus, so that not much, if any, genetic information is left for regulation machinery. Thus, any regulator has to be contrived extremely economically.

From these effects it was predicted[50] that picornaviruses code for a labile particle(s) with affinities for some ribosomal factor or components *and* for vRNA. To conserve genetic economy, there should be a single particle binding both to the small ribosomal subunit and to the 5′ end of vRNA, with a half-life of about 30 min. We have termed such a particle the "equestron," to denote a directive rider of both ribosomes and vRNA.

B. The Nature of the Proposed Regulator and Its Proposed Function

In the initial approach to the equestron idea,[50] various possible causes of each of the effects summarized above were examined, and the more complex and least all-embracing explanations were discarded. Rather than repeat the exercise here, we shall simply summarize the conclusion and illustrate it by a brief review of the growth process in terms of the proposed regulator. A crucial point is that any viral regulator system must be very economical genetically; therefore, it is expected to be multifunctional. It goes without saying that all of the very substantial number of anomalous but efficient facets of picornavirus replication have to be accounted for in some way.

The function of the hypothetical poliovirus regulator in relation to the growth process is outlined in Figure 4 and in the following paragraphs in conjunction with the flow sheet of Figure 5.

1. Establishment of Infection: 0 to 2 hr

The input vRNA must first join a ribosome, with an average random delay of 11 min, and an equestron is produced within a further 5 to 10 min. Soon after the first round or two of translation, the input vRNA must join with a replicase I molecule (liberated after its translation and cleavage near the 3′ end of vRNA, its site of action), making a cRNA strand in 1 to 2 min. The infection is then established. The equestrons form a reversible complex with small ribosomal subunits and/or vRNA, linking them specifically like an initiation factor and speeding up vRNA translation. More equestrons are then made in a self-accelerating process. The 45S-equestron complex

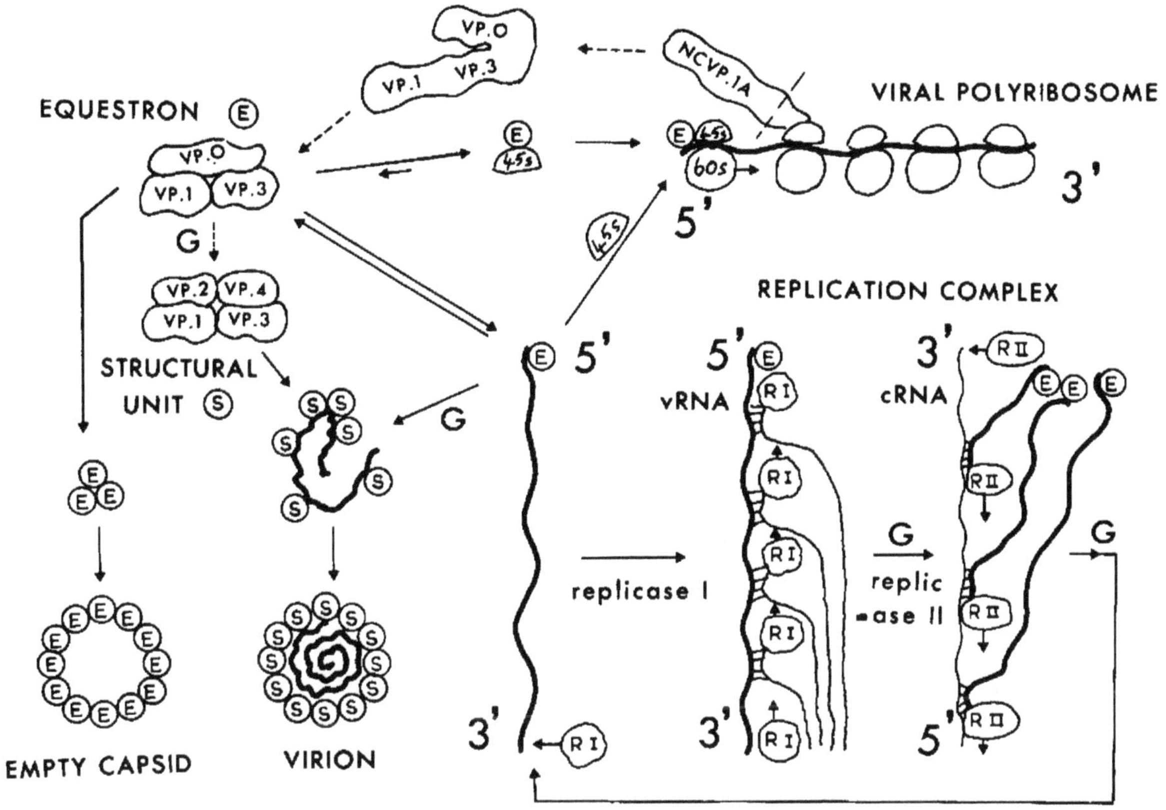

FIGURE 4. Detailed functions proposed for the hypothetical equestron regulator. The thick and thin wavy lines represent vRNA and cRNA, respectively; broken arrows represent cleavage steps; and G represents the steps blocked by guanidine. (From Cooper, P. D., Steiner-Pryor, A., and Wright, P. J., *Intervirology,* 1, 1, 1973. With permission.)

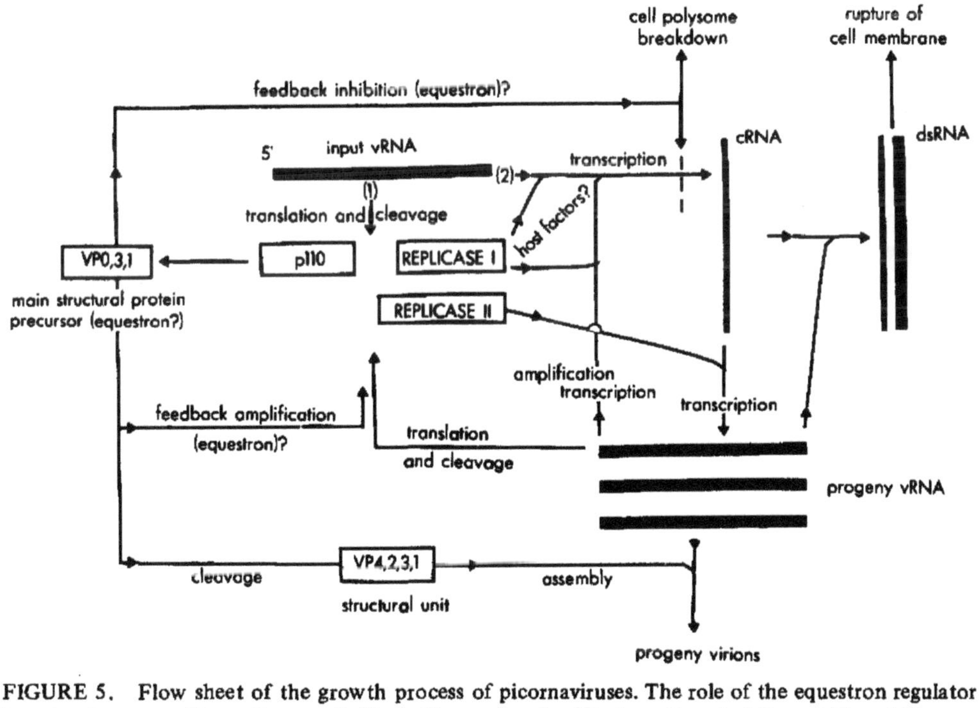

FIGURE 5. Flow sheet of the growth process of picornaviruses. The role of the equestron regulator is not yet proven. (From Cooper, P. D., in *Comprehensive Virology,* Fraenkel-Conrat, H. and Wagner, R. R., Eds., Plenum Press, New York, in press. With permission.)

coordinately blocks host mRNA association with ribosomes; which progressively become almost entirely available for viral polysome formation.

The affinity of equestrons for the small subunit must be high in order to be effective at such low initial concentrations and so the small subunits will preferentially bind the equestrons, which will therefore infrequently exist either free or on vRNA until the subunits approach saturation 2 to 3 hr after infection. The level of equestrons bound to 45S subunits shown in Figure 3 is based on the content of small subunits per cell (ca. 10^6). Progeny vRNA is transcribed from the first cRNA template by replicase II, and more cRNA from this vRNA is transcribed by replicase I. This process is exponentially self-increasing; it is probably unregulated at first and will at this time outstrip protein synthesis.

2. Polysome Development and Beginning of Regulation: 2 to 3.5 hr

The growing pool of equestron-primed 45S subunits takes more and more vRNA for polysomes, shifting the balance of the vRNA pool from replication to translation complexes. However, the amounts transferred are quantitatively insufficient to account for the reduction of the doubling time to 15 min. It is likely that the regulator comes into action at this time (see next section).

3. Restriction on Completion of cRNA Strands: 3.5 to 4.5 hr

When the number of equestrons becomes comparable to that of the small subunits, an excess becomes available to interact with vRNA. As an initiation factor, the equestron is proposed to bind specifically to the 5′ end, whose free time accordingly progressively decreases. Completion of the cRNA strands will gradually decrease in proportion (slowing up the exponential increase in vRNA) and ultimately will stop. When this happens, the cell makes steady use of those cRNA templates already completed, and the increase of vRNA (and protein) changes from exponential to linear (Figure 3).

4. Assembly Caused by Decay of Labile Intermediates: 3.5 to 6 hr

Much of the protein destined to enter virions has accumulated in the form of the structural unit precursor (VP0,1,3), since its half-life (20 to 40 min) is greater than that of the primary cleavage product NCVP1A (5 to 10 min). Its cleavage to the structural unit is presumably a first-order reaction, and the vRNA templates for its production have increased exponentially up to now: along with other viral protein, it has maintained a constant ratio to vRNA and will itself increase exponentially. However, when the vRNA increase slows to a constant rate, the increase of structural unit precursors also becomes linear; because their cleavage will remain exponential, their concentration will drop sharply about one half-life later (Figure 3), suddenly creating a large pool of mature structural units. The theoretical kinetics of Figure 3 closely resemble those for incorporation of soluble antigen (probably the VP0, VP1, VP3 complex) into virions.[46] At this time almost all the cell's ribosomes (3 to 4 × 10^6/cell) are locked up in viral polyribosomes, and most of the new vRNA is available to associate with the newly created structural units. Thus, the pool of vRNA shifts again, from translation complex to progeny virions. This corresponds with the end of the eclipse period (production of the first infectious progeny/cell).

5. Late Effects: 4 hr Onwards

The change of kinetics will also cause a drop, at about the same time and for the same reasons, in other labile components, one of which is the viral replicase. RNA production can be expected to tail off when replicase concentration, rather than available templates, becomes the rate-limiting factor (Figures 2 and 3). This sets a limit of vRNA molecules made per cell, as it is independent of input virus but depends on the change of kinetics (which is itself held to depend on the regulator) and on the half-life of replicase.

Finally, although the cRNA and vRNA "factories" are held in separate replication complexes, it must be expected that later in the cycle vRNA and cRNA molecules will now and again achieve full base pairing to dsRNA, which will not dissociate in vivo. Its final accumulation appears to be the cause of membrane rupture and release of virus,[31] signaling the abrupt end of the replication cycle.

C. The Candidate for the Equestron

Several lines of evidence, which all point in the same direction, have provided an idea of the candidate for the equestron. Prime among these is a search for polioviral proteins associated with the translation complex[51] showing that the smaller ribosomal subunit isolated from infected cells carried the structural unit precursor proteins (VP0, VP1, VP3) in equimolar ratio. These could not be ascribed to contamination with empty capsids or soluble protein. Disruption of the small subunit allowed these proteins to be isolated as a 6S particle. The larger (65S) subunit did not carry a label, but the VP0, VP1, VP3 complex was found associated with vRNA and with a complex of vRNA and a single ribosome; it was also associated with the membrane-free replication complex.[52] Accordingly, it is the candidate for the equestron.

This implication of structural protein in the regulation mechanism prompted a successful search for mutants defective in repression of host protein synthesis (one of the proposed functions of the equestron) in the light of known properties of mutants defective in structural protein. By this means, it was confirmed that the repression of host protein synthesis was dependent on the configuration of a product of the structural protein schizon.[27]

Thus, in regulating vRNA and cRNA production, the equestron is proposed to destroy itself in favor of the structural unit. Posttranslational cleavage is exploited as a regulation device. Saturation of the small subunit with equestrons, of ribosomes with vRNA, and of vRNA with equestrons leading to linear growth kinetics and the decay of equestrons to structural units are all stages with a fixed relation to each other, so that the system will be synchronized under a wide range of cultural conditions.

D. Genetic Evidence for the Nature of the Equestron

1. Guanidine

Although the poliovirus inhibitor guanidine has several effects on RNA synthesis, the genetic locus of resistance to guanidine lies in the structural protein gene.[8,9] The only change seen in viral proteins is a delayed cleavage of VP0,[53] implying that guanidine alters its conformation and therefore reacts directly with it. Guanidine sensitivity is dominant for vRNA and virus replication in mixed infection of sensitive and resistant strains plus guanidine,[54] implying that the sensitive strain makes a viral protein that slows all replicating centers equally. A single amino acid change neutralizes this effect in resistant strains, but is deduced to be in *structural* protein.[54] Guanidine is most effective 2 to 3 hr after infection;[55,56] if present only from 0 to 2 hr, it has much less effect on later growth, despite vigorous RNA replication at 0 to 2 hr in untreated cultures.[57] Thus, guanidine cannot appreciably interact with the simple replication process; it has no effect on replicase action in vitro, although it rapidly slows vRNA production in vivo, equalizing vRNA and cRNA synthesis.[15,58,59] It was concluded[50] that guanidine is likely to interact only when replication becomes *regulated.* This implies that guanidine primarily affects the regulator; therefore, from the genetic map, this implies that such a regulator involves structural protein.

2. Cystine

Many experiments have implicated a participation of cystine in the correct function of poliovirus structural protein. Apart from the effect of cystine on thermal stability of intact virions,[60] especially of some *ts* mutants,[16] some mutants are dependent on the presence of cystine at 37° without being either thermolabile or *ts*.[10,12] One particular *ts* mutant with a single defect in structural protein (*ts*-89) is thermostable and cystine dependent at 37°, yet forms only empty capsids at 39.5°.[16] The presence of cystine at 39.5° greatly lessens its *ts* defect (unpublished). Cystine dependence maps in the structural protein region of the genetic map[9] and changes in cystine dependence are often covariant with changes in guanidine sensitivity.[54] However, the *ts* defect of several cystine-dependent structural protein mutants also involves a defect in the synthesis of vRNA.[15] In summary, the absence of cystine has much in common with the presence of guanidine for poliovirus growth.

3. Interfering ts Mutants

In "interference-without-multiplication,"[61] an interference that can be even more effective than interference by multiplying strains can be induced by certain poliovirus strains under restrictive conditions in which multiplication of the interfering virus and its RNA is largely prevented. Thus, interference is not due to competition in vRNA replication. The restrictive conditions are either high growth temperature (interference induced by certain *ts* mutants with ts^+ as challenge) or the presence of guanidine (interference induced by guanidine-sensitive strains, with guanidine-resistant strains as challenge). Again, this is not a cell-surface phenomenon. A further study[54] showed that it was the *RNA replication* of the challenge that was blocked, but that the temperature-induced interference property was restricted to those mutants that had been shown to have a single *ts* defect in structural protein. However, only certain *ts* structural protein mutants had this interfering property. It was concluded that vRNA replication depended on the configuration of structural protein, and this was another instance of a genetically dominant inhibition of RNA replication by a misfolded product of the structural protein schizon. This conclusion has been interpreted to demonstrate the involvement of a product of the structural protein schizon in the regulation of poliovirus RNA replication.[50]

IX. A POSSIBLE SECOND CISTRON

The sizes of poliovirus translation and cleavage products were carefully determined by gel electrophoresis in order to attempt a definitive balance sheet between the coding potential of poliovirus RNA and its translation products.[25] It was found that its primary translation product, the polyprotein NCVP00 or p210, was not larger than 210,000 daltons. The two major primary cleavage products, NCVP1A (p110) and NCVP1B (p90), were 110,000 and 90,000 daltons, respectively, thus fully accounting in size for p210. The other ostensible primary cleavage product, NCVPX (p31), was 31,000 daltons, too large to be accounted for by error in the foregoing determinations and therefore implied *not* to be a cleavage product of p210.

The distinct nature of p31 was confirmed by tryptic peptide maps[62] showing that the amino acid sequences of p110 and p90 were fully contained in, and completely account for, p210. The sequences of p31 were quite different and thus are not contained in p210. These data accordingly show that p31 is not a cleavage product of p210 and imply, therefore, that it is specified by a distinct small cistron.

Several in vitro studies using picornavirus RNA as message and using $tRNA_f$ charged with ^{35}S-formylmethionine have led to the conclusion that there is only one tryptic peptide labeled with fmet and, therefore, only one initiation site for translation of picornavirus RNA.[63-66] However, in several cases[63-65] a small amount of a second peptide was detected. Celma and Ehrenfeld[67,68] have recently shown that there are in fact two distinct initiation sites for in vitro translation, where availability is markedly affected by Mg^{2+} ions. Only in a very narrow range of Mg^{2+} concentration are both sites used, which may account for their earlier oversight. This finding strongly suggests the idea of a second cistron, although it has not yet been demonstrated that the second initiation site is functional in vivo or that its product is p31. The possible function of the second cistron or p31 is not known, although p31 is generally found associated with the replication complex.[52,69]

X. SUMMARY OF THE GENE FUNCTIONS OF PICORNAVIRUSES

Using poliovirus as an example, the gene functions of picornaviruses as revealed by the mutant studies are summarized in the lower portion of Figure 6 in terms of their relation to the genome (top portion) and their probable relation to the various products of translation and cleavage (middle portion). The sizes are to scale,

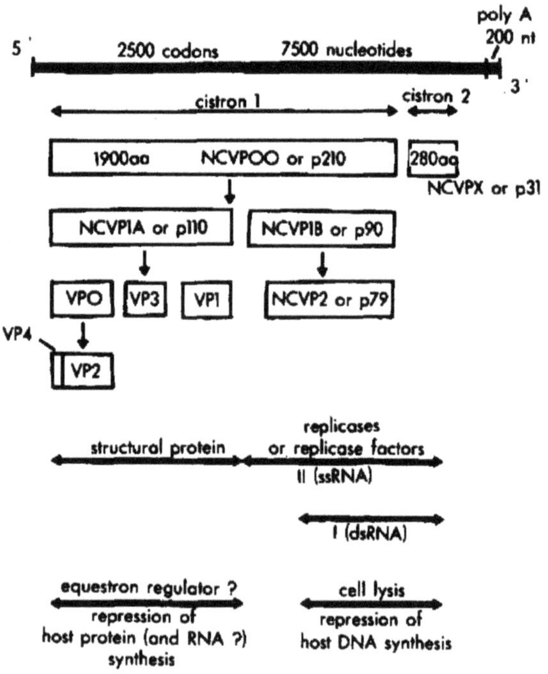

FIGURE 6. The quantitative relations of the poliovirus genome RNA (top) and its primary and secondary gene functions (bottom), with the main translation and cleavage products (center), as indicated by comparisons of tryptic peptide analysis. (From Cooper, P. D., in *Comprehensive Virology*, Fraenkel-Conrat, H. and Wagner, R. R., Eds., Plenum Press, New York, in press. With permission.)

and the genetic map is reversed here to conform to the convention of placing the 5′ end of messenger RNA to the left.

The orientation of the map in relation to the message has been determined by inhibition of initiation of translation (which occurs at the 5′ end) with pactamycin or high NaCl concentrations, followed by a pulse of amino acid label.[48, 70–74] The products more distant from the 5′ end are labeled more heavily. These studies show that the schizon for the structural protein precursor (p110) is nearest the 5′ end. Its relation to the virion proteins (VP1, VP2, VP3, and VP4 for poliovirus) has been established by tryptic and cyanogen bromide fragment analysis. From the genetic map, the replicase proteins are presumed to be those products translated from the other end, but the precise nature of the polypeptides that actually function as replicases has not yet been defined for any picornavirus. They are probably cleavage products of p90, possibly in association with some host protein (as core enzymes?) and may themselves be unstable to the cell's proteolytic cleavage system.

From its rate of labeling in the pactamycin studies, it has been assumed that p31 (NCVPX), or its analogue in other picornaviruses, is located in the middle of the map, i.e., between p110 and p90. However, the studies of Abraham and Cooper[25,62] indicate that this does not seem to be the case. The pactamycin studies make many assumptions,[6] paramount being that only one cistron is involved. For the time being, the second cistron option seems the more likely and is presented in Figure 6.

Figure 6 also shows the duplication of function of most stretches of the genome that has been indicated by the mutant studies. The structural protein schizon doubles as the schizon responsible for repression of host protein and perhaps RNA synthesis, as well as for control of vRNA synthesis, and accordingly as the candidate for the equestron regulator. All of the replicase schizon is responsible for replicase II activity, but most of it also serves as replicase I. This conclusion is derived from the observation that all mutants near the 3′ end (some of which can be demonstrated to be single mutants) are defective in both replicase I and replicase II activities. Finally, the replicase I schizon has yet a third function (deduced to be expressed via its product dsRNA) of finally rupturing the cell membrane and releasing progeny virus.

XI. POSSIBLE SITES OF ACTION OF PICORNAVIRUS INHIBITORS

Of all the steps in picornavirus replication that are conceivably open to chemotherapeutic interference, the hypothetical regulator function appears the most vulnerable. Accordingly, it seems highly significant that all of the poliovirus inhibitors whose resistance loci have been identified in the genetic map appear to act primarily on a product of the structural protein schizon (Figure 1). These include guanidine and cystine (as mentioned in Section VIII) and the synthetic compounds S-7 (ethyl-2-methylthio-4-methyl-5-pyrimidine carboxylate),[12] 2-(3-chloro-p-tolyl)-5-ethyl-1,3,4-oxadiazole,[11] and 5-methyl-5-3,4-dichlorophenylhydantoin.[85] The thiopyrimidine S-7 has recently been shown to interact with extracellular virus particles and thus to prevent their entry into the cell,[75] although the possibility of additional action sites inside the cell has not been eliminated. All the other drugs act on intracellular growth stages.

The equestron in its hypothetical role as a multifunctional regulator (and the candidate for the equestron in its role as precursor to the structural unit) must have several reactive sites on its surface. In addition, there are many findings to suggest that it is very susceptible to conformational changes. For example, the structural protein precursor particle [VP0, VP1, VP3] can self-assemble to form empty capsids, and in the unassembled form it binds strongly to vRNA. Despite these facts, vRNA is never enclosed in a [VP0, VP1, VP3] capsid, but the simple cleavage of VP0 to VP2 + VP4 is sufficient to allow the capsid to include vRNA in its assembly pattern and make the mature virion. Other findings include the large effect of a small change of temperature (from 37 to 39.5°) to allow a product of the structural protein schizon to inhibit vRNA synthesis[54] and the effect of a single amino acid substitution to change guanidine sensitivity to guanidine resistance. Thus, one would expect the equestron to be uniquely susceptible to perturbation by agents that may interact with it, change its conformation slightly, and thus upset one of its multiple activities. Its role as regulator is probably essential to efficient replication, but it is doubtful whether any of the secondary gene functions (e.g., the repression of host protein, RNA and DNA synthesis) are essential as such. Thus, they are probably side effects that do not represent targets for chemotherapy.

The precise action of guanidine to prevent virus growth is not known, but a change in conformation (caused by binding a few molecules of guanidine) that increases the affinity of the equestron for the 5′ end of vRNA, thus blocking the *completion* of cRNA strands, could account for all of the salient data on the mode of action of guanidine. An interesting chemical feature common to many synthetic inhibitors is that they tend to have flat, rather lipophilic structures without much ionization, suggesting some analogy with nucleotide bases. From their structure, their most likely interaction is to intercalate in hydrophobic peptide folds and thus disturb conformational patterns.

Nevertheless, the equestron is not the only site that in theory offers itself as a target for chemotherapy. The replicases represent an equivalent amount of genetic information, and their interactive sites must be equally complex; yet, very little is known either of the specific peptides involved or of their precise function or binding patterns. It can be inferred from the genetic map (Figures 1 and 6) that only a small amount of genetic information makes the difference between replicase I and replicase II activity. However, it is not known whether it does so by altering the cleavage pattern (thus making two quite distinct polypeptides or clusters of polypeptides), by acting as a separate cofactor, or by simply constituting a particular region of a single large molecule.

At all events, specific recognition sites on the replicase molecule are required for the 3′ end of vRNA, which consists of poly A sequences that are essential for infectivity.[76] Interestingly, these are not added posttranscriptionally, but are present in the input vRNA strand and are copied from poly U sequences at the 5′ end of cRNA.[77] Thus, replicase I must be able to distinguish (and copy) these poly A sequences from those at the 3′ end of cellular messenger RNA, which are not copied by the replicase. The 3′ end of cRNA does not have poly A and therefore requires a separate recognition site on the replicase II moiety. These considerations raise intriguing possibilities for the design of specific interference with such recognition functions. It can be seen from Figure 6 that, even allowing for the putative second cistron, more than 10% of the genome is not translated and presumably plays some part in other activities. The poly A sequence accounts for less than 2% of the genome, and the one or two ribosome binding sites probably do not account for more than another 1.5%.

The concept of the small second cistron is so new that its possible significance has not yet been considered experimentally. As mentioned, its candidate product p31 is usually found associated with the replication complex, but its function is not known. The marked effect of Mg^{2+} concentration on the utilization of the second site[67] suggests that RNA conformational factors are important, at least in vitro, and implies a possible mode of regulation of translation in vivo that comprises another theoretically sensitive locus for the action of inhibitory drugs.

Another vulnerable point is the posttranslational cleavage system itself. The cleavage enzymes are probably mostly cellular. Since the synthesis of some cellular proteins (e.g., insulin and chymotrypsin) depends on these enzymes, it may not be possible to interfere with them without affecting the host, and it is still possible that one cleavage enzyme is virus coded. However, there are several reports indicating that the cleavage process is very sensitive to conformation of the viral proteins, which may provide a drug-sensitive target in yet another mode. The inhibition by Zn^{2+} of cleavage of rhinovirus polypeptides[78] appears to provide an example of such drug action.

One final but important point deserves mention. The messenger RNAs of animal cells and of most viruses that grow in them are "capped." That is, the 5′ end is occupied by a 7-methylguanosine residue that is coupled posttranscriptionally with reversed polarity so that the actual 5′ terminus is a 3′ hydroxyl group. This appears to be essential for efficient initial binding of mRNA to the smaller ribosomal subunit in eukaryotic systems.[79] Such a requirement for a reversed polarity cap might well be expected to account for the rarity (or absence) of true polycistronic messengers in animal cells, since a nonterminal nucleotide of reversed polarity can be expected to raise transcriptional problems that are probably insuperable. It is a familiar concept that animal and plant RNA viruses appear to take an extraordinary variety of expedients to bypass the need for a polycistronic message, one of which is the use of a polyschizonic messenger RNA by picornaviruses.

However, poliovirus messenger RNA is not

capped[80,81] and may even contain an internal initiation site, which presumably cannot have reverse polarity. The RNA is also infective, which means that it is indeed capable of being translated in an unprimed ribosomal system. As mentioned above, the reason for the lack of a cap may well be that translation by the first invading picornavirus vRNA must be initiated *slowly*[50] (about 30 times slower than host mRNA initiation), so that the rapidly following and crucial next step of transcription (Figure 5) is not blocked by premature addition of ribosomes. Yet, translation later in the cycle needs to be (and is) initiated much faster,[50] and this is probably the heart of the requirement for the proposed initiation factor activity of the hypothetical equestron regulator. This reasoning should apply to any infective viral RNA; that is, special expedients are required to slow up the initial addition of ribosomes in order to allow the first round of transcription which results in the first complementary strand. It is interesting that the 49S (virion) RNA of Sindbis virus is infectious, and both it and the 26S messenger RNAs of Sindbis virus are capped.[82] Yet while the 49S component predominates in infected cells,[83] the proportion of Sindbis viral RNAs in polysomes is vastly in favor of the 26S RNA (the messenger for structural protein). Even early in the cycle the messenger activity of the 49S RNA is undetectable in comparison to that of the 26S, so that (as required by the above reasoning) for some reason it is a much less efficient messenger, despite its possession of a cap.

Whatever the significance of the uncapped 5′ terminus of picornavirus RNA, this unusual device means that its initiation of translation is likely to depend on ribosome binding sites or sequences in the RNA distinct from those of host mRNA. Thus, it represents a possible area for specific chemotherapeutic attack.

XII. CONCLUSIONS

As the complexities of the picornavirus replication process become clearer, the number of potential sites for selective interference with viral, but not cellular, metabolic machinery appears to increase. In fact, there is a large number of compounds that will block picornavirus growth without detectably interfering with the host cell, although none is ready for clinical application. Apart from problems of long-term toxicity, virus variants that are resistant to many of the drugs arise rather readily. Probably for similar reasons, many of the drugs are not "broad spectrum," that is, they do not inhibit all members of the Family Picornaviridae. The stringencies of clinical development are such that these are serious handicaps. However, the future practical outcome of this field of research at present seems rather encouraging.

REFERENCES

1. **Cooper, P. D., Summers, D. F., and Maizel, J. V.,** Evidence for ambiguity in the posttranslational cleavage of poliovirus proteins, *Virology,* 41, 408, 1970.
2. **Cooper, P. D.,** Towards a more profound basis for the classification of viruses, *Intervirology,* 4, 317, 1974.
3. **Melnick, J. L., Agol, V. I., Bachrach, H. R., Brown, F., Cooper, P. D., Fiers, W., Gard, S., Gear, J. H. S., Ghendon, Y., Kasza, L., LaPlaca, M., Mandel, B., McGregor, S., Mohanty, S. B., Plummer, G., Rueckert, R. R., Schaffer, F. L., Tagaya, I., Tyrrell, D. A. J., Voroshilova, M., and Wenner, H. A.,** Picornaviridae, *Intervirology,* 4, 303, 1974.
4. **Burroughs, J. N. and Brown, F.,** Physicochemical evidence for the reclassification of the caliciviruses, *J. Gen. Virol.,* 22, 281, 1974.
5. **Benzer, S.,** The elementary units of heredity, in *The Chemical Basis of Heredity,* McElroy, W. D. and Glass, B., Eds., Johns Hopkins Press, Baltimore, 1957, 70.
6. **Cooper, P. D.,** The genetics of picornaviruses, in *Comprehensive Virology,* Fraenkel-Conrat, H. and Wagner, R. R., Eds., Plenum Press, New York, in press.
7. **Cooper, P. D.,** Genetic analysis of animal viruses, *Br. Med. Bull.,* 23, 155, 1967.
8. **Cooper, P. D.,** A genetic map of poliovirus temperature sensitive mutants, *Virology,* 35, 584, 1968.
9. **Cooper, P. D.,** The genetic analysis of poliovirus, in *Biochemistry of Viruses,* Levy, H. B., Ed., Marcel Dekker, New York, 1969, 177.

10. **Cooper, P. D., Geissler, E., Scotti, P. D., and Tannock, G. A.,** Further characterisation of the genetic map of poliovirus temperature-sensitive mutants, in *Strategy of the Viral Genome,* Wolstenholme, G. E. W. and O'Connor, M., Eds., Churchill Livingstone, London, 1971, 75.
11. **Cooper, P. D., Steiner-Pryor, A., Scotti, P. D., and DeLong, D.,** On the nature of poliovirus genetic recombinants, *J. Gen. Virol.,* 23, 41, 1974.
12. **Cooper, P. D., Geissler, E., and Tannock, G. A.,** Attempts to extend the genetic map of poliovirus temperature-sensitive mutants, *J. Gen. Virol.,* 29, 109, 1975.
13. **MacKenzie, J. S., Slade, W. R., Lake, J., Priston, R. A. J., Bisby, J., Laing, S., and Newman, J.,** Temperature-sensitive mutants of foot-and-mouth disease virus: the isolation of mutants and observations on their properties and genetic recombination, *J. Gen. Virol.,* 27, 61, 1975.
14. **Lake, J. R., Priston, R. A. J., and Slade, W. R.,** A genetic recombination map of foot-and-mouth disease virus, *J. Gen. Virol.,* 27, 355, 1975.
15. **Cooper, P. D., Stancek, D., and Summers, D. F.,** Synthesis of double-stranded RNA by poliovirus temperature-sensitive mutants, *Virology,* 40, 971, 1970.
16. **McCahon, D. and Cooper, P. D.,** Identification of poliovirus temperature-sensitive mutants having defects in virus structural protein, *J. Gen. Virol.,* 6, 51, 1970.
17. **Wentworth, B. B., McCahon, D., and Cooper, P. D.,** Production of infectious RNA and serum-blocking antigen by poliovirus temperature-sensitive mutants, *J. Gen. Virol.,* 2, 297, 1968.
18. **Arlinghaus, R. B. and Polatnick, J.,** The isolation of two enzyme-RNA complexes involved in the synthesis of foot-and-mouth disease virus RNA, *Proc. Natl. Acad. Sci. U.S.A.,* 62, 822, 1969.
19. **Caliguiri, L. A.,** Analysis of RNA associated with the poliovirus RNA replication complexes, *Virology,* 58, 526, 1974.
20. **Cooper, P. D. and Bennett, D. J.,** Genetic and structural implications of tryptic peptide analysis of poliovirus structural protein, *J. Gen. Virol.,* 20, 151, 1973.
21. **Burrell, C. J. and Cooper, P. D.,** N-terminal aspartate, glycine and serine in poliovirus capsid protein, *J. Gen. Virol.,* 21, 443, 1973.
22. **Burrell, C. J., Cooper, P. D., and Swan, J. M.,** The use of p-iodophenyl [^{125}I] isothiocyanate for determination of N-terminal amino acids in nanomole quantities of protein, *Aust. J. Chem.,* 28, 2289, 1975.
23. **Granboulan, N. and Girard, M.,** Molecular weight of poliovirus ribonucleic acid, *J. Virol.,* 4, 475, 1969.
24. **Tannock, G. A., Gibbs, A. J., and Cooper, P. D.,** A re-examination of the molecular weight of poliovirus RNA, *Biochem. Biophys. Res. Commun.,* 38, 298, 1970.
25. **Abraham, G. and Cooper, P. D.,** Poliovirus polypeptides examined in more detail, *J. Gen. Virol.,* 29, 199, 1975.
26. **Leibowitz, R. and Penman, S.,** Regulation of protein synthesis in HeLa cells. III. Inhibition during poliovirus infection, *J. Virol.,* 8, 661, 1971.
27. **Steiner-Pryor, A. and Cooper, P. D.,** Temperature-sensitive poliovirus mutants defective in repression of host protein synthesis are also defective in structural protein, *J. Gen. Virol.,* 21, 215, 1973.
28. **Ehrenfeld, E. and Hunt, T.,** Double-stranded poliovirus RNA inhibits initiation of protein synthesis by reticulocyte lysate, *Proc. Natl. Acad. Sci. U.S.A.,* 68, 1075, 1971.
29. **Cordell-Stewart, B. and Taylor, M. W.,** Effect of viral double-stranded RNA on protein synthesis in intact cells, *J. Virol.,* 11, 232, 1973.
30. **Bablanian, R.,** Mechanisms of virus cytopathic effect, in *Microbial Pathogenicity in Man and Animals,* Pearce, J. H., Ed., 22nd Symp. Society for General Microbiology, Cambridge University Press, New York, 1972, 359.
31. **Garwes, D. J., Wright, P. J., and Cooper, P. D.,** Poliovirus temperature-sensitive mutants defective in cytopathic effects are also defective in synthesis of double-stranded RNA, *J. Gen. Virol.,* 27, 45, 1975.
32. **Fenwick, M. L.,** The influence of poliovirus infection on RNA synthesis in mammalian cells, *Virology,* 19, 241, 1963.
33. **Zimmerman, E. F., Heeter, M., and Darnell, J. E.,** RNA synthesis in poliovirus-infected cells, *Virology,* 19, 400, 1963.
34. **Holland, J. J.,** Depression of host-controlled RNA synthesis in human cells during poliovirus infection, *Proc. Natl. Acad. Sci. U.S.A.,* 49, 23, 1963.
35. **Cordell-Stewart, B. and Taylor, M. W.,** Effect of double-stranded viral RNA on mammalian cells in culture, *Proc. Natl. Acad. Sci. U.S.A.,* 68, 1326, 1971.
36. **Cordell-Stewart, B. and Taylor, M. W.,** Effect of viral double-stranded RNA on mammalian cells in culture: cytotoxicity under conditions preventing viral replication and protein synthesis, *J. Virol.,* 12, 360, 1973.
37. **Phillips, B. A., Summers, D. F., and Maizel, J. V.,** In vitro assembly of poliovirus-related particles, *Virology,* 35, 216, 1968.
38. **Phillips, B. A.,** In vitro assembly of polioviruses. I. Kinetics of the assembly of empty capsids and the role of extracts from infected cells, *Virology,* 39, 811, 1969.
39. **Phillips, B. A.,** In vitro assembly of polioviruses. II. Evidence for the self-assembly of 14S particles into empty capsids, *Virology,* 44, 307, 1971.
40. **Perlin, M. and Phillips, B. A.,** In vitro assembly of polioviruses. III. Assembly of 14S particles into empty capsids by poliovirus-infected HeLa cell membranes, *Virology,* 53, 107, 1973.

41. **Penman, S.**, Stimulation of the incorporation of choline in poliovirus-infected cells, *Virology*, 25, 148, 1965.
42. **Dales, S., Eggers, H. J., Tamm, I., and Palade, G. E.**, Electron microscopic study of the formation of poliovirus, *Virology*, 26, 379, 1965.
43. **Mosser, A. G., Caliguiri, L. A., Scheid, A. S., and Tamm, I.**, Chemical and enzymatic characteristics of cytoplasmic membranes in poliovirus-infected HeLa cells, *Virology*, 47, 30, 1972.
44. **Mosser, A. G., Caliguiri, L. A., and Tamm, I.**, Incorporation of lipid precursors into cytoplasmic membranes of poliovirus-infected HeLa cells, *Virology*, 47, 39, 1972.
45. **Amako, K. and Dales, S.**, Cytopathology of mengovirus. II. Proliferation of membranous cisternae, *Virology*, 32, 201, 1967.
46. **Scharff, M. D., Maizel, J. V., and Levintow, L.**, Physical and immunological properties of a soluble precursor of the poliovirus capsid, *Proc. Natl. Acad. Sci. U.S.A.*, 51, 329, 1964.
47. **Summers, D. F., Maizel, J. V., and Darnell, J. E.**, Evidence for virus-specific non-capsid proteins in poliovirus-infected HeLa cells, *Proc. Natl. Acad. Sci. U.S.A.*, 54, 505, 1965.
48. **Paucha, E., Seehaffer, J., and Colter, J. S.**, Synthesis of viral-specific polypeptides in mengovirus infected L-cells: evidence for asymmetric translation of the genome, *Virology*, 61, 315, 1974.
49. **Lucas-Lenard, J.**, Cleavage of mengovirus polyprotein in vivo, *J. Virol.*, 14, 261, 1975.
50. **Cooper, P. D., Steiner-Pryor, A., and Wright, P. J.**, A proposed regulator for poliovirus: the equestron, *Intervirology*, 1, 1, 1973.
51. **Wright, P. J. and Cooper, P. D.**, Poliovirus proteins associated with ribosomal structures in infected cells, *Virology*, 59, 1, 1974.
52. **Wright, P. J. and Cooper, P. D.**, Poliovirus proteins associated with the replication complex in infected cells, *J. Gen. Virol.*, 30, 63, 1976.
53. **Jacobson, M. F. and Baltimore, D.**, Morphogenesis of poliovirus. I. Association of the viral RNA with protein, *J. Mol. Biol.*, 33, 369, 1968.
54. **Cooper, P. D., Wentworth, B. B., and McCahon, D.**, Guanidine inhibition of poliovirus: a dependance of viral RNA synthesis on the configuration of structural protein, *Virology*, 40, 486, 1970.
55. **Lwoff, A. and Lwoff, M.**, L'action de la guanidine sur le développement du poliovirus, *C.R. Acad. Sci.*, 256, 5001, 1963.
56. **Koch, A. S., Eremenko, T., Benedetto, A., and Volpe, P.**, A guanidine-sensitive step of the poliovirus RNA replication cycle, *Intervirology*, 4, 221, 1974.
57. **Baltimore, D., Girard, M., and Darnell, J. E.**, Aspects of the synthesis of poliovirus RNA and the formation of virus particles, *Virology*, 29, 179, 1966.
58. **Caliguiri, L. A. and Tamm, I.**, Action of guanidine on the replication of poliovirus RNA, *Virology*, 35, 408, 1968.
59. **Baltimore, D.**, Inhibition poliovirus replication by guanidine, in *Medical and Applied Virology*, Sanders, M. and Lennette, E. H., Eds., Warren H. Green, St. Louis, 1968, 340.
60. **Pohjanpelto, P.**, Stabilisation of poliovirus by cystine, *Virology*, 6, 472, 1958.
61. **Pohjanpelto, P. and Cooper, P. D.**, Interference between polioviruses induced by strains that cannot multiply, *Virology*, 25, 350, 1965.
62. **Abraham, G. and Cooper, P. D.**, Relations between poliovirus polypeptides as shown by tryptic peptide analysis, *J. Gen. Virol.*, 29, 215, 1975.
63. **Smith, A. E.**, The initiation of protein synthesis directed by the RNA from encephalomyocarditis virus, *Eur. J. Biochem.*, 33, 301, 1973.
64. **Oberg, B. F. and Shatkin, A. J.**, Initiation of picornavirus protein synthesis in ascites cell extracts, *Proc. Natl. Acad. Sci. U.S.A.*, 69, 3589, 1972.
65. **Oberg, B. F. and Shatkin, A. J.**, Translation of mengovirus RNA in Ehrlich ascites cell extracts, *Biochem. Biophys. Res. Commun.*, 51, 1186, 1974.
66. **Villa-Komaroff, L., Guttman, N., Baltimore, D., and Lodish, H. F.**, Complete translation of poliovirus RNA in a eukaryotic cell-free system, *Proc. Natl. Acad. Sci. U.S.A.*, 72, 4157, 1975.
67. **Celma, M. L. and Ehrenfeld, E.**, Translation of poliovirus RNA *in vitro:* detection of two different initiation sites, *J. Mol. Biol.*, 98, 761, 1975.
68. **Celma, M. L. and Ehrenfeld, E.**, Detection of two initiation sites utilised during translation of poliovirus RNA *in vitro, Collog. Inst. Natl. Santé Med.*, 47, 245, 1975.
69. **Caliguiri, L. A. and Mosser, A. G.**, Proteins associated with the poliovirus RNA replication complex, *Virology*, 46, 375, 1971.
70. **Summers, D. F. and Maizel, J. V.**, Determination of the gene sequence of poliovirus with pactamycin, *Proc. Natl. Acad. Sci. U.S.A.*, 68, 2852, 1971.
71. **Taber, R., Rekosh, D., and Baltimore, D.**, Effect of pactamycin on synthesis of poliovirus proteins: a method for genetic mapping, *J. Virol.*, 8, 395, 1971.
72. **Butterworth, B. E. and Rueckert, R. R.**, Gene order of encephalomyocarditis virus as determined by studies with pactamycin, *J. Virol.*, 9, 823, 1972.
73. **Butterworth, B. E.**, A comparison of the virus-specific polypeptides of encephalomyocarditis virus, human rhinovirus and poliovirus, *Virology*, 56, 439, 1973.

74. **Saborio, J. L., Pong, S. S., and Koch, G.,** Selective and reversible inhibition of initiation of protein synthesis in mammalian cells, *J. Mol. Biol.*, 85, 195, 1974.
75. **Lonberg-Holm, K., Gosser, L. B., and Kauer, J. C.,** Early alterations of poliovirus in infected cells and its specific inhibition, *J. Gen. Virol.*, 27, 329, 1975.
76. **Spector, D. H. and Baltimore, D.,** Requirement of 3′ terminal poly (adenylic acid) for infectivity of poliovirus RNA, *Proc. Natl. Acad. Sci. U.S.A.*, 71, 2983, 1974.
77. **Yogo, Y., Teng, M., and Wimmer, E.,** Poly(U) in poliovirus minus RNA is 5′-terminal, *Biochem. Biophys. Res. Commun.*, 61, 1101, 1974.
78. **Butterworth, B. E. and Korant, B. D.,** Characterisation of the large picornaviral polypeptides produced in the presence of zinc ion, *J. Virol.*, 14, 282, 1974.
79. **Muthukrishnan, S., Both, G. W., Furuichi, Y., and Shatkin, A. J.,** 5′-Terminal 7-methylguanosine in eukaryotic mRNA is required for translation, *Nature*, 255, 33, 1975.
80. **Nomoto, A., Fon Lee, Y., and Wimmer, E.,** The 5′ end of poliovirus mRNA is not capped with $m^7G(5')ppp(5')Np$, *Proc. Natl. Acad. Sci. U.S.A.*, 73, 375, 1976.
81. **Hewlett, M. J., Rose, J. K., and Baltimore, D.,** 5′ terminal structure of poliovirus polyribosomal RNA is pUp, *Proc. Natl. Acad. Sci. U.S.A.*, 73, 327, 1976.
82. **Dubin, D. T. and Stollar, V.,** Methylation of Sindbis virus "26S" messenger RNA, *Biochem. Biophys. Res. Commun.*, 66, 1373, 1975.
83. **Simms, D. T. and Strauss, J. H.,** Replication of Sindbis virus. V. Polysomes and mRNA in infected cells, *J. Virol.*, 14, 552, 1974.
84. **Black, J. and Brown, F.,** personal communication.
85. **DeLong, D., Steiner-Pryor, A., and Cooper, P. D.,** unpublished results.

Chapter 5

THE CORONAVIRUSES

Albert Z. Kapikian

TABLE OF CONTENTS

I. INTRODUCTION

Coronaviruses are the most recently described group of viral agents to be associated with human respiratory tract illnesses. They are unique among agents causing human respiratory tract illness since members of this group are capable of causing a rather broad spectrum of illnesses ranging from respiratory disease in humans, chickens, and rats to diarrheal disease in pigs, calves, turkeys, dogs, and possibly man as well. In this review emphasis will be placed on the human coronaviruses. At present, no method of control exists, neither immunoprophylaxis nor chemoprophylaxis. It is hoped that more attention will be paid to this virus group in the future and that representative strains will be included in screening tests for antiviral agents.

II. HISTORY

In 1965 English workers reported the cultivation of a novel virus derived from a nasal swab and washing obtained from a boy with a common cold.[1] This agent, designated B814, could not be propagated in cell cultures, but induced a common cold-like illness in volunteers. It was successfully propagated in human embryonic tracheal organ cultures. In 1966 workers in the United States reported the recovery in cell cultures of fastidious respiratory tract agents which were isolated from five medical students, four of whom had a mild common cold-like illness.[2] These viruses were extremely difficult to propagate, having been recovered initially in a second blind passage in human embryonic kidney (HEK) cell cultures in which they induced a cytopathic effect (CPE). These agents were related serologically, and the prototype strain was designated 229E. Electron microscopic examination of organ culture (OC) harvests of the B814 and 229E viruses revealed that these agents were indistinguishable from the avian infectious bronchitis virus (IBV) by negative staining.[3] In 1967 the recovery in human embryonic tracheal OC of six IBV-like agents from adults with common cold-like illnesses was reported.[4] These agents [designated 501 (OC16), 663 (OC37), 664 (OC38), 690 (OC43), 691 (OC44), and 703 (OC48)][5] could not be propagated in cell cultures. In related studies mouse hepatitis virus (MHV) was found to be indistinguishable morphologically from the IBV and IBV-like agents.[6] It was then proposed that these agents (the avian IBV, MHV, and human IBV-like viruses), which shared morphological as well as other properties, be classified into a new group of viruses designated "coronaviruses."[7] This name was suggested since the surface projections on the envelope of these viruses were reminiscent of the solar corona. Other human coronavirus strains isolated include 18 229E-like or related viruses, strain EVS, and strain 692.[8-13] An additional 16 coronavirus strains (most of which are similar to 229E) have been recovered from adults with common cold-like illnesses. Details of this study have not yet been published.[14,15] Thus, since 1965 when the first of the human coronaviruses was reported, only 48 strains have been described. This is a result of the extreme difficulty in propagating these agents from clinical specimens, as serologic studies (discussed later) have revealed that coronavirus infections occur much more frequently than indicated by virus isolation.[14,16-20] With the exception of the 229E-like coronaviruses, all other human coronaviruses were first recovered in organ cultures.

III. DESCRIPTION OF THE AGENTS

Morphology – By negative staining, the coronaviruses are seen as enveloped particles 80 to 200 nm in diameter, appearing round or elliptical and moderately pleomorphic.[3,19] They possess relatively widely spaced club- or pear-shaped surface projections which are spaced uniformly on the circumference of the particle (Figures 1 and 2). The projections are about 20 nm in length, narrow at the base, and about 10 nm at the outer edge. The shape and distribution of these unique projections readily distinguishes the coronaviruses from the myxoviruses. In a negatively stained preparation in which no other marker is used, it is essential that these characteristic surface projections be visualized in order to tentatively designate an agent as a coronavirus.

Morphogenesis – Coronaviruses appear to develop in the cytoplasm by budding from membranes of the endoplasmic reticulum and cytoplasmic vesicles, thus incorporating cellular membrane material into the outer coat[6,19,21] (Figures 3 to 6). Particles have not been observed to bud from the plasma membrane. During the intracellular development of virus, the virions

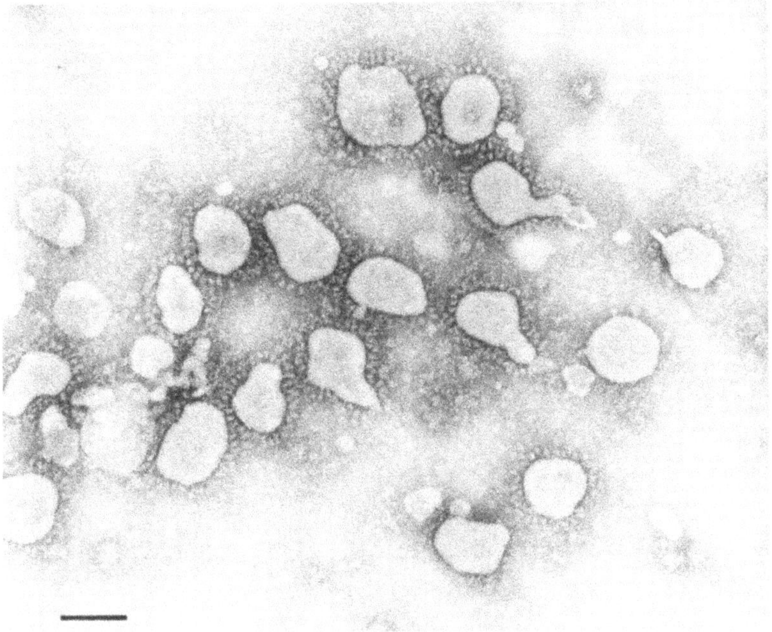

FIGURE 1. Coronavirus particles observed by electron microscopy in harvests of human embryonic intestine [HEI(MA-177)] cell cultures inoculated with strain 865 (a 229E-like strain). PTA, 2%. Details in Reference 10. (From Kapikian, A. Z., in *Diagnostic Procedures for Viral and Rickettsial Infections*, 4th ed., Lennette, E. H. and Schmidt, N. J., Eds., American Public Health Association, New York, 1969, 931. With permission.)

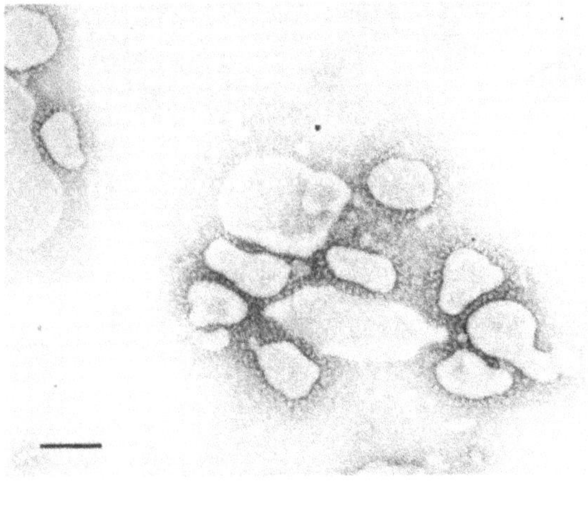

FIGURE 2. Coronavirus particles observed by electron microscopy in harvests of human embryonic intestine (HEI(MA-177)] cell cultures inoculated with strain 879 (a 229E-like strain); 2% PTA. Details are given in Reference 10.

appear to be uniform, with a round shape and a very limited size variation.

Nucleic acid composition – Coronaviruses possess RNA.[19] Recent reports indicate that IBV and a human coronavirus are single stranded.[22-24] In addition, the RNA of IBV was found to be comprised of one piece and have a sedimentation coefficient of 50S and a molecular weight of 9.0 × 10^6 daltons.[23]

Proteins – OC43 virus was found, by polyacrylamide gel electrophoresis, to have a minimum of six polypeptides with molecular weights ranging from 15,000 to 191,000.[25] A seventh band was occasionally found also. Four polypeptides were found to contain carbohydrate, and one polypeptide contained lipid. In a more recent communication, a human coronavirus has been reported as having 6 to 8 polypeptides with molecular weights ranging from 13,000 to 210,000;[24] IBV has been described as having 16 polypeptides with molecular weights ranging from 14,000 to 180,000.[24] Transmissible gastroenteritis virus, a porcine coronavirus, contains four major polypeptides and possibly two minor ones with molecular weights ranging from 28,500 to 200,000.[26]

Sensitivity to lipid solvents – Coronaviruses are ether and chloroform labile.[19]

Sensitivity to acid – Human coronaviruses are

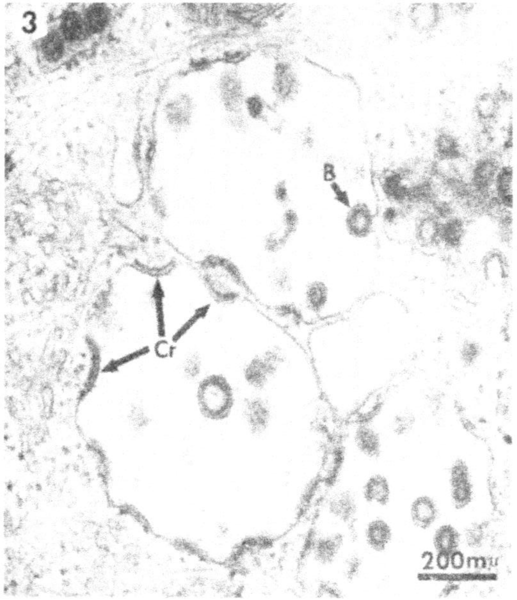

FIGURE 3. Strain 229E in WI-38 cells; this figure shows several cytoplasmic vesicles in which the process of budding is illustrated. The double vesicular membrane bulges inward to form a crescent (Cr); at the same time, a dense layer forms underneath, separated from the vesicular membrane by an electron-lucent space. A particle in a later stage of budding (B) shows the completion of the inner shell. (From Becker, W. B., McIntosh, K., Dees, J. H., and Chanock, R. M., *J. Virol.*, 1, 1019, 1967. With permission.)

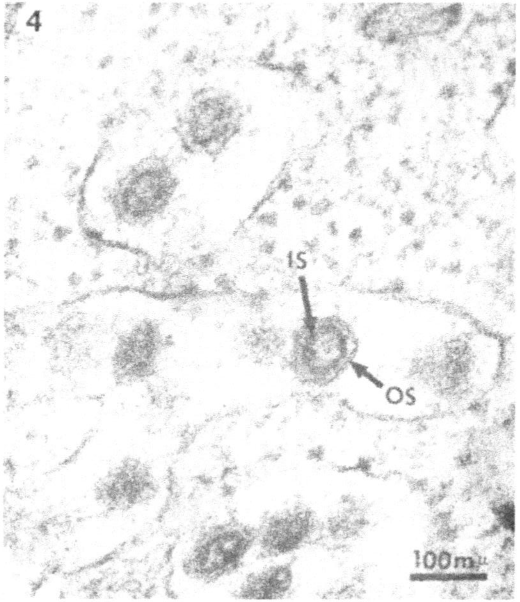

FIGURE 4. Strain 229E in WI-38 cells. A completed particle shows the double outer shell (OS), the inner shell (IS), and the electron-lucent zone between them. (From Becker, W. B., McIntosh, K., Dees, J. H., and Chanock, R. M., *J. Virol.*, 1, 1019, 1967. With permission.)

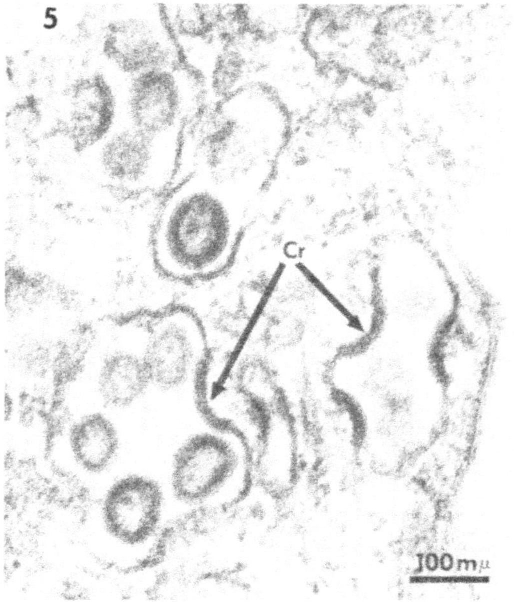

FIGURE 5. IBV in the chorioallantoic membrane. Crescents (CR) are formed by the bulging of cisternal membranes and the close apposition of a dense underlying layer. (From Becker, W. B., McIntosh, K., Dees, J. H., and Chanock, R. M., *J. Virol.*, 1, 1019, 1967. With permission.)

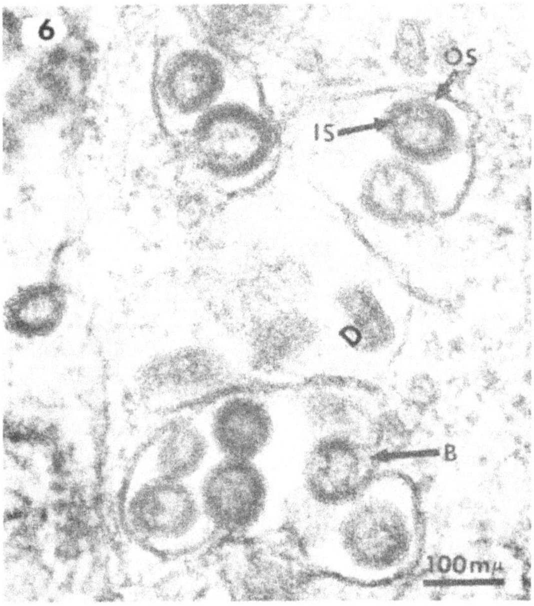

FIGURE 6. IBV in the chorioallantoic membrane. Continuity of the vesicular membrane with the outer shell of a budding particle (B) is visible. A completed particle shows the double outer shell (OS) and the dense inner shell (IS), surrounding a core of amorphous material. (Magnification: × 100,000.) (From Becker, W. B., McIntosh, K., Dees, J. H., and Chanock, R. M., *J. Virol.*, 1, 1019, 1967. With permission.)

acid labile (pH 3.0). However, among the animal coronaviruses this appears to be a variable characteristic, as some are acid resistant or relatively acid resistant and some are acid labile.[19]

Buoyant density – Coronaviruses have a density of 1.16 to 1.23 g/cm^3 in sucrose and 1.18/cm^3 in potassium tartrate; IBV has a density of 1:23 to 1.24 g/cm^3 in cesium chloride.[19,24]

Sedimentation coefficient – The sedimentation coefficient of OC43 virus is reported as 390 ± 16S[25] and 374 to 416S;[24] the sedimentation coefficient of 229E virus is reported as 378 to 400S;[24] the sedimentation coefficient of IBV is reported as 330S.[24]

Molecular weight – The molecular weight of the OC43 virion was calculated to be 112 ± 5 × 10^6 daltons;[25] the molecular weight of IBV was calculated as 9 × 10^6.[23]

Effect of Inhibitors – Coronaviruses are described as sensitive to 6-azauracil and virazole and insensitive to 5′-iododeoxyuridine, 5′-bromodeoxyuridine, 5′-fluorodeoxyuridine, cytosine arabinoside, aminopterin, and actinomycin D.[19,24] Obviously, much work needs to be done in this area.

Classification – A coronavirus study group has proposed that the coronaviruses are sufficiently distinct from other viruses to constitute a family, Coronaviridae, with only one genus, Coronavirus.[24] Agents in this family share a similar morphologic appearance, possess RNA, are ether labile, replicate in cytoplasmic vesicles by budding from membranes of the endoplasmic reticulum, and have a similar buoyant density.

IV. ANTIGENIC RELATIONSHIPS AMONG THE CORONAVIRUSES

The fastidious nature of these organisms has made studies of the antigenic relationships among the human coronaviruses very difficult, especially since certain strains have never been adapted to any cell culture or animal model system. A comparison of certain coronaviruses cultivable in cell cultures or animal models was made by two groups of investigators. Studies in the United States by complement fixation (CF) with OC38, OC43, MHV (A-59), 229E, and IBV-42 revealed that

1. The OC38 and OC43 viruses were identical.
2. The OC38-OC43 viruses fixed complement with polyvalent MHV antisera.
3. 229E virus, MHV-A59, and IBV-42 failed to react with heterologous antisera[5,27] (Table 1).

Neutralization tests confirmed the identity of strains OC38 and OC43 and, in addition, showed that there were no significant cross relationships among MHV, 229E, IBV-42, and OC38 viruses.[5] Studies in England demonstrated many of the above findings, but revealed a greater number of cross reactions.[29] In addition, they demonstrated that B814 virus was not related to IBV, MHV, 229E, OC43, and LP viruses and that LP virus was similar but not identical to 229E virus. The similarity of the OC38 and OC43 viruses has also been demonstrated[30,31] by HI and hemadsorption-neutralization studies. In addition, strain 692 has been shown to be distinct from OC43 and 229E by immune electron microscopy.[13]

A summary of described human coronavirus strains and their antigenic relationships is given in Table 2. For studies of antiviral screening,

TABLE 1

Coronavirus Antigenic Relationships Studied by Complement Fixation (CF) with Hyperimmune Animal Sera

			Reciprocal of antibody titer in animal serum prepared against designated virus[a]			
Virus strain	Source of CF antigen	Units used in test	OC38 (mouse)	OC43 (mouse)	MHV polyvalent (mouse)[b]	IBV-42 (hamster ascites)[c]
OC38	Suckling mouse brain	4–8	160	320	40	<10
OC43	Suckling mouse brain	4–8	160	320	80	<10
MHV (A-59)	Mouse liver tissue culture[d]	4	<10	<10	≥160	<10
229E	WI-38 tissue culture	8[e]	<10	<10	<10	<10
IBV-42	Allantoic fluid	4	<10	<10	<10	160

[a]Control CF antigens did not fix complement with these sera at a 1:10 dilution.
[b]Serum prepared by Dr. J. Parker against MHV strains A-59, JHM, MHV-1, and MHV-S.
[c]See text (p. 5).
[d]Tissue culture line NCTC-1469.[28]
[e]Measured with a convalescent human serum.

From McIntosh K., Kapikian, A. Z., Hardison, K. A., Hartley, J. W., and Chanock, R. M., *J. Immunol.*, 102, 1109, 1969. With permission.

TABLE 2

Antigenic Relationships Among Described Human Coronavirus Strains

Human coronavirus strain	Major antigenic relationships to other human coronaviruses	Reference
B814	Distinct	1, 29, 32
229E	Distinct	2, 5, 27, 29
241B, 243D, 276D, 299G	Similar to 229E	
OC38 (664)	Distinct	4, 5, 27, 29–31,
OC43 (690)	Identical to OC38[a]	33, 34
OC44 (691)	Related to OC38 and OC43	
OC16 (501), OC37 (663), OC48 (703)	Not definite	
489, 511, 515, 840, 844, 862, 865, 868, 879	Similar to 229E	10
LP	Related to 229E	9, 29, 35
EVS	Not determined	
Linder	Similar to 229E	11
220, 227, 297, 0661, 0744, 0765, 0312	Similar to 229E	12
692	Not related to 229E or OC43	13

[a]OC38 and OC43 demonstrate one-way serologic relationship with certain MHV strains.[5,27,29,34]

After Kapikian.[14]

selection of a suitable serotype may be important. It is apparent that there are at least three distinct human coronaviruses: B814, 229E, and OC38 or OC43. Although these agents appear to be antigenically distinct, it should be noted that heterologous CF or HI responses occur quite frequently. For example, 14 of 70 volunteers who developed illness following challenge with coronaviruses other than OC38 and OC43 developed a seroresponse by HI to the OC43 virus; 6 of 13 challenged with OC38 or OC43 viruses developed the homologous response.[36] In additional volunteer studies, 10 of 63 volunteers who developed illness following challenge with coronaviruses other than 229E and a related strain LP developed seroresponses to the 229E antigen; also, 18 of 83 volunteers who developed illness following challenge with various human coronaviruses developed CF antibody responses with MHV_3 antigen.[36] Studies of naturally occurring respiratory illnesses in children and adults have also revealed the presence of dual responses with certain coronaviruses.[34] The largest percentage of dual responses was found with the OC38 and OC43 viruses in children; sharing of responses was also observed between MHV and the OC38 and/or OC43 viruses, and only a few or no dual responses were observed with 229E and MHV, and 229E and the OC38 and/or OC43 viruses. Thus, although there are at present three known distinct coronavirus serotypes, it is clear that antibody responses observed in seroepidemiologic studies to a specific human coronavirus antigen may represent infection with an identical agent, a closely related agent, or a more distantly related one.

V. DISEASES ASSOCIATED WITH CORONAVIRUSES IN HUMANS

A. Respiratory Tract Illnesses

The classic epidemiological method of studying shedding patterns of a virus in patients and controls (cross-sectional study) or in patients during periods of illness and well-being (horizontal or longitudinal study) has not been possible with the coronaviruses since relatively few of these agents have been isolated.[37,38] As shown in Table 3, 45 and 50% of volunteers who were inoculated intranasally with B814 or 229E viruses, respectively, developed common cold-like illnesses.[39] These illnesses had an incubation period of 2 to 5 days (mean just over 3 days) and a duration of 2 to 18 days (mean 6 or 7 days). Characteristic signs and symptoms are shown in Table 3. A comparison of the clinical manifestations observed in volunteers challenged with either coronaviruses or rhinoviruses is also shown in this table. The mean incubation period of colds due to coronaviruses was significantly longer (about 1 day), the duration of illness somewhat shorter, and the mean maximum number of paper tissues used per day was greater than seen in illnesses induced by the two rhinoviruses shown. Each of the six OC strains (OC16, OC37, OC38, OC43, OC44, and OC48) was also administered by the nasal route, and 35 of 49 volunteers developed common cold-like illnesses (Table 4).[36] Thus, the disease-producing ability of the human coronaviruses has been firmly established from these volunteer studies. The illnesses observed in the volunteers were similar to those in adults undergoing natural coronavirus infections.[10] In serologic studies coronaviruses have been associated with a portion of upper respiratory tract illness of infants and young children; however, these agents do not appear to be important causes of pediatric lower respiratory tract illness,[10,34] although recent reports indicate that they may not only be responsible for a small percentage of such illnesses but also for exacerbations of wheezing among asthmatic children.[40–42] Clinical features are discussed further in Chapter 1, Volume I.

B. Gastrointestinal Illnesses

The etiologic relationship of transmissible gastroenteritis virus (TGE), a porcine coronavirus with diarrheal disease, in newborn pigs has been well established;[43] in addition, coronaviruses have been implicated in diarrheal diseases of calves, turkeys, and dogs.[44–48] Although stools from animals had been examined by electron microscopy for the presence of virus particles, such studies had not been carried out with stools from humans until recently.[49–53] Thus, with the descriptions of the Norwalk agent of epidemic gastroenteritis and the reovirus-like agent of infantile gastroenteritis,[52,54,55] there has been renewed interest in human diarrheal disease. The former was visualized in human stool preparations and associated with an elementary school outbreak of gastroenteritis by immune electron microscopy; and the latter was visualized in duodenal biopsies and stool preparations from infants and children with gastroenteritis[52,54–61] (Figure 7). In

TABLE 3

Clinical Features of Colds Produced by Inoculating Four Viruses

	Avian-infectious bronchitis-like		Rhinoviruses	
	229E	B814[a]	Type 2 (HGP or PK)	DC[a]
No. of volunteers inoculated	26	75	213	251
No. getting colds	13 (50%)	34 (45%)	78 (37%)	77 (31%)
Incubation period (days)				
Mean	3.3	3.2	2.1	2.1
Range	2–4	2–5	1–5	1–4
Duration (days)				
mean	7	6	9	10
Range	3–18	2–17	3–19	2–26
Maximum no. of handkerchiefs used daily				
Mean	23	21	14	18
Range	8–105	8–120	3–38	3–60
Malaise (%)	46	47	28	25
Headache (%)	85	53	56	56
Chill (%)	31	18	28	15
Pyrexia (%)	23	21	14	18
Mucopurulent nasal discharge (%)	0	62	83	80
Sore throat (%)	54	79	87	73
Cough (%)	31	44	68	56
No. of volunteers with colds of indicated severity				
Mild	10 (77%)	24 (71%)	63 (80%)	36 (47%)
Moderate	2 (15%)	7 (20%)	12 (15%)	28 (36%)
Severe	1 (8%)	3 (9%)	4 (5%)	13 (17%)

[a]Virus was administered only as nasal washings.

The statistical probability of some of the features of the colds due to 229E virus in relation to colds due to other cold viruses is as follows. (1) t-test on incubation periods – 229E: B814, $0.3 > P > 0.25$; 229E: Rhino type 2, $0.005 > P > 0.001$; 229E: DC, $0.05 > P > 0.025$; DC: Rhino type 2, $0.498 > P > 0.495$. (2) t-test on duration of colds – 229E: B814, $0.15 > P > 0.10$; 229E: Rhino type 2, $0.10 > P > 0.05$; 229E: DC, $0.20 > P > 0.15$; B814: Rhino type 2, $0.10 > P > 0.05$; B814: DC, $0.01 > P > 0.005$; DC: Rhino type 2, $0.95 > P > 0.90$.

From Bradburne, A. F., Bynoe, M. L., and Tyrrell, D. A. J., *Br. Med. J.*, 3, 767, 1967. With permission.

addition, in other studies a particle similar morphologically to the Norwalk agent but antigenically distinct was visualized in stool preparations of patients with hepatitis A and associated with this disease by immune electron microscopy also[62] (Figure 7). Thus, it is not surprising (in view of the role of animal coronaviruses in gastrointestinal disease and the now frequent examination of appropriate human stool preparations by electron microscopy by research laboratories) that a recent report has suggested a possible role of a coronavirus in human gastroenteritis.[63]

Coronavirus-like particles were observed in stool preparations derived from three outbreaks of adult nonbacterial gastroenteritis. The coronavirus-like particles observed in a stool preparation from one of the outbreaks appears to have been propagated in human embryonic intestinal organ culture and human embryonic kidney cell cultures. In addition, in prevalence studies, neutralizing antibody to the neonatal calf diarrhea coronavirus (NCDC) has been detected in human sera, suggesting the existence of an enteric human coronavirus.[64] However, an alternate explanation

TABLE 4

Clinical Response of Volunteers Inoculated with Various Coronavirus "OC" Strains

Virus inoculum	No. of volunteers inoculated	No. of colds produced	Classification of cold		
			Mild	Moderate	Severe
690 (OC43)	6	3	2	1	0
664 (OC38)	7	7	1	5	1
691 (OC44)	9	6	5	0	1
703 (OC48)	7	7	2	4	1
501 (OC16)	8	2	1	1	0
663 (OC37)	12	10	7	2	1
Total	49	35 (71%)	18	13	4

After Bradburne and Somerset.[36]

for the presence of such antibody has been proposed[65] serologic studies have shown that OC43 virus and NCDC, although distinct, appear to have some antigenic relationship; of 37 children with upper respiratory illnesses who developed a serologic response to OC43 virus, 20 also developed a response to NCDC. In addition, four of five adults with respiratory illness associated with OC43 virus also developed a seroresponse to the NCDC.[65] Thus, the presence of NCDC antibody in humans may represent heterologous previous infection with a human coronavirus such as OC43 rather than infection with an enteric coronavirus. However, it is obvious that further studies are needed since the primary response could have also been due to an as yet unknown human enteric strain.[65] The possible role of coronaviruses in human enteric disease represents a fertile area for additional investigation.

VI. SEROEPIDEMIOLOGIC STUDIES IN HUMANS

Seroepidemiologic studies of coronaviruses are limited by the fact that suitable serologic tests for detecting coronavirus infection exist for only two of the human coronaviruses, 229E and OC43. For 229E virus the available tests include neutralization, complement fixation, and indirect hemagglutination inhibition, whereas for OC43 virus they include neutralization or neutralization hemadsorption, complement fixation, and hemagglutination inhibition. No practical serologic tests are available for use in seroepidemiologic studies for any of the other human coronaviruses described.

Table 5 summarizes seroepidemiological studies with 229E virus and OC43 or OC43-related agents (OC38 and/or MHV) in adults, infants, and young children. It is apparent from these data that coronavirus infections have been associated with an appreciable proportion of upper respiratory tract illness occurring over varying periods of time. They are second only to the rhinoviruses as etiologic agents of common cold-like illnesses of adults.[78-80] Their relative importance in upper respiratory illnesses in infants and young children has not been determined. However, these agents have not been implicated as important etiologic agents of lower respiratory tract illnesses in adult or pediatric populations. These aspects are discussed further in Chapter 1, Volume I. One facet of the epidemiology of the coronaviruses not reflected in the table is the temporal distribution of coronavirus infections. In many studies infection with these agents reaches a peak during the late fall, winter, and early spring periods—a time when rhinovirus infections occur notably infrequently.[10,14,20,34,68] Thus, the coronaviruses appear to be the major known agents to be associated with adult common cold-like illnesses during the late fall, winter, and early spring months. Indeed, prior to these seroepidemiologic observations with the coronaviruses, the winter months were fallow periods for the detection of infection with noninfluenzal respiratory viruses. In addition, analyses of several of these studies have revealed a cyclic behavior in the infection rates of

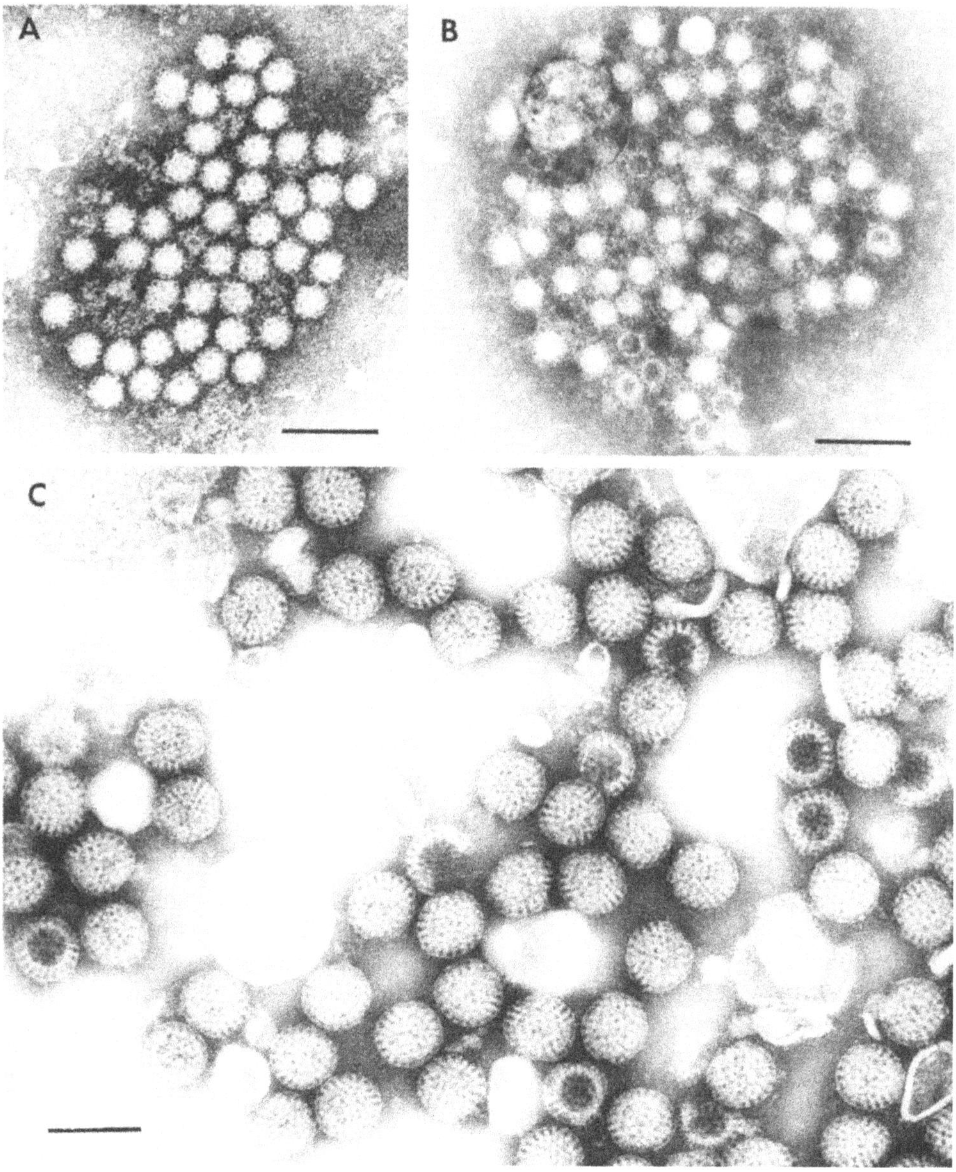

FIGURE 7. Particles observed by immune or conventional electron microscopy in filtrates made from a stool of an adult with gastroenteritis (A), an adult with hepatitis A (B), and a child with gastroenteritis (C). (A) The 27-nm Norwalk agent. (From Kapikian, A. Z., Wyatt, R. G., Dolin, R., Thornhill, T. S., Kalica, A. R., and Chanock, R. M., *J. Virol.*, 10, 1975, 1972. With permission.) (B) The 27-nm hepatitis A particle. (From Kapikian, A. Z., Feinstone, S. M., Purcell, R. H., Wyatt, R. G., Thornhill, T. S., Kalica, A. R., and Chanock, R. M., *Perspect. Virol.*, 9, 9, 1975. With permission.) (C) The 70-nm reovirus-like agent. (From Kapikian, A. Z., Kim, H. W., Wyatt, R. G., Rodriguez, W. T., Ross, S., Cline, W. L., Parrott, R. H., and Chanock, R. M., *Science*, 185, 1049, 1974. With permission. Copyright 1974 by the American Association for the Advancement of Science.) The bar = 100 nm in A, B, and C.

the coronaviruses 229E and OC43.[20] Figure 8 shows the cyclical activity of 229E and OC43 viruses in several studies. For 229E virus, infections appear to occur during the same years in the various locations, and a 2-to 3-year cycle of activity has been postulated. With OC43 virus, a 2-to 4-year cycling of infection was observed, but the cycles did not coincide with specific years in all locations. This may have been due to actual differences in the occurrence of infection or to detection of infection with heterologous coronaviruses by the OC43 antigen[20] which, as noted

TABLE 5

Incidence of Coronavirus Infections in Various Populations as Determined by Serologic Tests

Study designation	Reference	Study group	Major illness[a]	Location of study	Duration of study	No. of illnesses	No. under surveillance	No. with antibody increases[b] to indicated agent: 229E	No. with antibody increases[b] to indicated agent: OC43[c]
A	66	Military recruits	Unspecified	Camp Lejeune, N.C. during 3 years	Jan.–March each of 3 years	–	544	NT	65 (12%)
B	12	Medical students	URI	Chicago, Ill.	1961–1967[d]	–	937 sy[e]	133 (15%)	NT[e]
C	67	Hospital employees	URI	Chicago, Ill.	1967–1968	88	–	7 (8%)	NT
D	68	Insurance co. employees and their selected families	URI	Charlottesville, Va.	1963–1966[f]	592	–	10 (1.7%)	
					1968–1970[f]	620	–		15 (2.4%)
		Insurance co. employees			Surveillance 1 yr or longer		683 py[e]	51 (7%)	
							693 py		48 (7%)
E	69	5- to 19-yr-old church home residents	URI	Atlanta, Ga.	1960–1967	1328	–	NT	44 (3.3%)
	70				1960–1968	1477		63 (4.3%)	
F	71	Military group	URI	South Cerney, Gloucestershire England	1966–1967	91	–	0	0
G	72	Family study	URI	Tecumseh, Mich.	1966–1967	–	758 py	58 (8%)	NT
	73				1965–1969	–	910 py	NT	156 (17%)
H	40	Outpatient infants and children	RI	Sao Paulo, Brazil	July 1969	36	–	4 (11%)	NT
I	74	Adults	P or ARTI	Krakow, Poland	–	59		0	NT
J	34	Hospitalized infants and children	LRTI	Children's Hospital, Washington, D.C.	1965–1966	567	–	–	20 (3.5%)
	10				1966–1967	567	–	–	20(3.5%)
	10				1962–1965	892	–	1	NT
					1966–1967	222	–	0	NT

TABLE 5 (continued)

Incidence of Coronavirus Infections in Various Populations as Determined by Serologic Tests

Study designation	Reference	Study group	Major illness[a]	Location of study	Duration of study	No. of illnesses	No. under surveillance	No. with antibody increases[b] to indicated agent: 229E	No. with antibody increases[b] to indicated agent: OC43[c]
K	10	Infants and children	LRTI	WHO collaborative study[g]	1964–1965	261	–	0	NT
L	10, 34	NIH employees	URI	Bethesda, Md.	1962–1964	256	–	5 (2%)	4 (1.6%)
	10, 34				1965–1967	317	–	24 (7%)	23 (7%)
	15				1965–1970	541	–	32 (6%)	31 (6%)
M	75, 76	College students	URI	College Park, Md.	1960–1961	110	–	12 (11%)	7 (6%)
N	77	Military recruits	ARD	Parris Island, S.C. and Camp Lejeune, N.C.	1970	75	–	4 (5%)	2 (3%)
					1971	73	–	0	37 (51%)
					1972 Jan.–March each year	–	225	5 (2%)	1 (0.4%)

[a]URI, upper respiratory tract illness; RI, respiratory tract illness; P, pneumonia; ARTI, acute respiratory tract infection; LRTI, lower respiratory tract illness; ARD, acute respiratory disease.
[b]Antibody increases for 229E virus determined CF test in all studies except E, in which an indirect hemagglutination test was used; Antibody increases for OC43 virus (or group) determined by CF test in studies A, J, L, and M, by HI test in study D E, and N, and by HI and CF tests in study G. Percentages below 4 rounded to nearest 0.1, otherwise rounded to nearest whole number.
[c]OC43 or OC43 group (OC43 and/or OC38 and/or MHV A-59).
[d]Academic years.
[e]sy, student years; py, person years; NT, not tested.
[f]No sera available 1967, and only 11 pairs studied from winter and spring of both 1966 and 1968.
[g]Sera from: Cairo, Egypt; Hong Kong; New Delhi, India; Jamaica; Singapore; and Trinidad.

After Kapikian.[14]

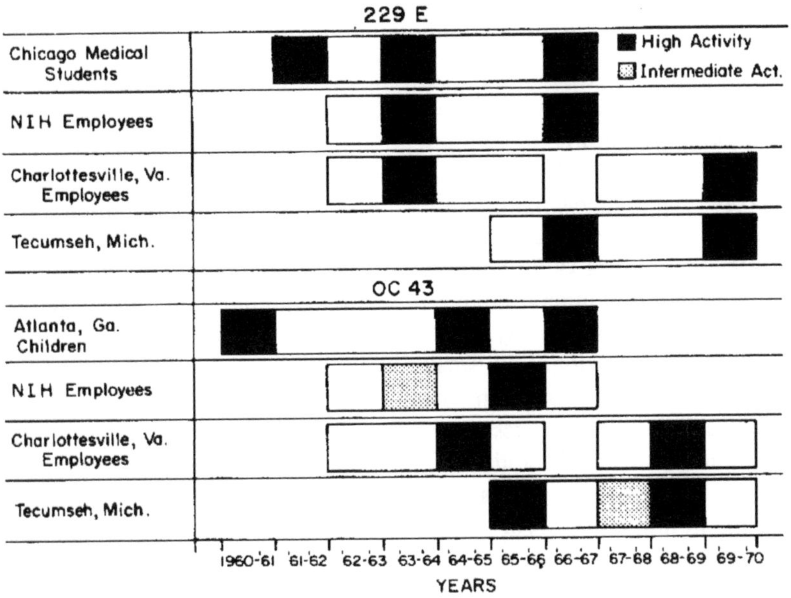

FIGURE 8. Cyclic behavior of 229E and OC43 viruses observed in five longitudinal studies. (From Monto, A. S., *Yale J. Biol. Med.*, 47, 234, 1974. With permission.)

above, cross reacts much more commonly with other coronaviruses than does 229E.

The epidemiologic surveys also demonstrated that reinfections with the coronaviruses occurred frequently (references in Table 5 and Reference 81). Thus, if vaccines are found to be necessary to prevent illness with these agents, it is possible that parenteral administration of such vaccines will not be beneficial since a vaccine administered by this route would stimulate the production of serum antibody predominantly and would not be efficient in inducing the production of nasal antibody.[82] Epidemiologic studies will be necessary to answer these important questions.

VII. LABORATORY DIAGNOSIS AND METHODOLOGY FOR ANTIVIRAL SCREENING

A. Virus Cultivation

1. In Animals

With the exception of OC43 (or OC38) viruses, none of the human coronaviruses has been reported to produce disease in an animal model. OC43 (or OC38) virus produces an encephalitic syndrome in the suckling mouse after IC or IP inoculation, but the IC route is more efficient.[27] Obviously, this would be an unsuitable model infection for screening or testing of antivirals. No data have been reported on the efficiency of this procedure for isolation of these viruses from clinical specimens. In the original description, suckling mice were inoculated with OC38 or OC43 virus passaged four to five times in human embryonic tracheal OC; illness began 11 to 15 days after inoculation and was characterized by tremors, rigidity, and lethargy. Only 2 of 8 mice inoculated with OC38 virus and 3 of 16 inoculated with OC43 virus developed illness. However, serial subpassage of mouse brain specimens resulted in increasingly shorter incubation periods, so that by four passages in suckling mice all inoculated animals died. The period between inoculation and death was then 48 to 60 hr. The encephalitic syndrome could be induced in mice 5 days of age or younger, whereas weanling mice did not develop illness although asymptomatic infection most likely occurred. Two blind passages of OC43 virus were needed in a repeat of the original experiment using the OC-passaged virus before the encephalitic syndrome occurred.

2. In Embryonated Eggs

Multiplication in embryonated eggs has not been reported for any of the human coronavirus strains.

3. In Tissue Cultures

229E virus may induce a CPE on primary passage of clinical specimens into tissue culture.[2,10] Semicontinuous human embryonic intestine (HEI) fibroblast cultures appeared to be more sensitive than WI38 cell cultures.[10] However, infection was detected more efficiently by CF studies in which pair era were examined for antibody rises to 229E virus.[10] The CPE in human embryonic lung or HEI cultures inoculated with positive clinical specimens develops slowly, commencing 5 or 6 days after inoculation, with small granular round cells appearing throughout the monolayer. The cell sheet is rarely destroyed completely. Three strains (B814, LP, and EVS) originally recovered in OC have been propagated from clinical specimens in human embryonic lung L-132 cells.[9,35] Attempts to propagate B814 virus, a distinct coronavirus serotype, in various cell cultures (including L-132) have been made by others but without success.[33] LP virus is related to 229E virus, whereas the antigenic relationship of LVS has not been reported.[9,35] Organ-culture and suckling-mouse grown OC43 and OC38 viruses were adapted successfully to primary monkey kidney cell cultures inducing a CPE on second or third passage.[33a] The monkey-kidney passaged virus was successfully adapted to BSC-1 cells in which it induced a CPE on first passage.[33] OC43 virus-infected BSC-1, rhesus monkey kidney, HEK, and WI38 cell cultures were found to hemadsorb rat and mouse erythrocytes, thus facilitating neutralization tests.[31] This might be suitable for antiviral screening. Attempts to adapt OC16, OC37, OC44, OC48, and 692 to monolayer cell cultures have been uniformly unsuccessful.[13,33]

4. In Organ Cultures

B814, OC16, OC37, OC38, OC43, OC44, OC48, LP, EVS, and 692 viruses were originally isolated in human embryonic tracheal OC.[1,4,8,9,13] Five of these (OC16, OC37, OC44, OC48, and 692) have yet to be isolated or adapted to any other system,[13,33] two (OC38 and OC43) have not yet been isolated from original specimens in any other system, and one (B814) has been grown in another system in only one laboratory.[9,35] Thus, the human embryonic tracheal OC system[83] has played a major role in the detection of human coronaviruses. The indicator system for detection of virus growth in OC may be (1) the comparison of ciliary activity in inoculated cultures with that in controls, (2) electron microscopic examination of concentrated OC harvests, or (3) virus interference. Ciliary immobilizing effect (CIE) is neither a reliable nor consistent indicator of infection since certain strains have no effect on ciliary activity, others have variable effects, and others induce CIE.[1,4] The most practical indicator for the detection of a human coronavirus in OC is the electron microscopic examination of OC harvests by the negative staining technique.[3,4] Such harvests may also be examined by immune electron microscopy.[13,84-86] In this technique the convalescent serum from a patient is added to the harvest from an OC inoculated with a specimen from the patient. The rationale for this method is that a low-titered coronavirus suspension not ordinarily detectable by conventional electron microscopy might be detected if the particles appeared in the form of antigen-antibody aggregates.[86] Following incubation, centrifugation, and further preparation for electron microscopy, coronavirus strain 692 was detected by immune electron microscopy[13] (Figures 9 and 10). In addition, it was also shown to be distinct from 229E and OC43 viruses, since antiserum for these two agents did not aggregate strain 692. A method utilizing human embryonic tracheal OC in test tube cultures has also been described.[87]

B. Identification of Isolates

An isolate may be identified as similar or identical to 229E virus by a conventional neutralization test in monolayer cell cultures with a specific 229E hyperimmune antiserum. Antiserum for 229E virus has been prepared in guinea pigs by various methods.[2,5,6,16] An isolate may be identified as similar or identical to OC38 or OC43 viruses by mouse or tissue culture neutralization tests, hemagglutination inhibition, and complement-fixation using specific hyperimmune serum or ascitic fluid. Antiserum and ascitic fluid for OC43 virus have been prepared in weanling mice.[5,27]

C. Serologic Methods

1. Neutralization

In cell culture – A neutralization test for 229E virus using absence of CPE as the indicator of virus neutralization has been used successfully in test tube cultures.[2,10,16] Plaque reduction tests have also been employed successfully.[5,21,27,34] For OC43 virus, a neutralization test using absence of

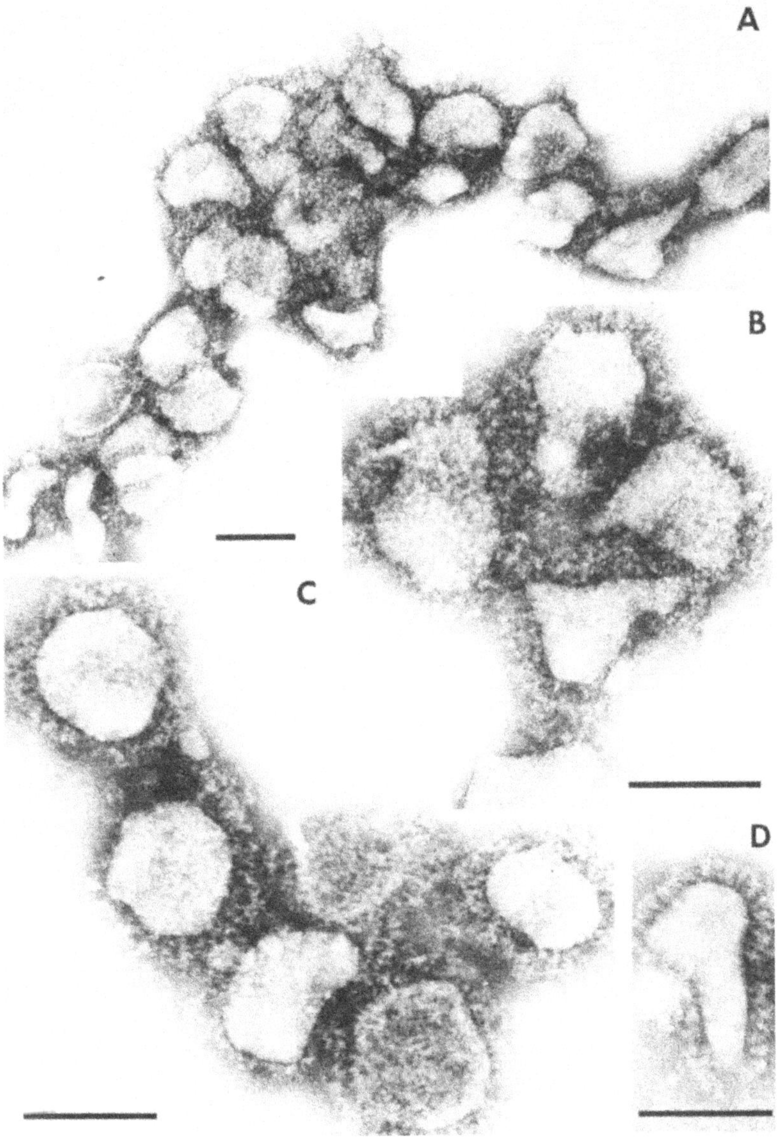

FIGURE 9. Large and small aggregates of coronavirus strain 692. (A, B, C) The HETOC harvest derived from a washing from a patient in the NIH common cold study was incubated with a 1:20 dilution of the patient's convalescent serum prior to further preparation for electron microscopy. The particles which were heavily coated with antibody were not randomly distributed but were present almost exclusively as groups which stood out clearly from the surrounding matter. The bar = 100 nm. (D) Prepared as Figure 9A to C. A single particle of coronavirus strain 692 was also found on the same grid on which particles in Figures 9A to C were observed. This particle appears to be covered with little, if any, antibody. The bar = 100 nm. (From Kapikian, A. Z., James, H. D., Jr., Kelly, S. J., and Vaughn, H. L., *Infect. Immun.*, 7, 111, 1973. With permission.)

CPE or absence of hemadsorption as the indicator of virus neutralization has been used successfully in test tube cultures.[31,33,73] Hemadsorption offers several advantages over CPE as the indicator system, since the former is considerably easier to read than the latter for OC43 virus; in addition, the hemadsorption pattern appears up to 1 week before the CPE, thus enabling more rapid completion of the test.[31] It is of interest that hemadsorption is also observed with the hemagglu-

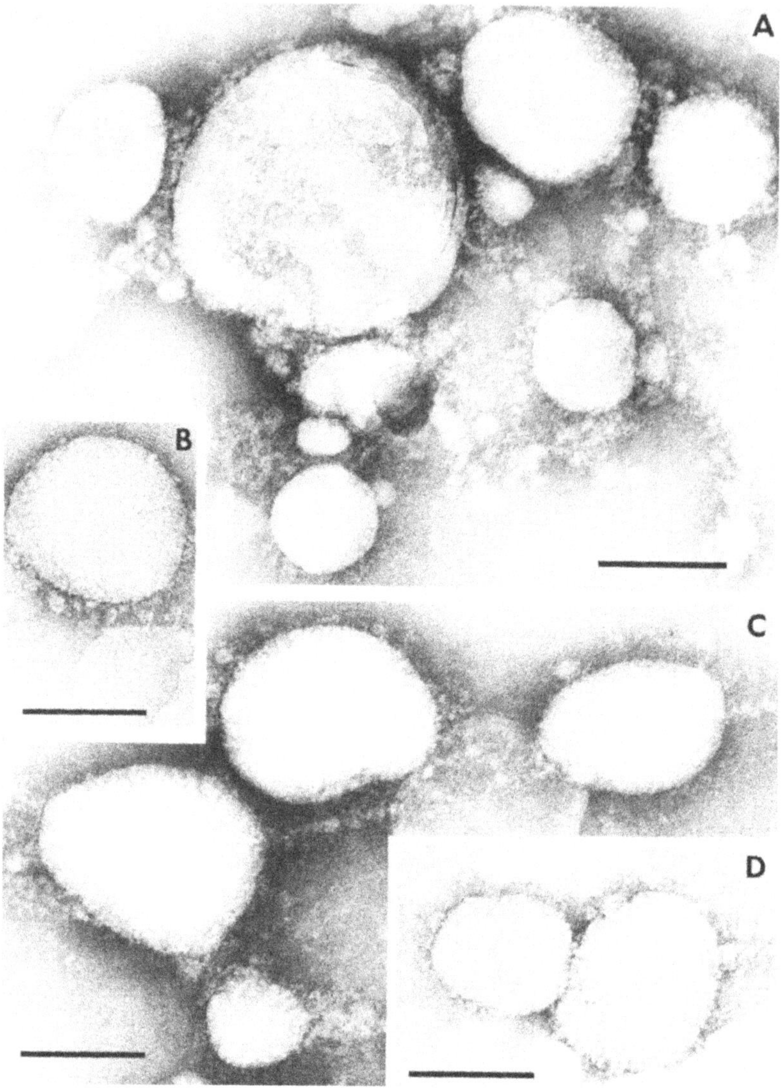

FIGURE 10. The same HETOC harvest from a washing from patient 692 was also incubated with phosphate buffered saline prior to further preparation for electron microscopy in the same experiment described in Figure 9. No coronavirus particles were observed, only predominantly round or oval-shaped objects, as shown in Figures 10A to D. The bar = 100 nm. (After Kapikian, A. Z., James, H. D., Jr., Kelly, S. J., and Vaughn, A. L., *Infect. Immun.*, 7, 111, 1973. With permission.)

tinating encephalomyelitis virus (HEV) of swine.[88-90]

In OC – A neutralization test for coronaviruses that grow only in OC has been described.[4,5] In this test neutralization is determined by the presence or absence of virus particles, as detected by electron microscopic examination, in harvests of organ cultures inoculated with appropriate virus-serum mixtures.

In suckling mice – A neutralization test for OC43 virus using the suckling mouse system has been described.[5,27] However, since this agent can now be grown in cell culture, the tests described above would be more practical.

2. Complement Fixation

CF antigen has been prepared for only two human coronaviruses – 229E, which grows in cell cultures, and OC43 (and OC38), which replicate in suckling mouse brain and monkey kidney cell cultures.[2,5,10,27] The CF antigen for 229E virus has been prepared by inoculating WI38 cells, allowing adsorption for 1 hr, and harvesting fluids after two freeze-thaw cycles about 2 to 3 days after inoculation.[2,10,21] Such early harvest is important since maximum virus titers have been found at 36 hr when CPE was minimal.[21] The harvest is clarified by low-speed centrifugation, and the supernate is used as antigen. CF antigen for OC43 (or OC38) virus consists of 10% suspensions of suckling mouse brain from animals which develop an encephalitic syndrome following inoculation with OC43 virus.[5,27] The suspensions are clarified by low-speed centrifugation, and the supernatate is used as antigen. CF antigen for OC43 virus has also been prepared in monkey kidney cell cultures.[33]

3. Direct Hemagglutination

The only human coronaviruses which have been shown to hemagglutinate are OC38 and OC43.[30] The hemagglutinating antigen[5,27,30] is prepared in the same manner as the CF antigen. Chicken, mouse, and rat cells are agglutinated by both OC38 and OC43 viruses at 4°C, room temperature, and 37°C. Human "O" and vervet cells agglutinated only at 4°C, whereas rhesus and guinea pig cells failed to agglutinate.[30] Treatment of chicken red blood cells with receptor destroying enzyme did not affect the agglutination of these cells by OC38 or OC43.[30,91] Hemagglutination of rat and mouse infected monkey kidney cell culture monolayers.[31] Hemagglutination has also been observed with the hemagglutinating encephalomyelitis virus (HEV) of swine and certain avian IBV strains.[19,88-90,92]

4. Indirect Hemagglutination

An indirect hemagglutination (IHA) test has been developed for detecting infection with the 229E virus.[69,93] Tannic acid-treated sheep erythrocytes were sensitized with 229E antigens. The IHA tests appears to be a rapid and sensitive method for serodiagnosis of 229E infections.

5. Immunofluorescence

229E virus-infected WI-38 cells and OC43 virus-infected BSC-1 cells showed specific cytoplasmic fluorescence when reacted with convalescent human serum and fluorescein conjugated antihuman globulin.[21,33] Fluorescence has also been demonstrated in L-132 cells infected with 229E or LP viruses, with convalescent human sera by the indirect method.[18] Tracheal epithelial cells from coronavirus-infected OC fragments have also been examined successfully by indirect immunofluorescence using specific mouse sera and fluorescein conjugated antimouse goat sera.[5]

6. Immunodiffusion

Gel diffusion studies with concentrated coronavirus tissue culture and mouse brain preparations and various antisera have been described.[29]

7. Immune Electron Microscopy

In this method[84] the presence or absence of antigen-antibody aggregates or particles heavily coated with antibody in absence of aggregation may be used as indicators for the study of coronaviruses that grow only in OC[13,86,94] (see above Section VII A4).

From the above descriptions, it is obvious that practical tests are available for the study of only two distinct coronaviruses, 229E and OC38 or OC43. Methods must be developed to study other coronaviruses not yet propagated in cell cultures or animal models.

VIII. CORONAVIRUSES OF ANIMALS

Coronaviruses cause a variety of clinical syndromes in animals, involving such diverse organs or organ systems as the brain, salivary glands, upper and lower respiratory tracts, kidney, genital tract, and intestine. A summary of known diseases associated with animal coronaviruses is shown in Table 6. A detailed comparative review of human and animal coronaviruses has recently been made.[19] It is possible that one of these natural animal models could find a use in antiviral screening.

IX. CONCLUSION

From this review, it is obvious that coronaviruses are unique in that they are capable of causing such a broad spectrum of illnesses. They also possess a rather broad host range which, at present, includes humans, chickens, pigs, mice, rats, calves, turkeys, and dogs. Studies of the human coronaviruses have been hampered by the fastidious growth requirements of these agents. However, despite only limited studies, coronaviruses have emerged as important etiologic agents of human respiratory disease. However, only after more efficient methods for cultivating or detecting

TABLE 6

Coronaviruses of Animals

Species type	Animal host	Serotypes or possible serotypes	Diseases associated with these agents	Reference
Infectious bronchitis virus (IBV)	Chicken	Massachusetts, Connecticut, Iowa 97, Iowa 609, Gray, Holte, JMK, SE17, Clarke 333, Delaware 2868, Delaware 2897, and others	Acute, highly contagious respiratory disease characterized by respiratory distress, tracheal rales, coughing, and in young chicks a nasal discharge also. Marked decrease in egg production and egg quality. Ovarian damage may occur. Mortality rate in very young chicks may reach 25%; in chickens >6 weeks of age mortality negligible. World-wide distribution. Inactivated and live attenuated virus vaccines available. Nephrosis and uremia recently reported.	95–100
Mouse hepatitis virus (MHV)	Mouse	JHM, MHV-1, MHV, (Pr), EHF-120, H747, MHV-3, A59, MHV-S, and others	Hepatitis, encephalitis; subclinical, endemic infections common.	5, 6, 28, 101–103
Transmissible gastroenteritis (TGE) virus	Pig	One type	Highly contagious, enteric disease with vomiting, severe diarrhea, dehydration, high mortality rate in piglets 2 weeks of age; mortality rate approaches 100% in newborn pigs. One of the major causes of illness and death in piglets. Adult pigs become infected also and develop a generally milder disease. Occurs predominantly in winter. Live attenuated vaccine available.	43, 104
Hemagglutinating encephalomyelitis virus (HEV) (vomiting and wasting disease virus)	Pig	All similar	Encephalomyelitis in piglets. Vomiting and wasting disease in piglets. High morbidity and mortality rates in newborn pigs. Piglets <2 weeks of age most susceptible to disease.	88–90, 105–107

TABLE 6 (continued)

Coronaviruses of Animals

Species type	Animal host	Serotypes or possible serotypes	Diseases associated with these agents	Reference
Rat coronavirus (RCV)	Rat	One strain described	Respiratory distress from pneumonitis in rats <48 hr of age. High morbidity rates in this group. Rats 7 and 14 days of age developed respiratory disease but recovered Rats > 21 days of age showed no outward sign of respiratory disease.	108
Neonatal calf diarrhea coronavirus (NCDC)	Calf	One strain described	Diarrhea in calves; may cause mortality in neonatal calf.	44, 45
Sialodacryoadenitis virus (SDAV)	Rat	One strain described	Acute disease of submaxillary and harderian glands; low mortality.	109
Transmissible enteritis virus	Turkey	One strain described	Outbreaks of enteritis in flocks; may be associated with high mortality rate. Associated with some cases of bluecomb disease.	46
Canine coronavirus	Dog	One strain described	Diarrheal disease in puppies.	47, 48

After Kapikian.[14]

these agents become available will the full extent of the role of coronaviruses in human disease be elucidated. Advances in methodology would be important and would facilitate the development of an easy and sensitive in vitro screening technique for antiviral studies with the group.

REFERENCES

1. **Tyrrell, D. A. J. and Bynoe, M. L.,** Cultivation of a novel type of common-cold virus in organ cultures, *Br. Med. J.,* 1, 1467, 1965.
2. **Hamre, D. and Procknow, J. J.,** A new virus isolated from the human respiratory tract, *Proc. Soc. Exp. Biol.,* 121, 190, 1966.
3. **Almeida, J. D. and Tyrrell, D. A. J.,** The morphology of three previously uncharacterized human respiratory viruses that grow in organ culture, *J. Gen Virol.,* 1, 175, 1967.
4. **McIntosh, K., Dees, J. H., Becker, W. B., Kapikian, A. Z., and Chanock, R. M.,** Recovery in tracheal organ cultures of novel viruses from patients with respiratory disease, *Proc. Natl. Acad. Sci. U.S.A.,* 57, 933, 1967.
5. **McIntosh, K., Kapikian, A. Z., Hardison, K. A., Hartley, J. W., and Chanock, R. M.,** Antigenic relationships among the coronaviruses of man and between human and animal coronaviruses, *J. Immunol.,* 102, 1109, 1969.
6. **Becker, W. B., McIntosh, K., Dees, J. H., and Chanock, R. M.,** Morphogenesis of avian infectious bronchitis virus and a related human virus (strain 229E), *J. Virol.,* 1, 1019, 1967.
7. Coronaviruses, *Nature,* 220, 650, 1968.
8. **Tyrrell, D. A. J., Bynoe, M. L., and Hoorn, B.,** Cultivation of "difficult" viruses from patients with common colds, *Br. Med. J.,* 1, 606, 1968.
9. **Bradburne, A. F.,** Sensitivity of L132 cells to some "new" respiratory viruses, *Nature,* 221, 85, 1969.
10. **Kapikian, A. Z., James, H. D., Jr., Kelley, S. J., Dees, J. H., Turner, H. C., McIntosh, K., Kim, H. W., Parrott, R. H., Vincent, M. M., and Chanock, R. M.,** Isolation from man of "avian infectious bronchitis virus-like" viruses (coronaviruses) similar to 229E virus, with some epidemiological observations, *J. Infect. Dis.,* 119, 282, 1969.
11. **Oshiro, L. S., Schieble, J. H., and Lennette, E. H.,** Electron microscopic studies of coronavirus, *J. Gen. Virol.,* 12, 161, 1971.
12. **Hamre, D. and Beem, M.,** Virologic studies of acute respiratory disease in young adults. V. Coronavirus 229E infections during six years of surveillance, *Am. J. Epidemiol.,* 96, 94, 1972.
13. **Kapikian, A. Z., James, H. D., Jr., Kelly, S. J., and Vaughn, A. L.,** Detection of coronavirus strain 692 by immune electron microscopy, *J. Gen. Virol.,* 7, 111, 1973.
14. **Kapikian, A. Z.,** The Coronaviruses, in *Proc. Int. Symp. on Immunity to Infections of the Respiratory System in Man and Animals,* Vol. 28, Regamey, R. H., Ed., Karger, Basel, 1975, 42.
15. **Kapikian, A. Z., James, H. D., Jr., Kelley, S. J., Vaughn, A. L., Cline, W. L., and Chanock, R. M.,** unpublished studies.
16. **Kapikian, A. Z.,** Coronaviruses, in *Diagnostic Procedures for Viral and Rickettsial Diseases,* 4th ed., by Lennette, F. H. and Schmidt, N. J., Eds., American Public Health Association, New York, 1969, 931.
17. **Tyrrell, D. A. J.,** Hunting common cold viruses by some new methods, *J. Infect. Dis.,* 121, 561, 1970.
18. **Bradburne, A. F. and Tyrrell, D. A. J.,** Coronaviruses of man, *Prog. Med. Virol.,* 13, 373, 1971.
19. **McIntosh, K.,** Coronaviruses: a comparative review, *Curr. Top. Microbiol. Immunol.,* 63, 85, 1974.
20. **Monto, A. S.,** Coronavirus, *Yale J. Biol. Med.,* 47, 234, 1974.
21. **Hamre, D., Kindig, D. A., and Mann, J.,** Growth and intracellular development of a new respiratory virus, *J. Virol.,* 1, 810, 1967.
22. **Tannock, G. A.,** The nucleic acid of infectious bronchitis virus, *Arch. Gesamte Virusforsch.,* 43, 259, 1973.
23. **Watkins, H., Reeve, P., and Alexander, D. J.,** The nucleic acid of infectious bronchitis virus, *Arch. Virol.,* 47, 279, 1975.
24. **Tyrrell, D. A. J., Almeida, J. D., Cunningham, C. H., Dowdle, W. R., Hofstad, M. S., McIntosh, K., Tajima, A., Zakstelskaya, L. Ya., Easterday, B. C., Kapikian, A., and Bingham, R. W.,** Coronaviridae, *Intervirology,* 5, 76, 1975.
25. **Hierholzer, J. C., Palmer, E. L., Whitfield, S. G., Kaye, H. S., and Dowdle, W. R.,** Protein composition of Coronavirus OC43, *Virology,* 48, 516, 1972.
26. **Garwes, D. J. and Pocock, D. H.,** The polypeptide structure of transmissible gastroenteritis virus, *J. Gen. Virol.,* 29, 25, 1975.
27. **McIntosh, K., Becker, W. B., and Chanock, R. M.,** Growth in suckling-mouse brain of "IBV-like" viruses from patients with upper respiratory tract disease, *Proc. Natl. Acad. Sci. U.S.A.,* 58, 2268, 1967.
28. **Gledhill, A. W. and Andrewes, C. H.,** A hepatitis virus of mice, *Br. J. Exp. Pathol.,* 32, 559, 1951.

61. **Kapikian, A. Z., Kim, H. W., Wyatt, R. G., Rodriguez, W. J., Ross, S., Cline, W. L., Parrott, R. H., and Chanock, R. M.,** Reoviruslike agent in stools: association with infantile diarrhea and development of serologic tests, *Science,* 185, 1049, 1974.
62. **Feinstone, S. M., Kapikian, A. Z., and Purcell, R. H.,** Hepatitis A detection by immune electron microscopy of a viruslike antigen associated with acute illness, *Science,* 182, 1026, 1973.
63. **Caul, F. O. and Clark, S. K. R.,** Coronavirus propagated from patient with nonbacterial gastroenteritis, *Lancet,* 2, 753, 1975.
64. **Sharpee, R. and Mebus, C. A.,** Rotaviruses of man and animals, *Lancet,* 1, 639, 1975.
65. **Kaye, H. S., Yarbrough, W. B., and Reed, C. J.,** Calf diarrhea coronavirus, *Lancet,* 2, 509, 1975.
66. **Hartley, J. W., Rowe, W. P., Bloom, H. H., and Turner, H. C.,** Antibodies to mouse hepatitis viruses in human sera, *Proc. Soc. Exp. Biol.,* 115, 414, 1964.
67. **Schaaf, D. M., Boswell, J., and Mufson, M. A.,** Viral and mycoplasmal infections in minor respiratory illness and nonstreptocococcal pharyngitis, *Am. Rev. Respir. Dis.,* 104, 41, 1971.
68. **Hendley, J. O., Fishburne, H. B., and Gwaltney, J. M., Jr.,** Coronavirus infections in working adults. Eight year study with 229E and OC43, *Am. Rev. Respir. Dis.,* 105, 805, 1972.
69. **Kaye, H. S., Marsh, H. B., and Dowdle, W. R.,** Seroepidemiologic survey of coronovirus (strain OC43) related infections in a children's population, *Am. J. Epidemiol.,* 94, 43, 1971.
70. **Kaye, H. S. and Dowdle, W. R.,** Seroepidemiologic survey of coronavirus (strain 229E) infections in a population of children, *Am. J. Epidemiol.,* 101, 238, 1975.
71. **Higgins, P. G., Ellis, E. M., and Wooley, D. A.,** Viruses associated with acute respiratory infections in Royal Air Force personnel, *J. Hyg.,* 68, 647, 1970.
72. **Cavallaro, J. J. and Monto, A. S.,** Community-wide outbreak of infection with a 229E-like coronavirus in Tecumseh, Michigan, *J. Infect. Dis.,* 122, 272, 1970.
73. **Monto, A. S. and Lim, S. K.,** The Tecumseh study of respiratory illness. VI. Frequency of and relationship between outbreaks of coronavirus infection, *J. Infect. Dis.,* 129, 271, 1974.
74. **Borysiewicz, J., Zgorniak-Nowosielska, I., Oleksy, J., Pastuszak, J., and Potec, Z.,** Adeno, parainfluenza, RS, and corona viruses and mycoplasma pneumoniae as etiologic agents of acute respiratory tract infections, *Acta Microbiol.,* 22, 102, 1973.
75. **Kapikian, A. Z., Cline, W. L., James, H. D., Jr., Vaughn, A. L., and Chanock, R. M.,** unpublished studies.
76. **Kapikian, A. Z., Johnson, K. M., Chanock, R. M., Bell, J. A., Dyke, L. M., and Iler, R. E.,** A pilot study of acute respiratory illness on a college campus, *Am. Rev. Respir. Dis.,* 90, 175, 1964.
77. **Wenzel, R. P., Hendley, J. O., Davies, J. A., and Gwaltney, J. M., Jr.,** Coronavirus infections in military recruits. Three-year study with coronavirus strains OC43 and 229E, *Am. Rev. Respir. Dis.,* 109, 621, 1974.
78. **Hamre, D.,** Rhinoviruses, *Monogr. Virol.,* 1, 1, 1968, Melnick, J. L , Ed., S. Karger, Basil.
79. **Tyrrell, D. A. J.,** Rhinoviruses, in *Virology Monographs,* Vol. 2, Gard, S., Hallauer, C., and Meyer, K. F., Eds., Springer-Verlag, New York, 1968, 68.
80. **Kapikian, A. Z.,** Rhinoviruses, in *Diagnostic Procedures for Viral and Rickettsial Diseases,* 4th ed., Lennette, F. H. and Schmidt, N. J., Eds., American Public Health Association, New York, 1969, 603.
81. **Zakstelskaya, L. Ya., Sheboldov, A. V., Vasilieva, V. I., and Alekseenkova, L. I.,** Occurrence of antibody to coronaviruses in sera of people living in the USSR, *Vopr. Virusol.,* 17, 161, 1972.
82. **Perkins, J. C., Tucker, D. N., Knopf, H. L. S., Wenzel, R. P., Kapikian, A. Z., and Chanock, R. M.,** Comparison of protective effect of neutralizing antibody in serum and nasal secretions in experimental rhinovirus type 13 illness, *Am. J. Epidemiol.,* 90, 519, 1969.
83. **Hoorn, B. and Tyrrell, D. A. J.,** On the growth of certain "newer" respiratory viruses in organ cultures, *Br. J. Exp. Pathol.,* 46, 109, 1965.
84. **Almeida, J. D. and Waterson, A. P.,** The morphology of virus-antibody reaction, *Adv. Virus Res.,* 15, 307, 1969.
85. **Kapikian, A. Z., Almeida, J. A., and Stott, E. J.,** Immune electron microscopy of rhinoviruses, *J. Virol.,* 10, 142, 1972.
86. **Kapikian, A. Z., Feinstone, S. M., Purcell, R. H., Wyatt, R. G., Thornhill, T. S., Kalica, A. R., and Chanock, R. M.,** Detection and identification by immune electron microscopy of fastidious agents associated with respiratory illness, acute nonbacterial gastroenteritis, and hepatitis A, *Perspect. Virol.,* 9, 9, 1975; *The Gustav Stern Symposium, Antiviral Mechanisms,* Pollard, M., Ed., Academic Press, New York, 1975.
87. **Harnett, G. B. and Hooper, W. L.,** Test-tube organ cultures of ciliated epithelium for the isolation of respiratory viruses, *Lancet,* 1, 339, 1968.
88. **Grieg, A. S. and Girard, A.,** Encephalomyelitis of swine caused by a hemagglutinating virus. II. Virological studies, *Res. Vet. Sci.,* 4, 511, 1963.
89. **Phillip, J. I. H., Cartwright, S. F., and Scott, A. C.,** The size and morphology of transmissible gastroenteritis and vomiting and wasting disease viruses of pigs, *Vet. Rec.,* 88, 311, 1971.
90. **Mengeling, W. L., Boothe, A. D., and Ritchie, A. E.,** Characteristics of a coronavirus (Strain 67N) in pigs, *Am. J. Vet. Res.,* 33, 297, 1972.
91. **Zakstelskaya, L. Ya., Sheboldov, A. V., and Molibog, E. V.,** Some aspects of interaction between coronaviruses OC38 and OC43 and erythrocytes, *Vestn. Akad Med. Nauk SSSR,* 27, 40, 1972.
92. **Bingham, R. W., Madge, M. H., and Tyrrell, D. A. J.,** Hemagglutination by avian infectious bronchitis virus-a coronavirus, *J. Gen. Virol.,* 28, 381, 1975.

29. **Bradburne, A. F.**, Antigenic relationships amongst coronaviruses, *Arch. Gesamte Virusforsch.*, 31, 352, 1970.
30. **Kaye, H. and Dowdle, W. R.**, Some characteristics of hemagglutination of certain strains of "IBV-like" virus, *J. Infect. Dis.*, 120, 576, 1969.
31. **Kapikian, A. Z., James, H. D., Jr., Kelly, S. J., King, L. M., Vaughn, A. L., and Chanock, R. M.**, Hemadsorption by coronavirus strain OC43, *Proc. Soc. Exp. Biol.*, 139, 179. 1972.
32. **Kendall, E. J. C., Bynoe, M. L., and Tyrrell, D. A. J.**, Virus isolations from common colds occurring in a residential school, *Br. Med. J.*, 2, 82, 1962.
33. **Bruckova, M., McIntosh, K., Kapikian, A. Z., and Chanock, R. M.**, The adaption of two human coronavirus strains (OC38 and OC43) to growth in cell monolayers, *Proc. Soc. Exp. Biol.*, 135, 431, 1970.
33a. **Bucknall, R. A., King, L., Kapikian, A. Z., James, H. D., Jr., and Chanock, R. M.**,
34. **McIntosh, K., Kapikian, A. Z., Turner, H. C., Hartley, J. W., Parrott, R. H., and Chanock, R. M.**, Seroepidemiologic studies of coronavirus infection in adults and children, *Am. J. Epidemiol.*, 91, 585, 1970.
35. **Bradburne, A. F. and Tyrrell, D. A. J.**, The propagation of "coronaviruses" in tissue culture, *Arch. Gesamte Virusforsch.*, 28, 133, 1969.
36. **Bradburne, A. F. and Somerset, B. A.**, Coronavirus antibody titers in sera of healthy adults and experimentally infected volunteers, *J. Hyg.*, 70, 235, 1972.
37. **Bell, J. A., Huebner, R. J., Rosen, L., Rowe, W. P., Cole, R. M., Mastrota, F. M., Floyd, T. M., Chanock, R. M., and Shvedoff, R. A.**, Illness and microbial experience of nursery children at Junior Village, *Am. J. Hyg.*, 74, 267, 1961.
38. **Kapikian, A. Z., Bell, J. A., Mastrota, F. M., Johnson, K. M., Huebner, R. J., and Chanock, R. M.**, An outbreak of febrile illness and pneumonia associated with respiratory syncytial virus infection, *Am. J. Hyg.*, 74, 234, 1961.
39. **Bradburne, A. F., Bynoe, M. L., and Tyrrell, D. A. J.**, Effects of a "new" human respiratory virus in volunteers, *Br. Med. J.*, 3, 767, 1967.
40. **Candeias, J. A. N., de S. Carvalho, R. P., and Antonacio, F.**, Seroepidemiologic study of coronavirus infection in Brazilian children and civilian adults, *Rev. Inst. Med. Trop. Sao Paulo,* 14, 121, 1972.
41. **McIntosh, K. M., Ellis, E. F., Hoffman, L. S., Lybass, T. C., Eller, J. J., and Fulginiti, V. A.**, The association of viral and bacterial respiratory infections with exacerbations of wheezing in young asthmatic children, *J. Pediatr.*, 82, 578, 1973.
42. Virus infection and asthma, *Med. J. Aust.*, 2, 304, 1973.
43. **Bohl, E. H.**, in *Transmissible Gastroenteritis in Diseases of Swine,* 3rd ed., Dunne, H. W., Ed., Iowa State University Press, Ames, 1970, 158.
44. **Stair, E. L., Rhodes, M. B., White, R. G., and Mebus, C. A.**, Neonatal calf diarrhea: purification and electron microscopy of a coronavirus-like agent, *Am. J. Vet. Res.*, 33, 1147, 1972.
45. **Mebus, C. A., Stair, E. L., Rhodes, M. B., and Twiehaus, M. J.**, Pathology of neonatal calf diarrhea induced by a coronavirus-like agent, *Vet. Pathol.*, 10, 45, 1973.
46. **Ritchie, A. E., Deshmukh, D. R., Larsen, C. T., and Pomeroy, B. S.**, Electron microscopy of coronavirus-like particles characteristic of turkey bluecomb disease, *Avian Dis.*, 17, 546, 1973.
47. **Binn, L. N., Lazar, E. C., Keenan, K. P., Huxsoll, D. L., Marchwicki, R. H., and Strano, A. J.**, Discovery and characterization of a coronavirus from military dogs with diarrhea, Proc. 78th Annu. Meet. U.S. Animal Health Association, 1974, 359.
48. **Keenan, K. P., Jervis, H. R., Marchwicki, R. H., and Binn, L. N.**, Intestinal infection of neonatal dogs with canine coronavirus 1-71: Studies by virologic, histologic, histochemical and immunofluorescent techniques, *Am. J. Vet. Res.*, 37, 247, 1976.
49. **Mebus, C. A., Underdahl, N. R., Rhodes, M. B., and Twiehaus, M. J.**, Calf diarrhea (scours): Reproduced with a virus from a field outbreak, *Res. Bull. Univ. Nebr. Coll. Agric. Home Econ.*, 233, 2, 1969.
50. **Fernelius, A. L., Ritchie, A. E., Classick, L. G., Norman, J. O., and Mebus, C. A.**, Cell culture adaption and propagation of a reovirus-like agent of calf diarrhea from a field outbreak in Nebraska, *Arch. Gesamte Virusforsch.*, 37, 114, 1972.
51. **Ferris, A. A., Kaldor, J., Gust, I. D., and Cross, G.**, Fecal antigen in viral hepatitis, *Lancet,* 2, 243, 1970.
52. **Kapikian, A. Z., Wyatt, R. G., Dolin, R., Thornhill, T. S., Kalica, A. R., and Chanock, R. M.**, Visualization by immune electron microscopy of a 27 nm particle associated with acute infectious nonbacterial gastroenteritis, *J. Virol.*, 10, 1075, 1972.
53. **Almeida, J. D.**, Visualization of fecal viruses (editorial), *N. Engl. J. Med.*, 292, 1403, 1975.
54. Rotaviruses of man and animals, *Lancet,* 1, 257, 1975.
55. Virus of infantile gastroenteritis, *Br. Med. J.*, 3, 555, 1975.
56. **Bishop, R. F., Davidson, G. P., Holmes, I. H., and Ruck, B. J.**, Virus particles in epithelial cells of duodenal mucosa from children with acute nonbacterial gastroenteritis, *Lancet,* 2, 1281, 1973.
57. **Flewett, T. H., Bryden, A. S., and Davies, H.**, Virus particles in gastroenteritis, *Lancet,* 2, 1497, 1973.
58. **Bishop, R. F., Davidson, G. P., Holmes, I. H., and Ruck, B. J.**, Detection of a new virus by electron microscopy of faecal extracts from children with acute gastroenteritis, *Lancet,* 1, 149, 1974.
59. **Bortolussi, R., Szymanski, M., Hamilton, R., and Middleton, P.**, Studies on the etiology of acute infantile diarrhea, *Pediatr. Res.*, 8, 379, 1974.
60. **Middleton, P. J., Szymanski, M. T., Abbott, G. D., Bortolussi, R., and Hamilton, J. R.**, Orbivirus acute gastroenteritis of infancy, *Lancet,* 1, 1241, 1974.

93. **Kaye, H. S., Ong, S. B., and Dowdle, W. R.,** Detection of coronavirus 229E antibody by indirect hemagglutination, *Appl. Microbiol.*, 24, 703, 1972.
94. **Kapikian, A. Z., Dienstag, J. L., and Purcell, R. H.,** Immune electron microscopy as a method for the detection, identification and characterization of agents not cultivable in an in vitro system, *Manual of Clinical Immunology*, in press.
95. **Berry, D. M., Cruickshank, J. G., Chu, H. P., and Wells, R. J. H.,** The structure of infectious bronchitis virus, *Virology*, 23, 403, 1964.
96. **Winterfield, R. W. and Hitchner, S. B.,** Etiology of an infectious nephritis-nephrosis syndrome in chickens, *Am. J. Vet Res.*, 23, 1273, 1962.
97. **Cowen, B. S., Hitchner, S. B., and Lucio, B.,** Characterization of a new infectious bronchitis virus isolate. I. Serological and pathogenicity studies of Clark 333, *Avian Dis.*, 15, 518, 1971.
98. **Hirai, K. and Shimakura, S.,** Isolation and characteristics of avian nephrosis-inducing infectious bronchitis virus (Coronavirus), *Jpn. J. Vet Sci.*, 33, 209, 1971.
99. **Hofstad, M.,** Infectious bronchitis, in *Diseases of Poultry*, 5th ed., Biester, H. E. and Schwarte, L. H., Eds., Iowa State University Press, Ames, 1965, 605.
100. **Cunningham, C. H.,** Avian infectious bronchitis, in *Advances in Veterinary Science and Comparative Medicine*, Vol. 14, Brandly, C. A. and Cornelius, C. E., Eds., Academic Press, New York, 1970, 105.
101. **Cheever, F. S., Daniels, J. B., Papfenhiemer, A. M., and Bailey, O. T.,** A murine virus (JHM) causing disseminated encephalomyelitis with extensive destruction of myelin. I. Isolation and biological properties of the virus, *J. Exp. Med.*, 90, 181, 1949.
102. **Rowe, W. P., Hartley, J. W., and Capps, W. I.,** Mouse hepatitis virus infections as a highly contageous, prevalent, enteric infection of mice, *Proc. Soc. Exp. Biol.*, 112, 161, 1963.
103. **Piazza, M.,** *Experimental Viral Hepatitis*, Charles C Thomas, Springfield, Ill., 1969.
104. **Tajima, M.,** Morphology of transmissible gastroenteritis virus of pigs. A possible member of coronaviruses, *Arch. Gesamte Virusforsch.*, 29, 105, 1970.
105. **Roe, C. K. and Alexander, T. J. L.,** A disease of nursing pigs previously unreported in Ontario, *Can. J. Comp. Med. Vet. Sci.*, 22, 305, 1958.
106. **Mitchell, D.,** Encephalomyelitis of swine caused by a hemagglutinating virus. I. Case histories, *Res. Vet. Sci.*, 4, 506, 1967.
107. **Cartwright, S. F. and Lucas, M.,** Vomiting and wasting disease in piglets. Virological and epidemiological studies, *Vet. Rec.*, 86, 278, 1970.
108. **Parker, J. C., Cross, S. S., and Rowe, W. P.,** Rat coronavirus (RCV): a prevalent naturally occurring pneumotropic virus of rats, *Arch. Gesamte Virusforsch.*, 31, 293, 1970.
109. **Bhatt, P. N., Percy, D. H., and Jonas, A. M.,** Characterization of the virus of sialodacryoadenitis of rats: a member of the coronavirus group, *J. Infect. Dis.*, 126, 123, 1972.

Chapter 6

THE DISCOVERY AND DEVELOPMENT OF ANTIVIRAL DRUGS – THE COMMERCIAL APPROACH

R. A. Bucknall and D. A. Rutty

TABLE OF CONTENTS

I. INTRODUCTION

The replicative mechanisms of viruses illustrate to what extreme lengths the parasitism of one organism may be carried by another – an intricate association at a molecular level which, in some cases, does not lead to the production of progeny viruses, but only to a change in the behavior of the host cell. Therefore, it is little wonder that the elucidation of even the simplest virus life cycle depended on the development of modern biochemical and microbiological technology. Although the intricacies of the life processes are steadily being revealed, we are still far from understanding the subtleties of many virus mechanisms.

However, a lack of information and full understanding of the nature of viruses did not deter pioneers from trying to prevent the diseases. Jenner, Pasteur, and Reed knew little or nothing of the organisms that caused smallpox, rabies, and yellow fever; yet, they devised approaches which are still used today to control these diseases. Their efforts well illustrate the principle that an extensive knowledge of a problem is often not necessary in order to arrive at a practical solution; nowhere is this truer than in the field of discovering antiviral drugs. Despite the heroic advances of the past 25 years toward the understanding of virus replicative processes, our knowledge is still not sufficiently detailed to enable predictions of entirely novel chemical structures which will intervene at specific points in virus-directed metabolism. No doubt, in time, increases in knowledge will permit a rational approach to the design of drugs, not only against virus diseases but also in other fields of medicine. For the moment, however, the discovery of new active chemicals, as distinct from the development of known ones, is still largely an empirical process. To some, the empirical approach is so intellectually unsatisfactory that they condemn it altogether as a futile and unproductive exercise, but the range of novel antiviral substances now available is largely the result of random screening. Although most are of no clinical value, they have been of considerable importance in the understanding of the nature of viruses, and undoubtedly one day they will also yield valuable medicines.

In the mid-1950s, when the search for antiviral drugs was beginning in earnest, the biological climate was very different from today. Techniques for growing viruses in tissue culture were just becoming routine. There were still many viruses that could not be cultured, and there were virtually no defined antiviral chemotherapeutic substances. There was also a feeling that potent, broad-spectrum drugs would be developed that would repeat the successes of the antibacterial antibiotics, but against virus diseases.

Twenty-five years have passed, and these early hopes have borne but little fruit. The antiviral drugs in current clinical use may be counted on the fingers of one hand, and the problem of finding useful antiviral drugs is now seen as much greater than at first realized.

Nevertheless, the few successes, and the many near misses, have shown that although the problem is formidable, it is not insurmountable. The search continues, but with a growing awareness that the virologist must now lift his eyes from the bench and consider more than just the disease organism, more than its destructive effects, and more than arresting it if he is to finally produce a worthwhile medicine which will find a use, and a market, in the world outside the laboratory walls. Some of this change in thinking is reflected in the reviews of the past decade; the reader is referred to Osdene,[1] McFazdean,[2] Goz and Prussoff,[3] Swallow,[4] and Bucknall.[5]

The great advances toward the understanding of viruses, their replicative processes, and their significance as disease organisms have inevitably influenced the approach to control of virus diseases. The modern virologist may feel he is now in a stronger position than ever before to tackle the development of antiviral drugs. However, virology has not advanced in isolation; there have been concurrent and equally immense developments in the science of drug safety, both at a practical and an administrative level. Stringent regulations now exist to which potential new drugs must conform before they can be sold to the public. Because of the importance of these statutory requirements, they will be dealt with in some detail in the later sections of this chapter, and the virologist who is intent on discovering an antiviral drug for human or veterinary use will find these very relevant to his task. With some knowledge of toxicological testing and drug registration requirements, the virologist is better able to guide his researches along lines that may eventually lead to discovery of useful medicines.

II. THE COMMERCIAL VIABILITY OF ANTIVIRAL DRUGS

In a commercial organization, the cash to finance its various activities ultimately comes from the sale of its products; it is surprising to many to learn just how large a burden a firm's annual sales must support in a modern manufacturing industry. Some of the internal organization of a not-too-hypothetical pharmaceutical manufacturing concern is shown in Figure 1, as well as the way in which the various departments work together to produce a range of saleable products.

The cost of discovering and developing a drug in the United Kingdom was £30 million in 1975; assuming an inflation rate of 10% over the next decade (and with increased complexity of work), by 1985 the cost will be £120 million. In other countries the cost is similar, except in the United States where it is 50% higher. Therefore, drug research is an expensive undertaking, and it is important to consider the possible commercial outcome of a research project before any work is begun. In the past, some projects have been undertaken without proper prior commercial analysis, and pharmaceutical firms have invested substantially in certain research areas only to find that there was no worthwhile financial return in the final products. In the case of virus diseases, the commercial viability of each disease must be fully analyzed before research is initiated; some of the more important considerations are given below.

A. Is the Disease Widespread and of Economic Importance?

If a drug is to be a commercial success, it must sell in sufficient quantities to pay for the research and development which have gone into producing it, as well as make a worthwhile financial contribution towards maintaining all the other essential aspects of the manufacturing concern. In theory, a drug could be a commercial success even if it sold only in small amounts, provided the price of each dose was very high. In practice, drug prices are relatively low, and drugs must therefore sell in large amounts if a worthwhile return can be

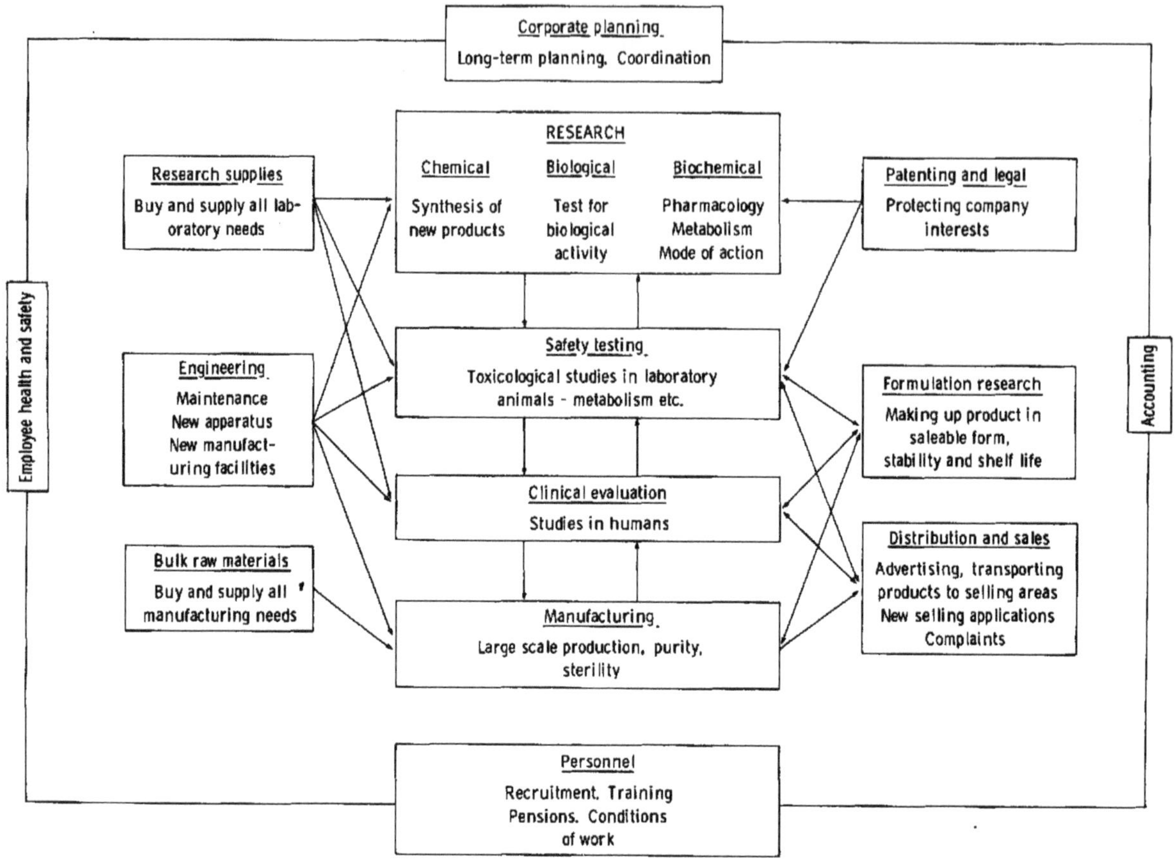

FIGURE 1. Chart showing some of the departmental organization of a pharmaceutical manufacturing concern. The arrows indicate the main flow of information, ideas, goods, etc. between departments. The whole is encompassed by the quartet of corporate planning, accounting, personnel, and employee health.

expected. This means that drugs are more likely to be developed against diseases that affect large numbers of individuals rather than against the less common ailments. However, no matter how common a disease may be, there will be no market for a drug unless the disease is worth curing. These two factors are critically important in deciding whether research into a particular drug area will be commercially viable: are there a large number of individuals involved, and is the disease serious enough to warrant treatment?

In the case of virus diseases, these principles can be illustrated by considering three diseases: rabies, rubella, and respiratory syncytial disease. Rabies is extremely serious for the infected individual, and an effective antirabies drug would be a godsend to patient and physician alike, but because the disease is relatively rare, the sales for an antirabies drug would be very small. With drug development costs being what they are, no firm is prepared to undertake work on an antirabies medicine.

Unlike rabies, rubella is an extremely common disease, no doubt due to its highly infectious nature, and many millions of people are attacked every year. However, a high proportion of cases are mild, or even subclinical, and at worst the infected individual is not at serious risk. There may be grave consequences for a fetus whose mother contracts the disease early in her pregnancy, but by and large the disease is not serious. Therefore, despite the fact that it is so common, there is no call for a specific drug to treat it.

Respiratory syncytial disease is widespread and serious enough (Chapter 1; Volume I) to warrant the use of a drug. Therefore, perhaps one day research into this disease may produce a medicine that will alleviate the suffering of thousands and pay its own way.

B. Can the Disease be Satisfactorily Controlled by Immunization?

Immunization is an excellent way of preventing infectious disease and has been widely and successfully used in the control of dangerous virus diseases such as smallpox, poliomyelitis, and yellow fever. Effective vaccines are also available against certain serious animal virus diseases. The most noticeable advances have been in the control of virus diseases of poultry, reared under modern intensive conditions, by the use of live-virus vaccines administered in the birds' drinking water. Diseases such as infectious bronchitis, Marek's disease, and Newcastle disease can be controlled with vaccines, and it is doubtful whether the present broiler industry could survive without them.

Where there are good vaccines, there is no need for antiviral drugs; however, despite the undeniable success of virus vaccines in the fields of both human and animal health, not all virus diseases can be controlled in this way. It is here that chemotherapeutic agents may find application.

There are a number of reasons why certain virus diseases cannot be controlled by vaccines. Some of these are technical and may be overcome in due course; others are more fundamental, and it is difficult to see how they might be tackled. The various aspects of controlling a disease by immunization are outlined below, and the scientist hoping to develop antiviral drugs must consider how each of these may affect the market for such a drug, should a good drug become available.

1. Does the Disease have a Wide Range of Causative Agents?

If a disease has a large number of serologically distinct causative agents, the prospects for developing an effective vaccine are poor. This is illustrated by the common cold, the chief causative agents of which are the rhinoviruses (Chapter 1, Volume I) and the coronaviruses. These two major groups of cold agents include a formidable array of serologically distinct agents. There are at least 89 rhinoviruses,[6] none of which falls into serologically related groups,[7,8] and a smaller but growing number of coronaviruses,[9] which again do not seem to share close serological determinants. Thus, it is obvious that the likelihood of developing a vaccine to give any practical immunity against the common cold is vanishingly small. Therefore, control of the common cold by antiviral drugs must be considered, although it is scarcely a less daunting task than the impossibility of immunological control.

2. Is the Causative Agent Prone to Antigenic Change?

Unless the causative agent(s) of any particular virus disease is antigenically stable, the clinical effectiveness of any vaccine will be seriously diminished. For example, vaccines against influenza A have been available for many years. In controlled clinical trials where volunteers are first immunized and then challenged with strictly homologous virus, a substantial degree of protection can be demonstrated (Chapter 4, Volume I); under practical conditions the wild influenza viruses are continually mutating. Once any substantial portion of the human population has become immune to a current influenza serotype, a mutant variant becomes dominant which, by virtue of its antigenic novelty, can overcome the population immunity and again cause disease. Thus, this type of antigenic mobility is a serious problem in developing an effective vaccine. In effect, it means that new vaccines based on the latest antigenic variants of the virus must be introduced for human use at regular intervals. In the case of influenza A, this is about every 2 years.

Three years ago, workers at the Pasteur Institute in Paris suggested that the current influenza A_2 Hong Kong could be experimentally developed to an ultimate mutant that would be antigenically stable and would confer protection against infection with any naturally occurring variant of A_2 Hong Kong. They produced laboratory evidence to support their claim, but no clinical trials in man had been done at that time, nor have reports of any appeared since.[10]

3. Is the Right Sort of Antibody Formed?

Following an intramuscular injection of a vaccine, the antibodies in the blood rise and may be readily detected in a serum sample by appropriate tests. In the case of diseases that have a viremic phase, such as poliomyelitis and yellow fever, these circulating antibodies quickly inactivate virus in the blood; this is why intramuscular vaccines give such good protection against this type of disease. However, circulating antibodies do not readily find their way into the mucus secretions of the respiratory and intestinal tracts; for this reason, intramuscular vaccines offer much less protection against diseases of the respiratory tract and the gut, particularly when only the superficial epithelium is involved, as in the case of influenza, the common cold, and virus gastroenteritis.[11-16] The same principle may also account for the recurrence of cutaneous and ocular herpes infections in human subjects, most of whom have very high levels of circulating antibody.[17,18] In the case of herpesvirus, the cellular immune function may be as important, if not more so, in limiting recurrent attacks of the disease.[19] In order to protect secretory epithelia, local secretory antibody must be stimulated which, in the case of viruses, means establishing an infection with attenuated live virus.

4. Does the Antibody Last?

A further problem with vaccination is the duration of immunity. Following immunization with an intramuscular vaccine at the time of peak antibody levels in the blood, sufficient circulating antibody finds its way into mucus secretions to give a certain degree of protection to the underlying epithelium. This is the principle by which intramuscular influenza vaccines work. However, when circulating antibody declines, the protection of superficial epithelia falls rapidly, which accounts for the relatively short and variable duration of protection offered by intramuscular influenza vaccines.[20] This would also be true of similar vaccines against infections of the respiratory or digestive tract epithelia. Attenuated live-virus vaccines stimulate local secretory antibody from the epithelium in which they grow, therefore giving very good local protection against infection with virulent virus. Here again the duration of protection is limited because when the vaccine virus infection is complete, the antibody-producing cells in the mucosa gradually disperse and the epithelium once more becomes vulnerable to infection. There is a recall mechanism so that local immunity can be reestablished quickly in the event of reinfection. Live-virus vaccines thus offer a better prospect for control of local infections of, say, the respiratory or gastrointestinal tract than intramuscular vaccines.

5. Will Immunization Worsen the Disease?

Leaving aside the purely technical aspects of the toxicology, local irritancy, and sensitization reactions associated with certain types of vaccine, there is at least one well-known example where the use of an experimental vaccine actually made the subsequent disease more severe rather than less so.[21-23] This unexpected anaphylactic response was seen in children vaccinated intramuscularly against respiratory syncytial virus. The vaccine was well tolerated with no serious side effects, and good levels of circulating antibody appeared. However, when the vaccinated children came into contact naturally with the disease, not only were they not protected from infection, but their symptoms were much more serious than they would ordinarily have been. The protracted bronchiolitis, which respiratory syncytial virus normally causes, was substantially increased, and there was serious involvement of pleural tissue, which is normally an exceptional occurrence in this disease. The underlying reason for the increased severity of the disease seems to be that in addition to the usual cytopathic effects of the virus on the lower respiratory mucosa, the newly formed virus combined with the circulating antibody, resulting from the vaccination, causing a serious allergic reaction with consequent worsening of the clinical symptoms. It has been suggested that the stimulation of local secretory antibody, rather than circulating serum antibody, may give adequate protection against respiratory syncytial virus infection without unwanted hyperreactivity,[24] but no clinical work has yet been done.

6. If a Vaccine is Available, Will it be Used?

This may seem a curious question, but the fact remains that although there are good vaccines against a number of infectious diseases (bacterial as well as viral), there is a definite reluctance on the part of the general public to use them. How many children today in Western Europe or North America are vaccinated against diphtheria or cholera? Even antitetanus vaccines are used only after injuries. It is true that the incidence of these serious diseases has declined of late and antibiotics now offer alternative protection, but the resistance to vaccination is also obvious in the case of virus diseases where no comparable alternative therapy exists; nor can it be that the virus diseases are less serious than bacterial ones. Smallpox vaccination is now seldom done routinely in areas where the disease is not endemic, and poliomyelitis vaccine again is not used as widely as might be expected.[25,26] The same is true in veterinary medicine. The scourge of poultry virus diseases can be kept in check by the excellent vaccines now available to the farmer; yet never a year passes without serious outbreaks of poultry diseases, always because vaccination has been omitted.

The reasons for this widespread reluctance to

use vaccines are many and varied – psychological, financial, and administrative. Collectively they will have an important bearing on the possible market for a potential antiviral drug.

The pros and cons of immunological control of virus diseases have been dealt with at length because vaccines offer the most serious challenge to the development of antiviral drugs. Therefore, the scientist wishing to work in the field of antiviral drugs must consider carefully those diseases that he thinks are worthwhile targets and assess whether or not they will, at some foreseeable date, be controlled by vaccines. This does not simply mean whether or not an effective vaccine will become available, but whether such a vaccine will find a widespread and effective practical use.

C. What Other Factors are Relevant?

In addition to considering the economic status of a virus disease, and also the possibility of its control by vaccines, there are other factors which also must be taken into account in deciding whether there would be a market for an antiviral drug. Each virus disease must be considered in detail, and the special factors associated with it must be analyzed to determine what bearing they may have on the commercial future of a potential drug. To a certain extent, each virus disease will have its own pecularities, but a discussion of some considerations of general relevance may be useful.

1. Is the Etiological Agent Properly Defined?

It may seem needless to point out that before undertaking research into developing a drug for a disease, one should be sure that the causative agent is clearly known. However, in diseases such as bovine pneumonia and human epidemic gastroenteritis, while many infectious agents have been implicated, there is as yet no real evidence as to the causative agent of these diseases. In the case of bovine pneumonia, it may well be that the disease syndrome is induced by a combination of agents, none of which could cause such a severe disease on its own. In human virus gastroenteritis, while the causative agent may be one of the fastidious organisms just now being characterized,[27-36] it may equally well be that again the disease is caused by an agent or agents yet unknown. In any event, it would be unwise to set up a screen against an organism which was presumed to be the causative agent of a poorly understood disease, only to subsequently find that another agent was more important.

2. How would an Antiviral Drug be Administered?

This aspect of the commercial analysis has too often been neglected, when in fact the route of dosing a potential drug may be of sufficient importance to decide its whole future. There are two considerations: the route and the timing. A few examples should help to stimulate thinking in this important area. First, what is the best route of administration? In the laboratory drugs are most often given to experimental animals by injection (intraperitoneally, intramuscularly, subcutaneously, and occasionally intravenously or intracerebrally) – the object being to deliver the drug quickly and conveniently to the appropriate place. In human medicine, and particularly with prescribable drugs, parenteral routes are used less frequently because oral administration is more convenient and more acceptable to the patient. A drug that is eventually envisaged as being off prescription (the so-called "over-the-counter" preparation) could only be given orally or perhaps intranasally. Thus, for instance, although an experimental compound given intraperitoneally to mice infected with influenza may protect the animals, unless the compound also shows protective activity when given orally, or possibly intranasally, its commercial prospects are nil. The same applies to any "common cold" drugs: only oral or intranasal dosing will be acceptable in any marketable preparation. This does not mean that oral activity will inevitably be necessary in all antiviral drugs. In many veterinary applications involving large animals, it is much easier to medicate by injection than to attempt oral dosing, as any veterinarian will testify. A further advantage of an injectable for veterinary application is that a long-lasting formulation could be given which would allow prolonged release of drug. This is an important consideration with diseases of herd animals where the labor of dosing is considerable.

Herpetic diseases present a range of options for dosing routes. At present, cutaneous and ocular herpes simplex are both normally treated topically with iododeoxyuridine in a bland solvent or ointment, but deep-seated infections do not respond well to topical application. It may well be that for any new drug to be successful in cutaneous herpetic disease, it is more important to devise a new method of delivery than to improve its intrinsic antiviral activity. Such was the case with Zoster: the success of topical therapy with iododeoxyuridine is largely due to the use of dimethylsulfoxide as a solvent and vehicle to carry the drug deep into the infected skin.[37-39]

Having discussed the question of the route of dosing, the timing must now be considered. Unless there is an opportunity to administer an antiviral drug to an infected individual, human or animal, in time for it to reduce the severity of the ensuing virus disease, the drug will have little clinical application. It is often supposed that drug treatment of the common cold will never be a practical proposition because by the time the main symptoms of the disease develop, much of the virus growth is over. Unfortunately, until good antirhinovirus compounds are available, this premise cannot be put to the test. However, data on the major virus growth phase of upper respiratory tract infections suggest that it is not excessively short, and under practical conditions there should be ample time for medication. For example, in a study of volunteers infected intranasally with Coxsackie A21 virus, Patel et al.[40] found that although the phase of maximum virus growth preceded the onset of symptoms by 24 hr, there was a considerable amount of virus growth concurrent with the severest symptoms over a period of 2 to 3 days, which should be amenable to treatment. Also, in a further human volunteer experiment, Douglas et al.[41] found that the period of rhinovirus growth following intranasal infection could be as long as 4 days before the severest symptoms developed, which again should give a useful period for treatment. Certainly in the case of influenza A_2, the course of the illness can be favorably modified by treatment with adamantanamine even after the onset of symptoms[42-45] (see Chapter 9 for further details).

Returning to the common cold, this disease raises special problems in making a common cold drug available to the general public. For such a drug to be a medical and a commercial success, it must be freely available through pharmacies and drugstores rather than be prescribable by doctors. This raises a host of toxicological and medico-legal questions which must be faced and resolved realistically if the public is ever to benefit from a potential cold cure.

One final point to be considered in assessing the prospects for any potential antiviral drug is the effect of changes in existing legislation regarding the control of particular diseases. For example, foot-and-mouth disease and rinderpest together cause enormous economic losses among livestock around the world, and an effective drug may find widespread use in certain areas. However, in some countries these diseases are controlled by a policy of slaughter and compensation; should this policy be extended, the potential market for an antiviral drug would be correspondingly reduced. The same is true of swine fever and as a testimony to the effectiveness of the slaughter policy, there has been no case of the disease in the United Kingdom for over 30 years.

To sum up, there are a number of important questions which must be answered when deciding on the commercial future of drugs against virus diseases and if research should be undertaken.

1. Is the disease sufficiently important to generate adequate sales of drug?
2. Can vaccines be developed which will eventually control the disease?
3. Is the etiological agent clearly identified?
4. Could a drug be administered successfully under practical circumstances?
5. Will changes in existing legislation have an adverse effect on the drug market?

III. THE DESIGN OF ANTIVIRAL SCREENS

A. Tissue Culture Screens

Most of the major pharmaceutical manufacturers have programs of antiviral research. At the conference on antiviral substances held under the auspices of the New York Academy of Science in 1965, many commercial firms disclosed for the first time the details of their test procedures.[46] This forum was so successful that a second was held in 1970 and a third in 1976. Collectively the published proceedings of these conferences give a good idea of how various workers have approached the problems of devising antiviral screens and how thinking and methods have altered over the past decade.[47,48] The worker in this field will be familiar with all of this, but to someone beginning in this type of work, these changing attitudes may be less obvious. This is partly because the earlier work tended to appear very fully detailed in print, while the later developments either have not appeared or have been presented with less of the authoritarian, even dogmatic, style which characterized earlier publications. Therefore, it is of value to review the methods which are available for in vitro screening and include comments, particularly on practical matters, which someone entering this field may find of value.

1. In Vitro versus In Vivo Screening

Before tissue culture techniques had been reduced to simple routines, antiviral testing was carried out in mice. This was an expensive and tedious procedure, but it had the advantage that if a compound showed activity, the problems of drug absorption, distribution, metabolism and so forth were all answered – at least in the mouse. Since the advent of tissue culture, virtually all antiviral testing has been done in vitro, and only when active compounds are discovered is animal testing used. It is worthwhile to consider the implications of this change, which was made not for any very rational considerations, but only to ease the burden of antiviral screening. In vitro screening can test many more compounds than the same effort put into an animal screen, and this important point is discussed in more detail later; an in vitro screen will not detect compounds that induce interferon formation or stimulate the host immune response, and these are two potentially valuable approaches to the control of virus disease. Also it is not likely that an in vitro screen would detect antiviral metabolites of test compounds which, in theory at least, is another approach to antiviral chemotherapy.

Therefore, one must bear in mind that although most workers are committed to large-scale tissue culture screens, these will generate no significant discoveries in the field of interferon induction or immune stimulation. If a worker has any serious interests in either of these areas, the only approaches are speculative testing in small laboratory animals or special in vitro tests designed to detect this type of activity.

2. Which Viruses?

Having decided, after a suitable commercial analysis, which diseases would be commercially viable drug markets, the first practical question is which virus or viruses should be used as the test organism? At one time it was thought that since the basic metabolic features of nucleic acid replication, protein synthesis, and assembly of compounds into virions were common to all viruses, more or less any convenient virus which grew well in the laboratory would be suitable as a test organism in an antiviral screen (even a phage). It is now clear that, despite these basic similarities, there are many other features which give each virus its own distinctive pattern of replication, and to ignore these as possible points of attack on the virus metabolism is simply to reduce one's chances of success.

Wherever possible, the virus that causes the clinical disease should be used as the test organism, except in the case of dangerous viruses such as smallpox or rabies, when attenuated or vaccine strains should be used instead. With certain test systems, notably the "zone-inhibition" and "pH-swing" tests (see below), the choice of test virus is often seriously restricted for technical reasons. Here there is a danger of regarding the smooth operation of the test system as an end in itself rather than as a means to discovering relevant antiviral activity. This tendency was most noticeable in some of the earlier test procedures where viruses that were easily manageable rather than clinically relevant were often used as the test organisms. For instance, tests have been described in which parainfluenza type III, measles, and polio,[49] and pseudorabies, adenovirus 3, and mouse hepatitis virus[50] were the test organisms. While such tests may well reveal interesting antiviral compounds, unless the antiviral activity extends to more important human or animal pathogens, they will be of very little use; leaving aside interferon inducers, it seems that antiviral compounds are rather restricted in the range of viruses they are active against.

The simple approach of using the disease-causing virus rather than some related but more convenient cousin gives a good guide to the choice of test viruses; in some instances, however, other factors must be taken into account. For example, in a rhinovirus screen, which of the 89 serotypes is the one to choose? The answer is probably to select any one which is easily grown because an important property of any clinically useful anti-rhinovirus drug is that it must be active against most, and preferably all, of the rhinovirus serotypes. Thus, the choice of a test virus is simplified.

In the case of influenza A, although the disease is well understood and the causative viruses are well characterized, choosing a strain for use in an antiviral screen presents certain problems. Unless recognized at the outset, these may lead to a drug which has no commercial future.

At present, the prevailing influenza A serotype

is A_2 Hong Kong (H3N2), which differs antigenically from the previous A_2 Asian (H2N2) and the preceding A_1 (H1N1), A_0 (H0N1), and A swine (H_{swine}1N1) influenzas which have successively infected the human population since the beginning of this century (see Chapter 3, Volume I). If one decided to set up a screen to look for activity against influenza A, which of these viruses should one choose? One possibility is A/Hong Kong/I/68 (H3N2) on the grounds that it is the current serotype, and any drug which was discovered would be immediately applicable to human medicine; since it may take up to a decade to discover, develop, and introduce a potential A_2 Hong Kong drug for clinical use, it is likely that A Hong Kong/1/68 (H3N2) will by then have given way to another serotype – maybe the long-awaited A_3. We have no means of knowing what A_3 will be like. It may resemble one of the previous serotypes, it may be a relative of the current strain; or it may be an entirely novel virus.[51,52] To the virologist trying to set up an anti-influenza screen, the consequences are much the same: he is faced with discovering a drug against a disease organism which does not yet exist – surely a paradox with no parallel in any other field of research.

The foregoing examples serve to illustrate that the choice of the test virus in an antiviral screen, although primarily dictated by which diseases are considered to be worthwhile targets, is often far from an easy decision. Nevertheless, since many years of effort follow the setting up of a screen, the choice of the test virus (or viruses) clearly warrants most careful consideration if this effort is to be worthwhile.

3. Which Cells?

Most of the published in vitro screening procedures make use of neoplastic cells, HeLa and KB as human cells and L cells, RK13, and BHK as animal cells. Neoplastic cells have a number of characteristics which make them particularly attractive for screening procedures: they are available in virtually unlimited quantities; they perpetuate their characteristic morphology and metabolism indefinitely; they will often survive well under agar overlay; and because of their high growth rate and metabolic capacity, they can be made to do useful tricks such as taking up vital dyes or radioactive tracers in large amounts or swinging the pH of their medium. Above all, they support the growth of a wide range of viruses. For these reasons, neoplastic cells have found widespread use in large antiviral screening programs. However, it must be remembered that all of these seemingly desirable features are a direct result of the basic abnormality of the cells, and to use a neoplastic rather than a normal cell is to make an already artificial test even more remote from the natural infection which one hopes ultimately to treat.

It may be that when the next antiviral drug comes along, it will have been discovered in a test using neoplastic cells. There is at least one example of a class of antirhinovirus compounds which, although they are highly active in normal diploid human cells, are virtually inactive in heteroploid cells and thus would have been missed in the majority of screens being run today.[53,54]

4. What Else?

There are two further important aspects of antiviral screening which must be mentioned. What scale of testing is required, and what sort of compounds should be tested? Both are difficult questions to answer, but some guidance can be given at least in the scale of testing.

In a screen that is nominally random (that is to say that the compounds tested are not selected with any preconceived notion of their being either active or inactive) the likelihood of discovering antiviral activity is directly proportional to the number of compounds tested. There is little published information on the percentage of randomly tested compounds which show activity. Buthala[55] quotes 6% and Miller et al.[83] quote 11%, but in our experience between 1 and 0.01%, depending on the test virus, is a more realistic estimate. Bauer[56] has suggested that the larger viruses should be intrinsically more susceptible to inhibition than smaller viruses because both their physical size and metabolic complexity should make them easier targets. It is certainly true that on a random screen antiviral activity against influenza viruses is more frequently encountered than activity against picornaviruses, but a contributory factor is that the route of entry of influenza virus into its host cell via a receptor on the cell surface is a particularly vulnerable function. In the case of herpes viruses, leaving aside inhibitors of DNA synthesis, in our experience it is rare to find antiviral activity in a random antiviral screen, and this is a virus of great size and complexity. Perhaps the most important factor in determining the susceptibility of any particular virus to inhibition is not so much the complexity of its metabolism, but rather whether its metabolic processes are very similar to those going on in the host cell where it will replicate. If there is a close similarity between the virus-coded and host cell-coded synthetic and metabolic processes, then there is little likelihood of interfering with viral mechanisms without at the same time disturbing the analogous host cell processes.

In any event, large or small, simple or complex, viruses are difficult to hit with drugs. To stand any real chance of success, a random antiviral screen must be capable of examining a large number of compounds each year. Given that between 1 and 0.01% of compounds will show some activity, depending on the test virus used, a screening rate of several thousand compounds annually will yield enough leads to be used for further evaluation.

There are two important points regarding the choice of compounds for testing. The first is to ensure that the selection taken from, say, an extensive collection of synthetic chemicals is random in the sense that it covers the widest possible range of diversity in chemical structures without any hidden biases. This is often difficult to achieve since most chemical collections contain large families of related compounds, and the selection procedure should be weighted to provide appropriately less emphasis on large groups of analogues and correspondingly more on chemical types which are less well represented. This means, in effect, that computerized structure-searching facilities are needed, and this service is not always available.

The second point is that, whenever possible, natural products such as fungal and bacterial metabolites should be included for screening, if they are available. Although their testing presents certain technical difficulties, natural products offer a wealth of structural diversity, and hence potential biological activity, which is not found in synthetic chemicals. As such, they should certainly be included for screening if they are available.

B. Published Test Procedures

The various methods used for antiviral screening will be given in chronological order rather than in any order of relative merit. In this way, the development of antiviral screening can be seen, from simple tests in mice used in the 1950s to the cell-free virus enzyme inhibition assays which are available today (see also Chapter 5, Volume I for examples of the latter). Many of the tests fail to meet one or more of the criteria outlined in the previous pages. However, since any in vitro test is a compromise between what is desirable and what is practical, the reader must judge for himself whether the deficiencies of any one test are sufficiently serious to preclude its use for any particular application.

There are two effects which an antiviral test must measure: the concentration of the test substance that is toxic to the host (be it animal, egg, or tissue culture cell) and the concentration that inhibits virus growth. The virus-inhibitory effects of test substances are easily measured by any one of a number of methods – infectivity titration, hemagglutinin production, cytopathic effects, fluorescent, or peroxidase-linked antibody microscopy, etc. What is more difficult to determine is the toxicity of a test substance, since the concept of toxicity changes according to circumstances. The measurement of toxicity will be dealt with in greater detail later since it is of prime importance in characterizing what is, and what is not, true antiviral activity.

1. Tests in Mice and Whole Eggs

Much of the early work on antiviral screening was done simply by injecting test compounds into mice (usually intraperitoneally) or fertile hens' eggs, infecting with a lethal virus, and looking for a lengthening of the survival time of the infected host. A refinement of this method was to harvest either some internal organs, in the case of mice, or allantoic fluid, in the case of eggs, and measure the amount of infectious virus produced.

The merits of in vivo screening procedures have been mentioned earlier, the chief one being that the pharmacology of the test compounds is investigated simultaneously with their antiviral properties. However, animal and whole egg screening procedures are very time consuming, and although they produced a few active leads,[57,58] they were eventually superseded by in vitro tests, which are faster and can test large numbers of compounds.

2. Tests on Chorioallantoic Membrane Pieces

While tissue culture techniques were being developed in the late 1940s and early 1950s, other workers turned their attention to making better use of the hen's egg. Whole, fertile hens' eggs had already proved to be very useful in the cultivation of influenza viruses in the laboratory, when it was realized that if the contents of the fertile egg were removed after 10 days or so of incubation, the shell of the egg carried the composite chorioallantoic membrane firmly attached to it on the inside. This membrane was an excellent tissue for supporting the growth of certain viruses, notably influenza and vaccinia.[59,60] The shell and its attached membrane could be conveniently cut up into small pieces about a ¼ in. square, and each piece could be maintained in a synthetic medium for several days (for technical details, see Appendix of Chapter 5, Volume I). Under these conditions the chorioallantoic membrane remained viable and could be infected with viruses. Using this system, Tamm and his colleagues developed an early antiviral screening technique and used it extensively to study what appeared to high antiviral activity in a series of chlorinated benzimidazoles and their ribosides.[61-63] The compounds were added in a series of concentrations to virus-infected chorioallantoic membrane pieces, and after a suitable period of virus growth the culture fluids were harvested and assayed for virus. The yield of virus was then plotted against the concentration of test compound, and a dose-response curve was obtained. This type of yield reduction assay for measuring antiviral activity is perfectly sound and is still in use today, but, retrospectively, it can now be seen that the chorioallantoic membrane test has a major defect which makes it unsuitable for general use as a screening procedure. The problem arises in assessing whether or not test substances are toxic to the cells of the chorioallantoic membrane. Tamm and his colleagues attempted to do this using pieces of membrane stripped from the shell.[64] They observed that under normal conditions stripped membrane pieces curled up during their culture due, no doubt, to the differential growth or fluid uptake of one side of the membrane. This curling was a characteristic feature of healthy membranes, and if they were treated with a toxic concentration of a test compound, the curling was prevented. Therefore, Tamm and his colleagues used this as an index of a test compound's toxicity, which it undoubtedly is. Unfortunately, it seems to be a particularly insensitive measure, and it is now clear that the series of halogenated benzimidazole ribosides which were studied extensively by this technique are not specific inhibitors of virus growth but are in fact general inhibitors of nucleic acid synthesis — RNA and DNA of both viruses and host cells.[65-68]

3. The pH-Swing Test

With improvements in tissue culture techniques and the availability of cells that would support the growth of a wide range of viruses, the first true tissue culture screening procedure was developed by Rightsel and his colleagues.[69] This was based on the observation that when cells are healthy, they swing the pH of their culture medium from alkaline to acid, principally by the production of carbon dioxide. The test was carried out in small vinyl plastic trays with moulded wells, which today would be called "disposables." Each well was seeded with 30,000 cells and infected with a standard challenge of virus. Test compounds were added in a series of concentrations to rows of wells and the plate sealed with a transparent adhesive tape. After incubation for 5 to 7 days, the wells with healthy cells showed a color change from pink to yellow (with phenol red as the indicator). The wells in which the color had not changed contained cells that had been killed either by the destructive effect of the challenge virus or by the toxic effects of the test compound. Thus, compounds with antiviral activity were readily detected by looking for wells in which the color had changed from pink to yellow. Of course, solutions of test compounds must be carefully adjusted to neutrality before being used in this type of test if reliable results are to be obtained, but with proper attention to technical details, the test works well and is capable of screening large numbers of compounds. Because of the small volumes involved, the test is not extravagant with test compounds or tissue culture cells. The original authors described how the incubation period of the test could be reduced by adding a thin layer of mineral oil to each well to trap the carbon dioxide produced. They demonstrated the versatility of the test by using it not only for antiviral drug assays, but also for virus infectivity and antibody titrations. A similar test has also been used for interferon assay.[70] Because the test relies on the destruction of host cells by the virus challenge, it may not give clear results with viruses that are not strongly cytopathic, for example, influenza or respiratory syncytial virus. Also, the host cells must be reasonably robust to survive without fresh medium for up to a week in order to swing the pH sufficiently. These two factors inevitably restrict the usefulness of the test somewhat, but otherwise the test is commendable in its simplicity, reliability, and high testing capacity. A demonstration of the pH-swing test is shown in Figure 2, which illustrates the ease with which antiviral compounds can be detected.

4. Zone-Inhibition Test

This test is a clever adaptation of the principle, well known in the detection of antibacterial activity, of allowing the test substance to diffuse from a small reservoir or paper disk placed on an agar culture, creating a zone of inhibition of bacterial growth. The radius of the zone gives a measure of the strength of the antibacterial activity. DeSomer and Prinzie[71] and Porterfield[72] described the technique for assaying poliovirus antibody and interferon, respectively, but it was Herrmann[73] and his colleagues who developed the technique for screening large numbers of substances for antiviral activity.

The procedure is not unduly complex from a technical point of view, as it is basically an adaptation of the plaque assay system. Cells are

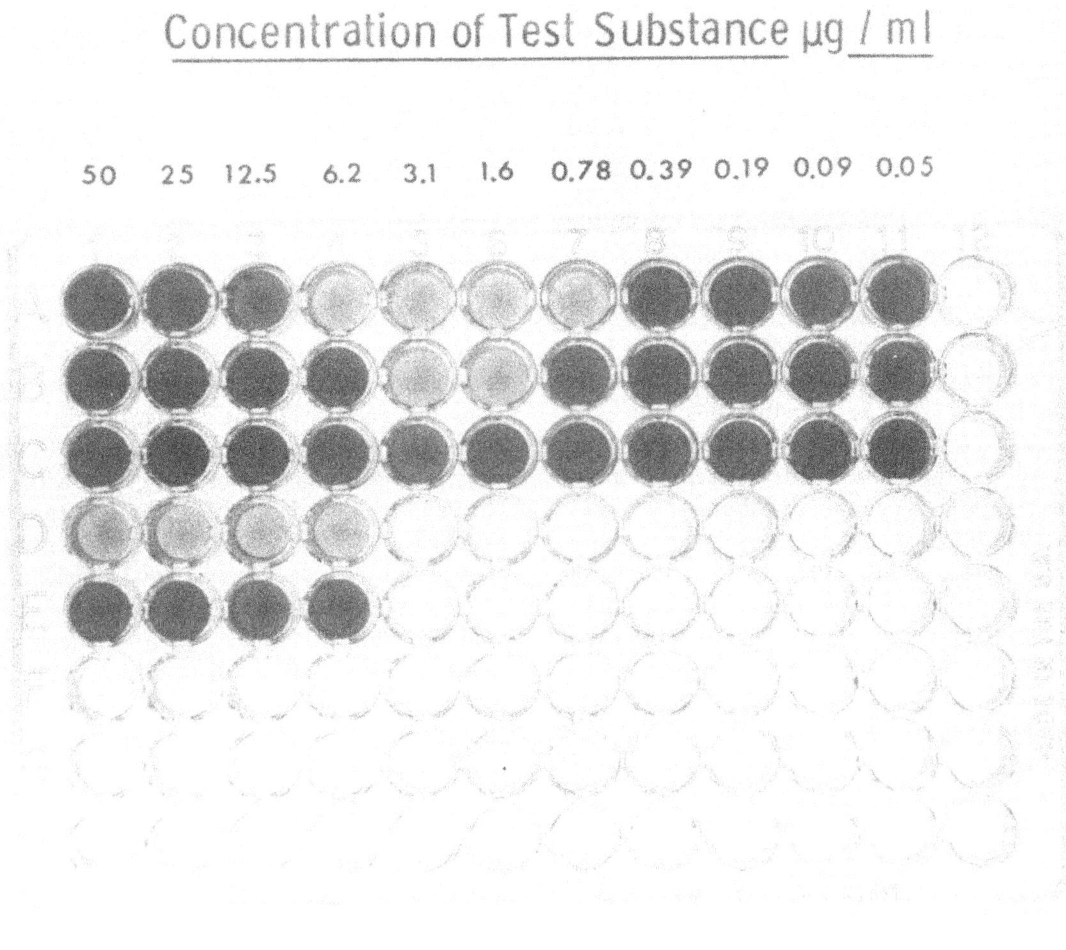

FIGURE 2. The "pH-swing" method for detecting antiviral compounds. The test is run in the first five rows of the plate, and the darker color represents a change of the indicator (phenol red) to rose pink. Row E was seeded with HeLa cells and infected with a cytopathic virus. After 4 days the cells were destroyed and the medium remained pink. Row D was seeded with HeLa cells, but not infected. The indicator has changed from pink to yellow. Row C carries serial twofold dilutions of a test compound together with HeLa cells and challenge virus. All the wells are pink, showing that the cells have died either from the toxic effects of the test compound or the destructive effects of the challenge virus. The compound thus shows no antiviral activity. Row B carries serial twofold dilutions of a compound with some antiviral activity. Down to 6.2 μg/ml, the compound is toxic. Between 3.1 and 1.6 μg/ml, the compound is not toxic, and it has protected the cells from destruction by the challenge virus. Below 1.6 μg/ml, the compound is too dilute to protect the virus-infected cells. Row A carries serial twofold dilutions of a better antiviral compound. It is less toxic and more active than Row B.

grown in petri dishes or other shallow containers; when confluent monolayers have formed the cell sheet is infected with the test virus. Enough virus is added to produce a large number of infectious foci. The infected sheet is then covered with fresh medium containing agar which is allowed to set. Test specimens are placed on the surface of the agar. These can be disks of filter paper, about ¼ in. in diameter, which have absorbed the test substance or small cylindrical reservoirs containing a solution of the test substance or even round holes punched in the agar to form wells.

When the test plate has been prepared, it is incubated, during which time the plaques of virus growth develop in the cell sheet. At the same time, the test substances diffuse out radially from where they have been placed on the agar, setting up a radial concentration gradient above the infected cell sheet. The test plate is then stained with either a tetrazolium stain or a vital dye, such as a neutral red, which shows up the living cells and reveals the virus plaques and dead cells as nonstaining areas. With an active antiviral substance, there will be two zones visible around the test disk or well. The inner zone will be dead cells killed by the high concentration of test substance in the immediate vicinity of the disk. Surrounding this will be a deep-staining zone of healthy cells which have been protected by the intermediate concentration of the test substance, and beyond, the cell sheet peppered with virus plaques which are too fai from the test substance to have been inhibited. The merit of the test is the ease with which large numbers of substances can be handled. Herrmann describes how up to 40 substances on filter paper disks can be tested against vaccinia in a single 10 × 6½ in. dish of chick embryo cells and how the test can be adapted for any virus which will produce plaques under agar.

The test has a few drawbacks, chief among these being that it only works well with viruses that give good plaques, which essentially means the more cytopathic viruses. The test is also relatively insensitive, and because of the difficulty in distinguishing small zones of virus inhibition, some of the less active leads may be missed. In addition to the difficulty of seeing small zones of stained or unstained cells, the radial gradient is not linear, an intrinsic property of any diffusion system. Thus, the biological effects of a test compound (toxicitiy or antiviral activity) are very rapidly lost with increasing distance from the reservoir. We have found, using the zone-inhibition test, that it is necessary to use compounds at concentrations roughly ten times greater in order to detect the minimum toxicity than is necessary in, for example, the neutral red uptake test or the CPE-inhibition test (see next section).

5. CPE-Inhibition Assays

Perhaps the simplest, and therefore the most reliable, form of antiviral assay is that in which one looks for a reduction of the cytopathic effects caused by the infecting virus. This particular form of test was selected for use in our own laboratory some years ago after experimenting with the assays previously described, and we have had no reason since to regret that choice.[5] Besides the merit of simplicity, the test is versatile and can be used with viruses which do not, for example, form plaques under agar and with cells that are fastidious in their culture requirements. It is economical in cells, test compound, and labor, and a single operator can easily screen between 100 and 200 compounds each week.

In our laboratory we use conventional tissue cultures grown in 3 × ½ in. glass tubes. Test compounds are made up and diluted in fourfold steps to give a range of concentrations that are added to the culture tubes: 1 ml of each dilution to a single tube. The tubes are then infected with virus and incubated for 48 hr, the challenge being sufficient to give good virus growth during this time. The tubes are then examined individually by a low-power microscope (100×) and scored for obvious cytopathic effect. With a little practice, the operator can readily distinguish between cytopathic effects due to virus growth and those due to the toxicity of the test compound, but this is hardly necessary, since the objective of the test is to find a tube of healthy looking cells which is affected by neither toxicity nor virus growth. The microscopic examination is thus only cursory, and a trained operator seldom spends more than 5 sec inspecting an individual tube. Figure 3 gives an indication of the range of cytopathic effects due to compound toxicity met with in a screen using human diploid lung cells and illustrates the ease with which they can be detected. In practice, it is best to do no more than about 1 hr of microscopic work at a time; in 1 hr, using four dilutions for each test compound and a simple manual recording system, a single operator can screen over 100 compounds. In the case of hemadsorbing

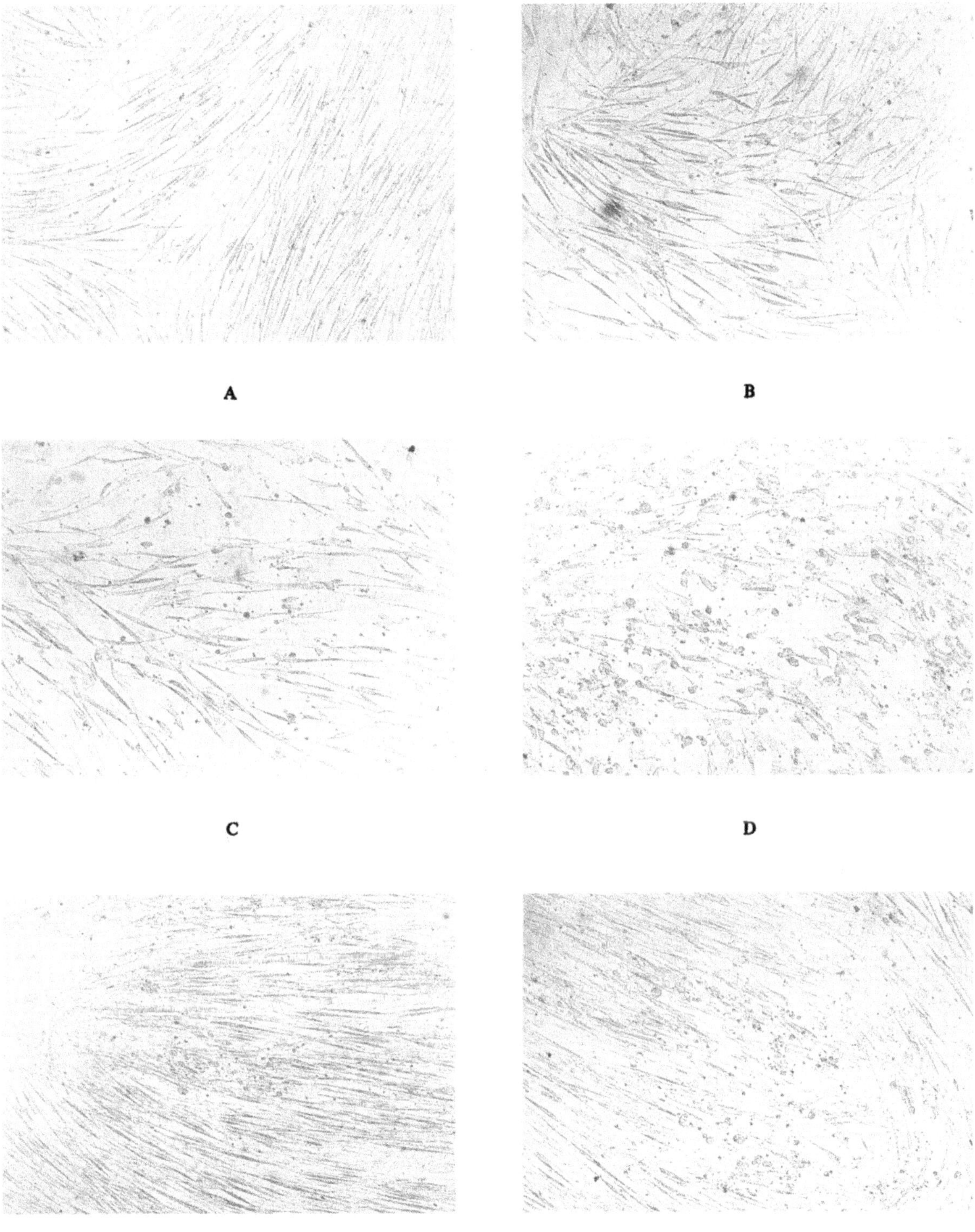

FIGURE 3. The microscope appearance of human diploid lung cells (a) before treatment and (b, c, and d) after 24 hr treatment with chemical compounds of increasing toxicity. The generalized cytotoxic effects of the compounds added to each culture can be clearly seen, even in the case of Figure 3b where the cytotoxic effect is slight and can be readily distinguished from the focal cytopathic effects which most viruses cause in the early stages of infection. (e and f) Rhinovirus type 2 cytopathic effects are surrounded by relatively normal cells, thus presenting a localized lesion in the cell sheet.

viruses, the reading of the test can be made easier by treating the tube cultures with red blood cells to make the virus growth more obvious.

The test is reproducible and precise enough for routine purposes. If the range of concentrations of test compounds is carefully chosen, the number of compounds which need retesting can be reduced to less than 10%. Apart from technical mishaps, the need for retesting arises when a compound is toxic at the lowest concentration tested. This means that a further series of dilutions and a repeat test are needed. An analysis of the data on the toxic concentrations obtained from several thousand random compounds is shown in Figure 4. It can be seen that virtually all compounds show toxic effects in a simple tissue culture test lasting 3 days at concentrations over 200 μg/ml. Thus, there is little point in using concentrations in routine screening higher than 200 μg/ml; 200 to 1 μg/ml gives the most efficient use of material and labor in a primary screen. In theory, the situation could arise where a compound had antiviral activity and was not toxic at, say, 1000 μg/ml, but was not active at 200 μg/ml; in an antiviral screen where the highest concentration tested was 200 μg/ml, such a compound would be missed. However, in practice, since the chance of encountering a compound which is not toxic at 200 μg/ml is about 0.50% and the chance of finding a compound with antiviral activity may be as low as 0.01%, the chance of these two properties occurring in a single compound is 1 in 2,000,000. Even if one tested 2 million compounds, there would still be only a 50% chance of finding this hypothetical compound. Therefore, one is faced with a choice. Does one put in the extra effort on each compound tested by using an extended range of concentrations in the hope of detecting this rarity, or does one use the effort to examine other compounds? The answer is fairly obviously the latter since by looking at random compounds, even with a truncated concentration range, one is faced with a 0.01% probability of success, whereas with the former the chance of success is 0.0000005%. If this does not persuade the skeptics, there is the further consideration that any antiviral compound that is active only at concentrations in excess of 200 μg/ml in vitro would present serious problems in achieving useful tissue levels in vivo. Assuming ideal absorption and distribution, and slow metabolism and excretion, in order to achieve 200 μg/ml or more in the blood and tissues, the whole body dose would be of the order of 2000 mg/kg, which makes a single human dose a daunting 14 g. The range of concentrations tested in a routine screen is a matter for the individual to decide, but factors such as those just discussed should be taken into account if best use is to be made of whatever facilities are available.

A CPE-inhibition assay system similar to that just described has been used by Walker et al.[74] to provide a clinical aid in the selection of antiherpes drugs. Using WI38, monkey kidney, or HeLa cells, they showed that it was possible to assess the drug sensitivity of strains of herpes viruses within 1 to 2 days and thus facilitate the choice of drug for clinical use. They used the inhibition of virus CPE as a measure to define the antiviral activity of iododeoxyuridine, fluorodeoxyuridine, and cytosine arabinoside, and used the cytopathic effects of the drugs as an index of their cytotoxicity. Unfortunately, with this particular class of drugs (which are all DNA inhibitors) the use of cytopathic effects as a measure of cellular toxicity can

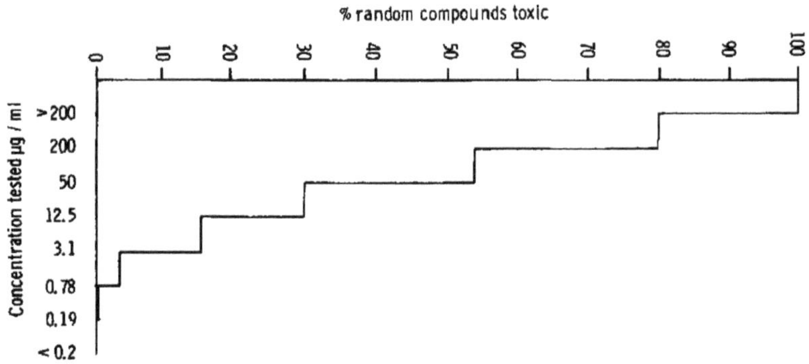

FIGURE 4. The toxicity, assessed in a CPE assay using human diploid lung cells, of 8000 randomly selected chemical compounds.

be misleading. Tissue culture cells are almost always confluent when used in viral tests, and cell division, and hence DNA synthesis, is virtually at a standstill. Inhibitors of DNA synthesis therefore have very little obvious effect on these cultures and appear to be relatively nontoxic, whereas if the inhibition of cell growth, or DNA synthesis, by such compounds is measured, very often it turns out that there is no selective effect on virus growth. For example, Bucknall et al.[75] showed that cytosine arabinoside was actually more inhibitory to the DNA synthesis of human diploid lung cells in culture than to the growth of herpes simplex virus.

A refinement of the CPE-inhibition assay was described by Finter who devised a method of quantitating the degree of CPE by measuring the uptake of the vital dye neutral red.[76] The test is essentially the same as the conventional CPE-reduction assay except that instead of reading the CPE microscopically, a solution of neutral red is added to all the tubes and they are incubated at 37° for 1 hr to allow the dye to be taken into the healthy cells. The unabsorbed dye is then rinsed off the cell sheets, and the intracellular dye is released by adding 70% alcohol. The concentration of the alcoholic dye solution is measured spectrophotometrically and is inversely proportional to the amount of cytopathic effect. The method will measure CPE due to the toxicity of test compounds and CPE due to virus growth, but to be successful as a virus assay the virus used must produce a pronounced destructive effect on the host cells. The method is very suitable for highly lytic viruses such as arborviruses, enteroviruses, and herpesviruses, but less so for myxo- and paramyxoviruses. Finter also described a method for quantitating hemadsorption caused by myxoviruses,[77] and this method is readily combined with the dye-uptake method to give a fully quantitative technique for measuring both compound toxicity and myxovirus growth.[78] It was with this method that the highly active anti-influenza compounds, the transdecalins, were discovered.[79] More recently, a quantitative microscopic hemadsorption method has been described which involves infecting cultures of 5C4 cells with hemadsorbing viruses and counting the number of hemadsorbing cells which subsequently develop.[80] The authors claim that the method is precise and reproducible, but it seems unlikely that it could be adopted for antiviral screening because of the labor involved in each assay.

A typical assay of an anti-influenzal compound done by Finter's quantitative techniques[78] is given in Figure 5. The toxic effects of the compound are measured by neutral red uptake, and the inhibition of two serotypes of influenza, one sensitive to the compound and one not, are measured by quantitative hemadsorption. (For further technical details, see Appendix of Chapter 5, Volume I).

The dye-uptake method, used either alone or in conjunction with quantitative hemadsorption, combines the advantages of simple CPE inhibition with reproducible and reliable quantification. Naturally, this demands more time, but when accuracy of measurement is important, the extra time involved is not excessive.

One aspect of the quantitative hemadsorption method which should be mentioned in connection with antiviral screening is the amount of challenge virus used. Both Finter[78] and Hahon et al.[80] recommend a high level of challenge virus to give single-cycle growth conditions and thus enable the results of each test to be obtained in as short a time as possible. However, the use of high virus challenges may give misleading results if certain types of antiviral activity are encountered. If compounds are screened which reduce the infectivity of the newly formed virus progeny but do not reduce the amount of hemagglutinin formed, such an antiviral action would be missed in a test using single-cycle growth conditions. This criticism is entirely hypothetical since no compounds are known which have this type of antiviral action. Nevertheless, it remains a real possibility, and in an antiviral screen it would be better to use a moderate virus challenge and allow several cycles of virus replication to occur before reading the test. In this way, the test will encompass all the aspects of the virus replicative processes and be correspondingly more comprehensive in its ability to detect antiviral activity.

It is not difficult to devise antiviral screening procedures based on histochemical methods combined with microscopic examination; indeed, such a system has been recently described by Boyd and Somerville.[81] They used a high challenge of virus to give single-cycle growth conditions so as to obtain results within a few hours (which is open to the same criticisms mentioned earlier in connection with the quantitative hemadsorption test). Virus growth was detected by indirect staining with fluoroscein-conjugated antibody and was roughly quantitated by eye. Compound toxicity was assessed by cytopathic effects. The method

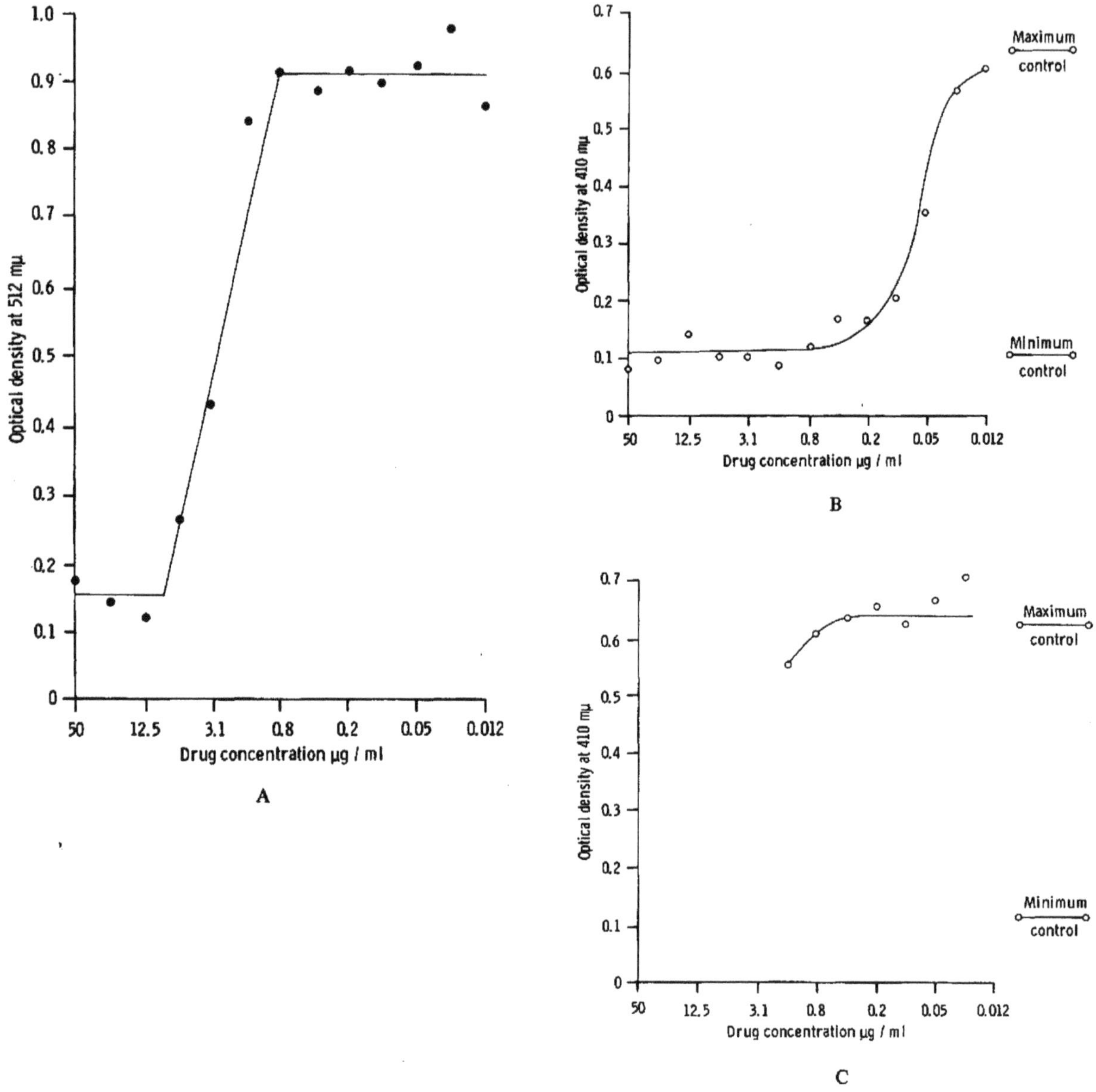

FIGURE 5. The assay of toxicity by neutral red uptake (5a) and anti-influenza activity by quantitative hemadsorption of an antiviral compound in calf kidney cells (b and c). The compound was in contact with the test cells for 48 hr before the measurements were made. (a) The 50% reduction of neutral red uptake, measured as optical density units at 512 nm, is taken to be half way between the maximum and minimum controls. The maximum control is obtained from tube cultures which have not been treated with the antiviral compound, but have been allowed to take up neutral red from a standard solution of 1.5×10^{-5} g/ml for 1 hr at 37°C. The minimum control is similarly derived, but from tube cultures which are first heated to 100°C for 10 min to kill the cells. The 50% neutral red uptake point of the test compound occurs at 3 μg/ml. (b) The 50% reduction of hemadsorption of guinea pig red blood cells is taken as half way between the maximum and minimum controls. The maximum control is obtained from cells infected with virus (influenza A strain DSP, HON1), but not treated with the antiviral test compound, and allowed to adsorb guinea pig red blood cells from a standard suspension of 0.1% for 1 hr at room temperature. The minimum control is obtained from cells neither treated with nor infected antiviral compound, but allowed to adsorb guinea pig red blood cells. The 50% reduction of hemadsorption by the test compound occurs at 0.05 μg/ml. Thus, there is a ratio of 60 between this figure and the neutral red 50% toxicity concentration. (c) Effect on the growth of influenza A/Hong Kong/1/68 (H3N2) of the test substance. It can be seen that no inhibition of virus growth occurs until concentrations of 3 μg/ml or greater are reached. Since these are toxic concentrations, as indicated by neutral red uptake in Figure 5a, the inhibition of virus growth must be a nonspecific effect due to compound toxicity.

could probably be made more quantitative if necessary by the application of Koerting's quantitative fluorescent microscopic assay of viruses,[82] but in a screening procedure this degree of refinement is probably superfluous. The authors claim that the method is rapid and reliable, but the labor involved in a histological method such as this is probably more than in most of the methods previously described.

6. *Radiochemical Assays*

In radiochemical assays the uptake of radiolabeled metabolic precursors by infected and noninfected tissue culture cells is used as a measure of both the toxic effects of compounds on the host cells and the inhibition of virus growth. The use of radiolabeled nucleic acid precursors to measure the effects of so-called antiviral substances on the metabolism of the host cells was described by Bucknall[68] in 1967. The technique has proven useful in distinguishing between substances that specifically inhibit a virus-coded process, and which are therefore truly antiviral, and those that interfere with the host cell metabolism, thus leading to a reduction in virus growth.[68] These types of substances are not and should not be referred to as antivirals. They are simply metabolic inhibitors which reduce the cell's capacity to support virus growth or interfere indiscriminately with some area of metabolism which is common to both cell and virus, such as nucleic acid or protein synthesis.

An antiviral screening test using radiolabeled nucleic acid precursors has been described and used by Miller and his colleagues.[83] This test measures the toxic effects of compounds on the metabolism of HeLa cells by the inhibition of the uptake and fixation of H^3-uridine, which is a measure of cellular RNA synthesis. The growth of RNA viruses (with the exception of influenza, which is itself inhibited by actinomycin D) is measured by infecting the cells, treating them with actinomycin D to suppress cellular RNA, and then measuring the incorporation of H^3-uridine at a time when the rate of virus RNA synthesis is maximal. If a test compound inhibits virus RNA synthesis more than cellular RNA synthesis, this is taken to be a specific antiviral effect. The authors describe how they have used the test to screen 1500 filtrates from cultures of microorganisms and have successfully detected selective effects on RNA virus growth in 11% of these. A ratio of 2.5 between the concentration of test substance causing 50% inhibition of virus RNA synthesis and the concentration causing 50% inhibition of cellular RNA synthesis was taken as the minimum indication of a specific antiviral effect.

The test can be adapted for use with DNA viruses by using H^3-thymidine uptake as a measure of DNA synthesis. To distinguish between cellular and viral DNA synthesis, virus-infected cells which have been exposed to H^3-thymidine are disrupted and the acid precipitable radioactivity is measured. This gives the combined virus and cell DNA synthesis. A sample is then treated with deoxyribonuclease which solubilizes cellular and free virus DNA, but does not affect encapsulated virus DNA. A further count of acid-insoluble radioactivity thus gives a measure of virus-coded DNA synthesis and allows a virus-specific effect of test compounds to be distinguished.

The authors say that their test can operate with a range of viruses, some of which are important disease organisms; of course, it would be difficult to use with influenza as it is an RNA virus which is sensitive to actinomycin.[84-86] The authors claim that the test is rapid and reliable, and certainly it has the merit of using a sensitive metabolic measure for determining whether or not test compounds are toxic to the test cells. However, since nucleic acid synthesis is the only measure of virus growth used, this could in theory mean that certain types of antiviral activity may be missed. For example, compounds that did not reduce virus nucleic acid synthesis but reduced its infectivity by affecting its secondary structure or sequencing would not show activity in this test. Also, the test may be rather insensitive since a large virus challenge must be used to ensure that all cells are infected simultaneously.

7. *In Vitro Virus Enzyme Assays*

The enzymes coded for by the virus genome fall into two classes: those that appear in the mature virion as part of its normal structure and those that are produced during the virus replicative cycle and function and remain in the host cell. The earliest virion-associated active proteins to be studied were the hemagglutinin and neuraminidase of the myxoviruses. It was hoped that an inhibitor of one or the other of these might find application in the control of myxovirus diseases. In light of later work, this seems still to be a reasonable approach to myxovirus chemotherapy, since hemagglutinin provides the mechanism by which infecting myxoviruses attach to the surface of their host cell[87] and neuraminidase may be important in the subsequent release of virus progeny[88-92] (see Chapter 5, Volume I). However, although inhibitors of hemagglutinin and neuraminidase are known, they have not proved sufficiently promising to be taken on for clinical trials in man.[93]

Enzymes are now known to form part of the mature virion of many viruses. Influenza,[94] reoviruses,[95] vesicular stomatitis virus,[96] Newcastle disease virus,[97] and Sendai virus[98] all carry an RNA-dependent RNA polymerase in the virion. The RNA tumor viruses carry reverse transcriptase[99-101] (which should more properly be called an RNA- or DNA-dependent DNA polymerase); in addition, certain RNA tumor viruses also carry protein kinases.[102,103] However, by far the greatest range of virion-associated enzymes which have been characterized to date have been found in the poxviruses. These include at least two nucleotide polymerases,[104-106] nucleotide phosphohydrolases,[107,108] RNAase and DNAases,[109-111] and protein kinases.[112,113]

Since presumably all of these enzymes have a function in the replication of the various viruses with which they are associated, identification of specific inhibitors of these enzymes would provide another approach to the suppression of the virus replicative process.

Similarly, the virus-coded enzymes which remain in the host cell, and collectively synthesize the various constituents of the virion, are different structurally and in their functional requirements from their cellular counterparts. Theoretically, they should also be susceptible to specific inhibition.

Virus-specific enzymes present opportunity for testing substances directly on a virus-coded process in vitro, free of the complications of cellular metabolism. It should be easy enough to automate an enzyme assay and screen large numbers of compounds against, say, an influenza RNA-dependent RNA polymerase (see Appendix to Chapter 5, Volume I for technical details) or a tumor virus "reverse transcriptase," as has already been done in the case of influenza virus neuraminidases[114] (see Appendix to Chapter 5, Volume I for technical details and Chapter 6, Volume I for a detailed discussion). In doing so, however, it should be recognized that the particular enzymic step on which attention is being focused is only a small part of the total replicative process, and by using that as the sole test criterion, compounds that act elsewhere in the virus replicative cycle may be missed. As an example one may cite adamantanamine which would show no activity in an "influenza polymerase" screen (see Chapter 5, Volume I).

Therefore, enzyme screens present a dilemma. Should one take advantage of their ability to screen large numbers of compounds in the hope of finding an inhibitor of a specific virus-coded function and then try to solve all the subsequent problems of its absorption, metabolism, toxicity, etc., or should one opt for a tissue culture screen which, although it may be prone to giving spurious "positives," nevertheless embraces all the aspects of the virus replicative process and therefore should detect compounds which act on any stage? Enzyme screens could have certain applications in, for example, correlating structure/activity relationships in a series of antiviral compounds whose mode of action is known or in detecting anti-RNA tumor virus compounds by utilizing the isolated reverse transcriptase. At present, however, it is probably better to use conventional tissue culture screens for the primary detection of antiviral activity and supplement these with enzyme screens wherever it seems desirable. For further discussion of this problem, see Chapters 5, Volume I and 8, Volume II. The salient features of the various kinds of antiviral tests just described are summarized for comparison in Table 1.

IV. THE USE OF ORGAN CULTURES

The term "organ culture," as is commonly known, is a misnomer. It is seldom if ever that a whole organ is cultured. Usually just a piece of tissue is encouraged to retain the morphology and the function it possessed in vivo. A technique of growing viruses in "organ culture" in vitro, as distinct from tissue explants which rapidly dedifferentiate into primitive fibroblasts, was described by Hoorn and Tyrrell[115] who grew rhinoviruses, coronaviruses, and myxoviruses in fragments of human tracheal epithelium. This system, sometimes modified for ease of manipulation, has since been extensively used in experimental and epidemiological studies of human respiratory disease.[14,117-127] Other human organ cultures used in virus investigations include gut,[13,15,16,128,129,131] liver,[130,131] nervous tissue,[132,133] fallopian

TABLE 1

A Comparison of Some of the Features of Antiviral Screening Procedures

Type of assay	Reference	Type of cell	Type of virus	Capacity to test number of compounds	Time taken	Reliability, reproducibility, accuracy	Comments
Chorioallantoic membrane pieces	59, 60	Chick	Very limited	+	4–5 days	Adequate	Poor in assessing toxicity
pH swing	69, 70	Any	Cytopathic	++	5–7 days	Good	Takes the longest to obtain results
Zone inhibition	71–73	Limited	Cytopathic	+++	4–5 days	Good	Limited to plaque-producing systems
CPE inhibition							
Simple	5, 74	Any	Any	+++	2–3 days	Good	Straightforward and versatile
Quantitative	76–78	Any	Cytopathic	++	1–2 days	Very good	Can be used with quantitative hemadsorption to extend range of viruses
Histochemical	81, 82	Any	Any	+	1–2 days	Good	Too laborious for use on a large scale
Radiochemical	68, 83	Most	Many	+++	1 day	Good	Combines advantages of CPE assays with quantification
Enzyme assays	See text	–	Many	+++	1 day	Good	Highly specific for virus-coded functions but may miss other aspects of antiviral action

tubes,[134] conjunctiva,[135] spleen, and blood vessels.[131]

Animal tissues, particularly respiratory tissues, have also been used to study the growth and pathogenic effects of respiratory viruses,[125,131,136–139] as well as the combined effects of viruses and other respiratory tract pathogens.[140] Cutaneous virus infections have also been studied in organ cultures of skin.[141]

Thus, the use of cultured fragments of animal tissue or organs is well established in virological research. With improvements in the methods of culture[142,143] and the possibility of storing fragments of tissue frozen in a viable form, analogous to liquid nitrogen storage of conventional tissue cultures,[144] there is no reason why organ culture should not become a useful adjunct to the evaluation of antiviral substances. However, relatively little use seems to have been made of organ cultures for this purpose, which is perhaps surprising since they offer an opportunity to measure the antiviral and toxic effects of experimental compounds free from the complications attendant on experiments with whole animals. With organ cultures, the concentration of experimental drug required to produce the desired degree of virus inhibition can be determined. The duration of contact with the drug and the effectiveness of intermittent dosing schedules can also be determined. In this way, one can build a picture of what requirements must be met if a therapeutic effect is to be achieved in the whole animal or man.

Of the few studies published, those of Coombes et al.,[145] Larin et al.,[146] and Herbst-Laier[125,147] deal with the evaluation of the antiviral effects of experimental drugs in tracheal organ cultures. Smorodintsev[148] and Albright et al.[129] have studied the antiviral effects of interferon. We have used tracheal cultures for a number of years in the evaluation of antiviral substances against respiratory viruses.[5,149] Compounds that show interesting activity against influenza in conventional tissue culture cells are tested in organ cultures of ferret trachea before being used for testing in whole animals.

Cultures are prepared by removing the trachea aseptically from a freshly killed ferret and cutting transverse rings. The rings are conveniently cut between the annular cartilages and from a mature ferret, this gives rings approximately 1 mm deep and 4 mm in internal diameter. The area of ciliated epithelium is thus approximately 18 mm^2. Each ring is cultured in 1 ml of Eagle's medium without serum but containing 0.2% bovine plasma albumin (Armour fraction V) in a stoppered 3 × ½ in. tube. Ciliary activity is assessed subjectively by microscopic examination, and only rings showing vigorous ciliary action over 75% or more of the inner circumference are used in tests. Groups of three tubes are treated with a range of concentrations of an experimental antiviral drug and then infected by adding approximately 100 TCD_{50} of influenza virus to each tube. The infected cultures in their drug-containing medium are incubated at 37°C on a roller machine, and each day they are inspected and scored for ciliary activity. Also, the medium is removed and replaced with fresh samples containing the appropriate concentration of drug. The removed medium is assayed for infectious virus by any suitable method (tissue culture infectivity, plaque assay, hemagglutination), and this procedure is followed for several days. The final results are displayed graphically as shown in Figure 6. Figure 6a shows the results of an active anti-influenza drug. At the highest dose tested (0.8 μg/ml) there is a gradual diminution of ciliary activity and no detectable virus growth over the 4 days of the experiment. We would interpret this as a true antiviral effect, but that the concentration of compound was just beyond the limit tolerated with no toxic (that is, ciliary inhibitory) effect. At one fourth of that concentration (0.2 μg/ml) there is no toxic effect and still a good inhibition of virus growth. At 0.05 μg/ml there is no toxicity, as expected, but virus growth is more pronounced, although still well below that in the control cultures on each of the 4 days. Figure 6b shows results from a compound that appeared to be a good anti-influenza drug when tested in conventional tissue cultures of calf kidney cells. In calf kidney cells there were no obvious morphological changes and no influenza growth in cells exposed to concentrations of the drug up to 50 μg/ml for 3 to 4 days. However, in ferret tracheal organ culture the drug is clearly toxic both at 50 and at 12.5 μg/ml, and may show slight toxic effects at 3.1 μg/ml after 4 days; only at the toxic concentration is there any inhibitory effect on influenza growth. Thus, the results from calf kidney cells are spurious and result from the fact that cytopathic changes are not always a reliable measure of a cytotoxic effect.

A further useful observation from tracheal

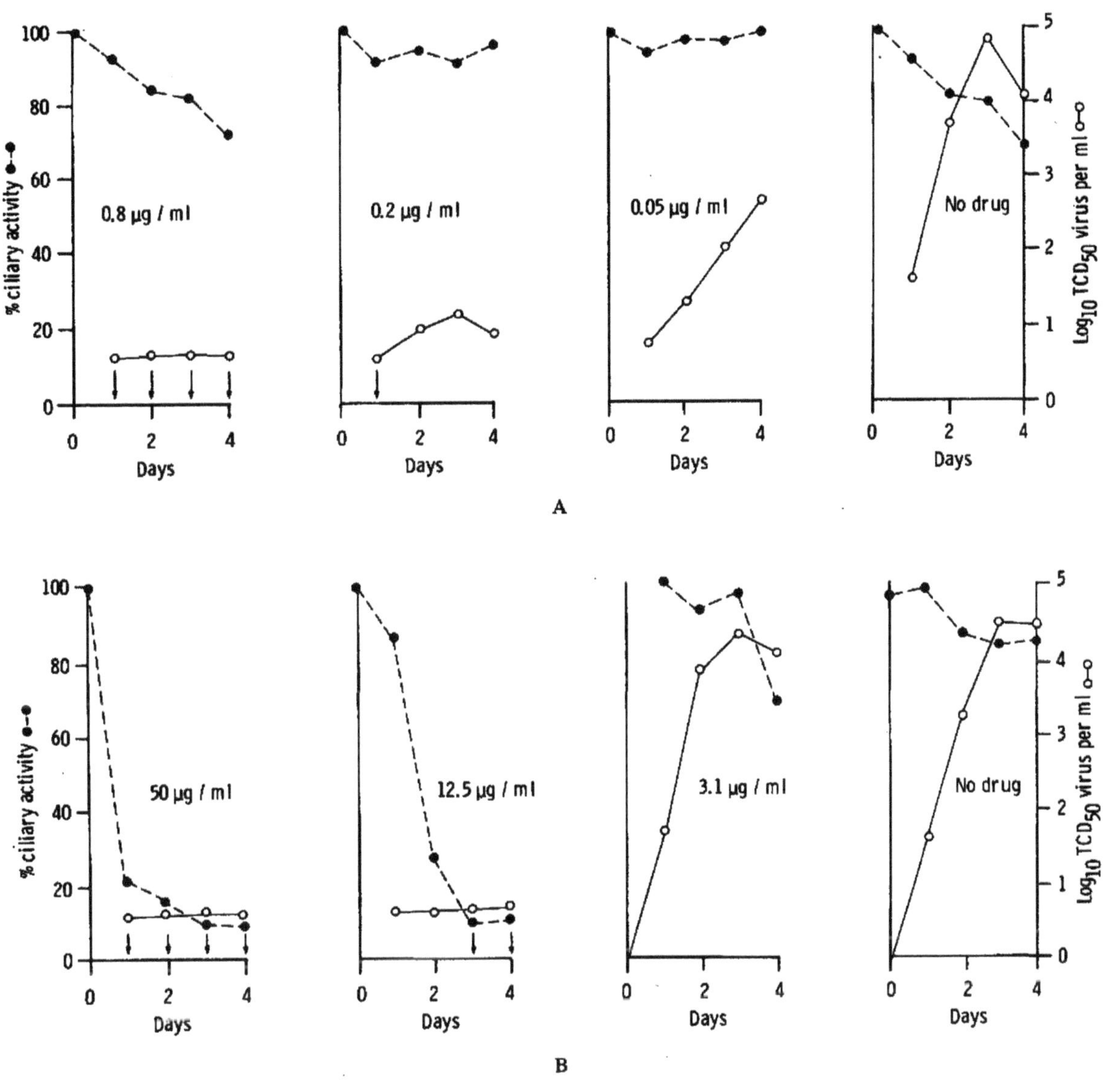

FIGURE 6. (a) The effect of an anti-influenzal compound on the growth of influenza A (H3N2) in ferret tracheal cultures. Virus growth is measured by assaying the amount of infectious virus in the culture fluid each day (o——o). At all concentrations of compound tested, there is inhibition of virus growth. The concentration of compound in each culture, assessed on a subjective basis and referred to as a percentage, is also shown. Only at the highest concentrations of compound tested is there any noticeable diminution of ciliary activity (●----●). The fall in ciliary activity in infected cultures which have not been treated with the anti-influenzal compound is due to the destruction of the ciliated epithelium by the growing virus. (b) Results with a compound that has no anti-influenzal activity. At concentrations down to 12.5 μg/ml, there is a serious toxic effect on ciliary activity. The absence of virus growth is therefore not surprising. At concentrations of 3.1 μg/ml or less, there is little or no effect on ciliary activity, denoting that these concentrations are not toxic, but there is also full virus growth, when compared to the controls with no test compound. The test compound thus shows no antiviral activity.

culture tests is that often one finds that considerably higher concentrations of test drugs are necessary to produce antiviral effects than in conventional tissue culture cells. Presumably this is due to differences in the permeability of the two types of cells. Paradoxically, tracheal cultures are almost invariably more sensitive to the toxic effects of applied compounds than are conventional tissue culture cells; this can only be because ciliary activity is a more sensitive measure of toxic effects than cytological changes. The net result of this increase in sensitivity to toxic effects and decrease in sensitivity to antiviral effects in tracheal cultures is to reduce the therapeutic ratio of test compounds (that is, the ratio of the active concentration to the toxic concentration). Often, what was an "active" compound by the criteria of conventional tissue culture testing turns out to be a "negative" in tracheal culture tests. Since the tracheal culture test is more relevant to the whole animal, these are results which must be considered as indicative of the test compound's potential performance in an animal model.

V. ANIMAL MODELS OF VIRUS DISEASES

In assessing the therapeutic potential of a novel antiviral substance, it is valuable to have an animal model of the virus disease; this is particularly true of human virus diseases. Unfortunately, the animal models of human diseases are often not as representative of the human condition as one might like and this, together with the fact that the fate of a drug given to an animal may be vastly different from its fate in a human, makes it impossible to do more than offer the most general comments on animal model diseases. Various aspects are discussed further in Chapters 2 and 6, Volume I.

In using animal models to evaluate the clinical potential of antiviral substances, a positive result (i.e., a curative effect) is usually accepted as a sign that the test compound is worth continuing with, and a negative result indicates that it is not. This approach tacitly assumes that the absorption, distribution, and metabolism of the test compound in the test animal will be much the same as that in the human being – which very often it is not so. Therefore, it is necessary to have some information on the pharmacology of a test compound in order to interpret the results of animal protection experiments, particularly if no therapeutic activity has been seen. For example, if a compound which is dosed orally to a test animal is rapidly destroyed by the enzymes of the animal's liver, then it is unlikely to achieve any useful concentration in the blood; the chances of it reaching, say, an influenza infection of the lungs or an herpetic infection of the brain are remote. However, it would be unwise to abandon that compound as having no future in man until it is known whether or not the human liver destroys it in the same way as the animal liver did.

A review by Hucker[150] is worthwhile noting in this connection. He points out that, given the pharmacological behavior of an experimental drug in one animal species, it is simply not possible to predict how the drug will behave even in a closely related species, let alone man. The variations between species in absorption, metabolism, excretion, and intrinsic sensitivity are too great to allow meaningful extrapolations. Even Old World and New World monkeys differ from each other pharmacologically. Of course, if an experimental compound shows a consistent pattern in several animal species, this would give one some confidence of its possible performance in man, but this type of behavior is the exception rather than the rule.

Then there is the question of how great a therapeutic effect one should expect in an animal model before proceeding to clinical trial in man. This again is virtually an unanswerable question, and until more antiviral drugs have been evaluated clinically, no parallels can be drawn.

A further important consideration in animal protection experiments is the intrinsic ability of any particular animal/virus system to reveal potential antiviral activity in an experimental antiviral drug. The essential aspects of devising animal test systems for both the initial detection of antiviral activity and for distinguishing fine differences in activity between related members of series of antiviral compounds have been concisely reviewed by Bauer.[151] There are broadly two approaches: the quantal assay and the continuous response method.

In a quantal assay a group of, say, mice is infected with a lethal virus, and half the group is treated with the test drug. In due course, the number of dead animals is scored, and the number of survivors in the treated group is expressed as a percentage of the number in the untreated group. Provided that the dose of virus is high enough to infect most of the animals, but not so high as to swamp any antiviral effects of a test compound, the quantal assay can give satisfactory results. Since quantal assays rely solely on a positive or negative response, they are less sensitive and less versatile than continuous dose-response assays which take into account all the intermediate possibilities between a full positive and a complete negative result. If, instead of simply recording how many mice die in the treated and the untreated groups, one also records the length of time each animal survives, this can give a measure of an antiviral effect which, although not great enough to prevent death, nevertheless increases the length of time an infected animal survives. Herein lies the improved sensitivity of the continuous dose-response assay over the simple quantal assay (see also Chapter 6, Volume I).

It is unfortunate for the prospects of potential antiviral drugs that animal models are such a poor guide to a drug's performance in man, because it is at this stage that a decision has to be made whether to continue with the next phase of development (formal toxicological studies often costing many thousands of pounds) or whether to abandon the drug. Therefore, it behoves the virologist and his colleagues to bring all their expertise to bear on this critical aspect of any new antiviral drug so that if it is decided not to continue, then a potentially valuable drug is not being missed.

VI. MEASURING THE TOXICITY OF TEST COMPOUNDS IN VITRO

A. In Vitro Studies

The assessment of the toxic effects of experimental drugs in vitro has already been dealt with in the preceding pages. However, this is such an important aspect of antiviral chemotherapy and so often is not done properly that it will be reviewed here in more detail.

All compounds that are screened for antiviral activity will be toxic to the cells used in the test system if the concentration is high enough, and meaningful antiviral tests can only be made when each test compound is diluted to below this critical level. If test compounds are used at or near concentrations that are toxic to infected cells, there is usually a reduction in virus growth. This is simply a consequence of the impairment of cellular metabolism and is not a specific action of the test compound on a virus-coded process; as such, it cannot be considered a true antiviral effect. Certainly no drug with this type of action should be countenanced for use in human medicine.

The correct assessment of a compound's toxicity is therefore of prime importance at all stages of antiviral testing – in the test tube, in animals, and in the clinic – and a common error is that insufficient attention is paid to the evaluation of a compound's toxicity from the outset of its career. All too often, routine screening procedures underestimate some intrinsic toxic property of a test compound and because of the resulting inhibition of virus growth, the compound is then credited with antiviral properties which it does not actually possess. Much time and effort are then spent in trying to demonstrate its supposed curative effects in endless laboratory and sometimes even clinical tests. Sometimes the confidence which workers place in a compound's supposed properties will carry their experimentation beyond the bounds of both reasonableness and ethical practice.

Before discussing how the toxic effects of test compounds might be assessed experimentally in vitro, it is necessary to define what these effects might be. At a cellular level, any disturbance of the metabolic processes caused by an extraneous compound must be considered a toxic effect, and it would be a daunting task to devise comprehensive tests which would pinpoint exactly where a compound was interfering with the normal running of the cell. Fortunately, it is not necessary to go to such lengths, and quite simple tests, properly applied, will give a good guide as to whether or not a test compound is worthy of more elaborate examinations.

The use of radiolabeled precursors to study effects on macromolecular synthesis has already been discussed.[68,83] In our laboratory we find this to be a rapid and informative method of revealing undesirable metabolic effects in potential antiviral compounds. Even when the test compound is not itself a direct inhibitor of nucleic acid or protein synthesis, its effects are rapidly reflected in these important areas of cellular activity. If a compound's antiviral effects are expressed only at concentrations which reduce cellular macromolecular synthesis, then the antiviral effect is likely to be a spurious one.

Some work of ours with Virazole (ribavirin), a compound claimed to be a specific inhibitor of influenza virus growth,[152-154] will serve to illustrate this point. When tested in calf kidney cells against influenza A_0, Virazole showed no obvious toxic effects at concentrations up to 50 μg/ml (the highest concentration tested) and suppressed virus growth at concentrations as low as 1 μg/ml but when the nucleic acid and protein metabolism of uninfected calf kidney cells were studied, however, Virazole was found to inhibit DNA synthesis profoundly at concentrations above 0.5 μg/ml. Leaving aside for the moment the mechanism of how Virazole inhibits influenza virus growth (see Chapter 5, Volume I for further discussion), it is clear that in addition to suppressing virus growth the drug has serious metabolic effects in cells, which must cast grave doubts on its future as a human medicine. It is now known that Virazole inhibits the synthesis of guanine nucleosides,[155,156] and it seems that its action is not specific to virus-coded enzymes, if indeed such enzymes exist.

The application of biochemical tests to determine whether or not a potentially antiviral substance affects some aspect of the host cell metabolism is an essential step towards establishing its bona fide antiviral status. Only in very few cases have such tests shown that there is a clear margin between the antiviral and the antimetabolic concentrations of so-called antiviral drugs. Some of the early work on guanidine and 2-(α-hydroxybenzyl)benzimidazole[157-160] went some way to prove that these substances genuinely have a selective effect on virus-coded processes. It is a pity that their action is not selective enough to make them suitable for use in man.

The use of organ cultures, and in particular tracheal ciliated epithelium, in detecting toxic effects has already been mentioned. This system is simple and relatively rapid and usually reveals toxic effects of test compounds at considerably lower concentrations than conventional tissue culture cells. Table 2 gives a series of compounds with their toxicity determinations in both conventional calf kidney cell cultures and ferret tracheal cultures. Both were assessed after 2 days' contact with the test compound. In the case of the calf kidney cells, the "toxic concentration" is that which causes 50% of the cell sheet to show morphological abnormalities; in the case of the tracheal cultures, it is that which reduces ciliary activity by 50%. The extra sensitivity of the tracheal cultures is obvious. In this particular case, the compounds are all uncouplers of oxidative phosphorylation; therefore, the energetic process of ciliary action provides a very sensitive system for detecting this type of toxic effect. We have also found tracheal cultures useful for detecting undesirable effects in a wide range of chemical compounds with a diversity of biological actions.

Perhaps the most stringent test of a toxicity in vitro is to grow cells in the presence of a compound and to measure effects on their growth rate. If cultured cells divide and grow at their normal rate in the presence of a test compound, then clearly the compound cannot be interfering with any of the cellular metabolic processes. We find that few so-called antiviral compounds pass this test. In our laboratory the test is done as follows. A series of concentrations of a test compound in Eagle's medium with 8% serum is prepared, and an inoculum of 10^6 cells is added to each. The mixtures are put into culture bottles and incubated for 3 days. The cell type used is that in which the antiviral activity has been measured. After 3 days the cells are trypsinized and counted, and an aliquot of 10^6 is added to a fresh sample of the appropriate concentration of the test compound in Eagle's medium, which is again cultured for 3 days. By this method cells can be kept in constant division, and any effects of the test

TABLE 2

"Toxic" Concentrations of a Series of Compounds Determined by Two Cytological Methods

Compound code number	Toxic concentration in calf kidney cultures, μg/ml[a]	Toxic concentration in ferret tracheal cultures, μg/ml[b]	Minimum concentration causing uncoupling of oxidative phosphorylation, μg/ml[c]
99,600	44	2	<0.33
98,173	>44	0.4	<0.33
93,249	5	0.02	<0.33
97,947	>44	0.01	<0.33
95,007	44	0.4	1–2
85,258	44	2	1–2
96,126	25	0.4	5–10
84,448	44	2.5	5–10
90,868	>44	20	10–20

[a]Concentration that causes 50% of cultured calf kidney cells to appear morphologically abnormal.
[b]Concentration that reduces ciliary activity of cultured ferret tracheal epithelium by 50%.
[c]Minimum concentration of compound that causes uncoupling of oxidative phosphorylation in isolated rat liver mitochondria.

compound on cell growth gradually become apparent when the cumulated cell yield is plotted over the course of a week or two. Some typical results from this type of test are shown in Figure 7.

The in vitro tests outlined above serve two purposes. First, they help to establish the validity of the initial observation on a test compound – that it possesses genuine antiviral properties. Second, they prepare the way for more detailed toxicological studies on the compound in vivo, which is a vast subject in itself (dealt with in outline in the next section).

VII. INTRODUCTION TO THE SAFETY EVALUATION OF ANTIVIRAL COMPOUNDS

During the last 20 years there have been considerable changes in attitudes concerning the toxicity testing of new drugs, and legislation has been introduced to regulate the development process. With modern concepts of human safety and with compounds of high biological activity, we can no longer call for volunteers to determine the efficacy and safety of drugs without prior recourse to thorough studies in the laboratory. When a new antiviral medicine is licensed so that it may be prescribed for the general public, it has been thoroughly tested in both animals and humans under strictly controlled circumstances and the physician is given full information on its use.

The laboratory tests for discovering and evaluating new antiviral substances described in the foregoing papers can detect only a relatively limited range of the undesirable or toxic properties which any potential antiviral drug may possess. Therefore, it is necessary to perform specific studies to detect the unwanted side effects of the new medicine. No laboratory test can establish absolute safety, but it is possible for the experienced toxicologist to do a number of tests in laboratory animals and in vitro that should predict whether an unacceptable level of undesirable effects would occur if the compound were given to man.

Toxicity testing involves identification of altered body functions or morphology after short-term and long-term administration of the compound. It is also necessary to do special tests for specific hazards including effects on reproduction,

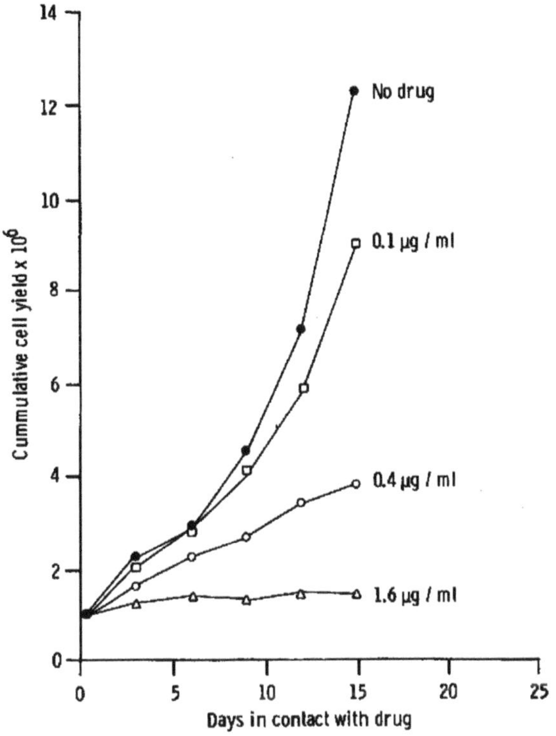

FIGURE 7. The growth of human embryonic diploid lung cells in the continuous presence of a test compound. The cells are kept in virtually continuous growth by trypsinization and subculturing at intervals of 3 days. The yield of cells at each subculture is used to calculate the theoretical total yield of cells from the initial inoculum of 10^6 cells seeded on day 0. Thus, even small differences in the growth rate are revealed with the passing of time. In this instance, 1.6 µg/ml reduces growth to a level which barely maintains the cell yield at 10^6. At 0.4 µg/ml the growth rate is reduced, while at 0.1 µg/ml it approaches normal. Concentrations of less than 0.1 µg/ml could be safely assumed to be nontoxic to cells cultured under these conditions. If an antiviral effect could be demonstrated at less than 0.1 µg/ml, one could be confident that this was a specific effect on a virus-coded process.

immunological function, and genetic material. It is also essential to study the pharmacokinetics (absorption, distribution, and excretion) and metabolism of the drug in animals and in volunteers.

It is very probable that unwanted effects will be seen during the safety evaluation of a substance, although these may only occur at very high doses. This is embodied in the principle enunciated by Paracelsus (Theophrastus Bombastus von Hohenheim, 1493 to 1541) that "Poison is in everything, and nothing is without poison," but he added a second principle of cardinal importance to the toxicologist: "It is the dosage which makes it either poison or remedy." It is only by an experienced and full review of all the known facts (physical, chemical, and biological), including the pharmacology and toxicological findings, together with a consideration of the compound's intended use that the therapeutic index and risk-to-benefit ratio may be estimated. Various aspects are discussed further in Chapters 2 and 6, Volume I.

The development of a new medicine is therefore a protracted and complex process. It is usual in most countries for the investigational work in man or animals to be done in several discreet stages.

It is desirable that the new medicine be given to man at the earliest safe time in the development program. The scientists will want to know whether the compound has the required activity, so that extension of its use may be considered or the search for substances with greater potency or with wider activity may be encouraged. The toxicologist will want to study the compound's metabolism and disposition in man to ensure that future studies are done with those species which will best predict mans' responses. If the compound proves to be inactive in man, then much time and money (which otherwise would have to be paid for, either directly or indirectly, by the public) can be saved.

Over the years, a general development scheme has evolved in which the initial clinical studies commence only after certain preclinical toxicology studies are satisfactorily completed. Then, both clinical and toxicological work advance together, with certain essential units of work completed by physicians and toxicologists before the next stage of the clinical studies begins.

In most countries there are government agencies that function to secure adequate safety testing before marketing and early detection of adverse drug reactions following marketing.

In the United States there is the Food and Drug Administration (FDA), which has over 70 years of legal, regulatory, and scientific experience beginning with the Pure Food and Drugs Act of 1906. This prohibited the use in drugs of an "ingredient deleterious or detrimental to health." It was consolidated and extended by the Food, Drugs, and Cosmetics Act of 1938 and again by the Kefauver-Harris 1962 Amendments calling for substantial evidence of a drug's effectiveness before it could be sold in the U.S.

In the United Kingdom at the time of the thalidomide disaster, the only medicines that were controlled (apart from the drugs of addiction) were those under the Therapeutic Substances Act of 1956, i.e., biologicals, injectable antibiotics, and certain hormones and enzymes. In 1962 the Medical Advisory Committee on the Safety of Drugs, under the chairmanship of Sir Derek Dunlop, was introduced together with the Advisory Committee on Pesticides and Toxic Chemicals and the Veterinary Products Safety Precautions Scheme. At this time the submission of toxicity data to these regulatory bodies by the manufacturer of new drugs was voluntary. In 1968 the Medicines Act established the Medicines Commission with the Committee on the Safety of Medicines (CSM) and the Veterinary Products Committee. Manufacturers of new products must satisfy these committees of the safety and efficacy of their products before a license will be granted to permit marketing. Many other countries have similar regulatory agencies, for example, the National Drugs Advisory Board of Eire, The Pharmaceutical Affairs Bureau of the Japanese Ministry of Health and Welfare, The Food and Drug Directorate of Canada, The Drug Committee of the National Board of Health and Welfare of Sweden, and many more. There are also advisory bodies, such as the World Health Organization scientific groups.[161]

In Europe the European Economic Community (EEC) began the harmonization of pharmaceutical legislation of member nations in 1965[162] and in 1975[163] issued directions on what toxicity tests would be required for new drugs. Member nations have agreed to implement the latter directive within 18 months of the document date. The authors understand that professional toxicologists employed by pharmaceutical companies will be considered to have "expertise" in their particular discipline; this term is used in France for a panel of personnel recognized as possessing an expert understanding of aspects of the work and who must be prepared to endorse a particular submission.

The whole system of control is based on a series of certificates and licenses without which products may not be tested or marketed in human or veterinary medicine. In the United Kingdom there are two major stages. First, a Clinical Trial Certificate is required before a new drug may be given to patients for the first time, and a Product License is required before marketing. In the U.S.

the two stages are similar and are called a Notice of Claimed Investigational Exemption for a New Drug (IND) and a New Drug Application (NDA), respectively. A similar two-stage process exists for licensing veterinary medicines, the Animal Test Certificate and Product License in the U.K. and INAD and ANDA in the U.S., respectively. In human studies where the participants are volunteers, no clinical trial certificate is required when the work is done in the U.K., whereas in the U.S. no human studies may be undertaken without first obtaining an IND.

In the U.S. the human investigational work is done in three tiers or phases, as follows.

Phase I – clinical pharmacology. A small number of subjects, usually healthy volunteers, receive graded doses of the compound, beginning with a dose about one tenth of what might be expected to be useful. This is done to study absorption, general tolerance to the drug, and metabolism. Occasionally, early efficacy studies with patients are also done.

Phase II – Early, rigidly controlled clinical trials in limited numbers of patients (100 to 200) are done to demonstrate efficacy and safety. There are early open trials where the clinicians know what medication the patients are receiving, and later double-blind comparisons where the identity of the medication (drug or placebo) is not revealed until the end of the trial. Phase II may start when Phase I has been successfully completed and if animal teratology studies have been done.

Phase III – This embraces expanded controlled studies for further evidence of efficacy and long-term safety, and for determining the optimum dosage for treating specific diseases.

The first phase is preceded by the preclinical laboratory studies, detailed below, but is generally shorter and less extensive than those required for the later stages of the clinical studies. The application for a certificate to test a drug includes the details of its chemistry and pharmacy, the results of experimental biological studies, the toxicology data, anticipated clinical use, and what further research is planned with the drug. The application also requires the qualifications of the investigators involved in the work; to further add to the protection of the participating volunteers, there is provision for their informed consent and approval of the sponsor for the work.[164]

As the drug proceeds through the phases, there must be regular assessment of its progress, and decisions must be made as to whether the program will continue. This is formalized by the FDA, which holds three "critical phase" conferences with the sponsor.

The application for marketing must contain all information on all human and animal studies, both favorable and unfavorable.[165,166] Each submission consists of up to 400 volumes of raw data and a summary which itself may run to 15 volumes. The data are reviewed by physicians, pharmacists, chemists, and other professional staff of the regulatory agency, assisted where necessary by experts from outside the agency. The key question is whether the benefits of the new drug, when used properly, will outweigh the risks of adverse reactions.[167] For example, if there were an antiviral compound that would successfully treat a terminal case of rabies, then a relatively high risk of adverse reactions would be tolerated, whereas a much lesser degree of risk would be acceptable for an antiviral compound to treat a common cold. If a license to market the product is granted, the evaluation does not stop there, because following approval it is necessary to compile records relating to the drug's safety and effectiveness in man.

As mentioned above most countries have their own regulatory agencies, and each has its own views on the type and quantity of toxicity studies that should be done. It is necessary to consider the specific demands of every country where the compound may be used.[168–180] Many regulatory authorities, national bodies, and international agencies issue guidelines, but it is of great importance to *distinguish between the advice (guidelines) and the demands (regulations).* Furthermore, each new drug must be treated individually. The toxicity work must be designed as a research program that will satisfy all scientific and ethical considerations and at the same time meet the minimum requirements of the relevant regulatory agencies.

Although guidelines are extremely valuable for the above reasons, *it is strongly advised* that if there are any doubts concerning the specific requirements of a regulatory authority for a particular compound, the authority be consulted before the trials begin. In general, all licensing authorities have similar basic requirements.

VIII. PRECLINICAL TOXICITY TESTS

A. Acute Toxicity Tests

These are to determine the amount of compound which is lethal or otherwise toxic to animals given a single dose. At least two species of animals and both sexes must be used, as well as at least two routes of administration. One route must be that proposed for human treatment. Details are required of the numbers of animals dying in each group and the mode of death, over both a 24-hr period and a 7- or 14-day period. A convenient way of expressing the results is as the median lethal dose – LD_{50}.

It is very important to understand that the LD_{50} is not an absolute value, but depends on many factors[181] including the species, strain, age, and sex of the animals; the route, rate, and vehicle of administration; and environmental conditions, e.g., temperature, number of animals per cage, nutritional state, season, etc.

The LD_{50} is useful, however, because it gives an index for the drug's acute toxicity which has some predictive value for the nature and degree of the response to be expected if there is inadvertent administration of a high dose. It may also be used with the median effective dose (ED_{50}) of the therapeutic action to obtain a therapeutic index for single doses. For example, rolicin (an antibiotic with in vitro activity against foot-and-mouth disease) had insufficient margin between the lethal and sublethal dose,[182] and gliotoxin (an antiviral compound with activity against type 2 polio virus) had insufficient therapeutic margin;[183] no extended development was done with them. The LD_{50} will sometimes give information on absorption by comparing the value obtained after administration by a parenteral route with that for oral dosing. Similarly, differences between sexes or species may be detected. It must be stressed that if comparisons are to be made, they should be done within a single experiment.

When estimating the LD_{50}, the aim should be to use the minimum number of animals consistent with the calculation of the required confidence limits; it is unnecessary to use large numbers of animals in the vain endeavor to obtain more "accurate" values. Whatever method is used it should provide a dose-response regression curve from which the median dose may be obtained. Several doses should be used and they should be spaced logarithmically, the highest killing all or nearly all animals and the lowest producing no effects. The LD_{50} may be calculated with sufficient precision by the method of Litchfield and Wilcoxon,[184] but methods using logit analysis are used by laboratories with computer facilities.

In some instances, particularly with non-rodents, it is sufficient to find the maximum tolerated dose by making incremental increases in the dose administered every few days until toxic effects are seen ("target organ" toxicity). In the authors' opinion, this is an opportunity to use clinical pathology in its most predictive mode.

B. Range Finding Studies

The doses for long-term trials are best selected after experience has been gained using several dose levels in short-term studies.

C. Repeated Dose Studies (Subacute or Chronic Studies)

These tests are designed to determine any effects that will arise following repeated daily doses of the compound for long periods. Most authorities agree on the following basic points:

1. The test compound should be of a known high standard of purity.
2. The compound should be administered by the route(s) proposed for man.
3. The compound should be tested in both male and female animals.
4. It is common knowledge that one species will often respond to a drug in a different manner than another, e.g., pencillin is lethal to guinea pigs. The difference in response may be due to a difference in the disposition of the drug in the two species or because of inherent differences in their pharmacological or toxicological responses.

For the last reason, at least two species are used, one rodent, e.g., dog, subhuman primate, or pig. These species are commonly used because they have one or more of the following advantages. They are readily available or easily bred; they can be obtained with known age and health status; their anatomy, pathology, and physiological responses are well known; they are inexpensive to house and maintain in sufficient numbers; and they are not difficult to dose and examine.

Primates are often demanded because of a misplaced anthropomorphic view, and this is only justified in specific instances, e.g., the placentation or endocrine status may present a more appropriate testing model. Ideally, the species selected should show similar drug disposition, metabolism, (both qualitatively and quantitatively) pharmacology, and biological response to man. Of course, this cannot be known until some human studies have been done; therefore, this type of species selection must be reserved for later toxicity studies. Drug disposition and metabolism studies should nevertheless be commenced with the animal species, before the first investigation in man.

5. The number of animals required varies, but there should be at least ten rodents of each sex in each group and three (U.S. IND) to five (U.K.) nonrodents of each sex in each group (see Table 3).

6. The duration of the trial should bear some relationship to the proposed duration of dosing for man.

7. The compound should be tested at at least three dose levels, and there should be untreated control animals. Occasionally it will also be necessary to have a group receiving the vehicle or placebo. The dose should be expressed in terms of the base, not the salt, and as the units of dose per kilogram of body weight.

8. The three dose levels are usually selected as follows:

a. High dose – The high dose is detected so that some overt toxic signs will be seen. If a compound is very bland, it may be impractical to give a toxic dose.

b. Mid-dose – This dose assists with determination of the therapeutic index. It is usually many times higher than the human exposure. A high multiple of the human dose is used to increase the confidence in a safety factor obtained with animals as they may be less responsive to toxic effects than humans either because of interspecies differences in drug deposition or because of intrinsic differences in response. In most toxicity studies the number of animals used is very small compared to the large human population which may receive the compound; accordingly, increased confidence in extrapolating from animals to man is given by a very high therapeutic index.

c. Low dose – This is selected to provide evidence that there are no toxic effects at a dose that is three to five times the human dose. The low dose or preferably the mid-dose should be the "no effect" dose, providing a "safety margin" of at least three to five times the human dose in the most responsive species.

9. Whenever possible and appropriate, the drug should be administered at least once daily 7 days/week.

10. Before the trial the animals should be allocated to the several groups in a random fashion, and subsequent sampling should be done using subgroups that are truly representative of the total group (replicates).

11. Before and during the trial the animals should be examined at least once daily for altered behavior[185] or condition and clinical signs. Food consumption, body weight, and food conversion should be observed and ophthalmological studies done. It is often advisable to estimate water consumption.

12. Clinical pathology (Appendix A), (hematology,[186] blood chemistry, urinalysis,[187] blood coagulation studies) should be done before the trial begins and at appropriate times throughout the study.

13. Clinical pharmacology measurements of heart rate, blood pressure, and respiration rate and an electrocardiogram should be done on the nonrodent species.

14. A full necropsy should be done on all animals whether or not they survive the full period of the study. The animals should be killed by an overdose of anesthetic or some other appropriate method, and major organs should be weighed[188,189] for histology before fixation. Samples from all organs (Appendix B) and tissues should be fixed for subsequent histopathological examination, although it may only be necessary to examine the animals receiving the top dose and the control animals if no adverse effects are found.

15. If effects are found, they should be further studied by inclusion of a reversibility group in which animals are allowed to survive after the full period of drug administration and then are killed at a suitable period of time after drug withdrawal.

16. The duration of the repeated dose studies varies considerably between countries. For example, in the U.S. it depends on the route and duration of the proposed trials in man. In the U.K. preclinical studies are usually of 3 months duration, although a shorter period may be appropriate if the drug is to be administered to man only on one or two occasions. The guidelines for the U.S. have been set out by Goldenthal[190] and are summarized in Tables 4 and 5. The requirements of the FDD are outlined in Table 6. The EEC calls

TABLE 3

Duration and Number of Animals Required for Toxicity Trials

Country	Proposed duration of clinical use	Duration of toxicity trial	Number of rodents per sex/group	Number of nonrodents per sex/group	Minimum number of species for acute studies
United Kingdom	A few doses	2–4 weeks	10	5	2
	Clinical Trial Certificate	3 months			
	Product Licence	6 months			
United States	See Tables 2 and 3		10	3–5	3
United States PMA/FDA	U.S.		10–25	2–3	Not specified
Sweden	One dose	2 weeks	10	3	
	1 week – several doses	1–3 months	10	3–5	2
	Long term	6 months	20	5	
European Economic Community	Preclinical	2–4 weeks	Not specified	Not specified	2
	Marketing	3–6 months			
World Health Organization	Single – four doses	2 weeks			
	1 month	13–26 weeks	Not specified	Not specified	Not specified
	More than 1 month	At least 26 weeks			

TABLE 4

Guidelines for Toxicity Studies for Registration in the United States, Part I

Route	Duration of human administration	Phase	Subacute or chronic toxicity	Special studies	Observations
Oral or parenteral[a]	Several days	I, II, III, NDA	2 species; 2 weeks	For parenterally administered drugs: irritation studies, compatibility with blood where applicable	Body weights, food consumption, behavior, hemogram, coagulation tests, liver and kidney function tests, fasting blood sugar, ophthalmological examination, metabolic studies, gross and microscopic examination, other as appropriate
	Up to 2 weeks	I	2 species; 2 weeks		
		II	2 species; up to 4 weeks		
		III, NDA	2 species; up to 14 weeks[b]		
	Up to 3 months	I, II	2 species; 4 weeks		
		III, NDA	2 species; 3 months		
		III, NDA	2 species; up to 6 months		
	6 months to unlimited	I, II	2 species; 14 weeks[b]		
		III	2 species; 6 months or longer		
		NDA	2 species; 12 months (nonrodent), 18 months (rodent) or 12-month rat plus a 2-year mouse carcinogenesis study		
Anesthetic inhalation		I, II, III, NDA	4 species; 5 days (3 hr/day)		
Dermal	Single application	I	1 species; single 24-hr exposure followed by 2-week observation		
	Single or short-term application	II	1 species; 20-day repeated exposure (intact and abraded skin)		
	Short-term application	III	As above		
	Unlimited application	NDA	As above, but intact skin study to be extended up to 6 months		

Note: Subacute or chronic studies should be by route to be used clinically.

[a]Where toxicity data are available on each drug individually.
[b]12 weeks in Goldenthal recommendations modified by FDA/PMA Committee.

TABLE 5

Guidelines for Toxicity Studies for Registration in the United States, Part II

Route	Duration of human administration	Phase	Subacute or chronic toxicity	Special studies	Observations
Ophthalmic	Single application	I		Eye irritation tests graded doses	Body weights, food consumption, behavior, hemogram, coagulation tests, liver and kidney function tests, fasting blood sugar, ophthalmological examination, metabolic studies, gross and microscopic examination, other as appropriate
	Multiple application	I, II, III	1 species; 3weeks daily application, as in clinical use		
		NDA	1 species; duration commensurate with period of drug administration		
Vaginal or rectal	Single application	I		Local and systematic toxicity after vaginal or rectal application in 2 species	
	Multiple application	I, II, III, NDA	2 species; duration and number of applications determined by proposed use		
Drug combination[a]		I			
		II, III, NDA	2 species; up to 3 months		

Note: Subacute or chronic studies should be by route to be used clinically.

[a]Where toxicity data are available on each drug individually.

TABLE 6

Guidelines for Toxicity Studies for Registration in Canada

Category	Duration of human administration	Species	Duration	Route of administration in animals	Special investigations
Oral	Not exceeding 1 month	2 (e.g., rat and dog)	2–8 weeks	Oral in feed or by stomach tube	
	Exceeding 1 month	2 (e.g., rat and dog)	6–12 weeks		
Parenteral	Not exceeding 1 month	2 (rat and dog)	2–8 weeks	Parenteral route to be recommended clinically	Site of injection should be examined for signs of irritation
	Exceeding 1 month	2 (rat and dog)	6–12 weeks		
Topical (dermal ophthalmic, nasal, rectal, etc.)	Not exceeding 1 month	2 (rabbit may be used instead of a nonrodent species)	2–8 weeks	Parenteral, if possible, and by method to be used clinically	Site of injection should be examined for signs of irritation
	Exceeding 1 month	2 (rabbit may be used instead of a nonrodent species)	6–12 weeks	Parenteral, if possible, and by method to be used clinically	
General anesthetics		At least 2, 1 a nonrodent	2 weeks for 3 hr/day	By inhalation to surgical plane of anesthesia	
Aerosols	Not exceeding 1 month	2 (1 nonrodent)	2–8 weeks, 3 hr/day	By inhalation; note: doses to be larger than clinically recommended	
	Exceeding 1 month	2 (1 nonrodent)	6–12 weeks, 3 hr/day		
Combinations of drugs	Where biological data are available on each active ingredient, then subacute toxicity studies are not required for preclinical submissions unless evidence of potentiation is evident from the acute toxicity studies. If required, 2 species should be studied for 2 to 12 weeks, depending on duration of human administration.				

for two tests, one lasting 2 to 4 weeks ("In respect of proprietary medicinal products to be administered once only to humans a single test lasting two to four weeks shall be performed") and the other long term, normally 3 to 6 months long.

Examples of toxicity testing of antiviral drugs are as follows.

1. An extensive report on amatadine hydrochloride was published by Vernier et al.[191] Clinical signs, clinical pathology, histopathological findings, reproductive studies, and pharmacology were reported. At high doses, central nervous system stimulation and cardiac arrhythmias were observed.

2. Systemic toxicity of polyinosinic acid: polycytidylic acid copolymer in rodents and dogs,[192] monkeys,[193] and mice[194] has been reported. Decreased hematopoiesis was seen in all three species.

3. Cyclocytidine has been shown to cause severe leukopenia, bone marrow depression, and elevation of transaminases in dogs receiving high doses.

4. Dayon[196] reported jaundice in a subject after dosing with iododeoxyuridine.

D. Teratology

Tests for effects of the compound on the developing fetus should be conducted (see details below).

E. Allergenicity

Tests for sensitising potential[197] should be done with guinea pigs. The Magnusson and Kligman "maximization" test[198] has received[199] acceptance in many laboratories. Some drugs have the potential to cause autoimmune reactions.[200]

F. Phototoxicity

The advice of a person with knowledge of photochemistry is very valuable in deciding what tests should be done. The erythrocyte hemolysis test, macrophage test, Aspergillus test, and/or animal studies should be done. The hemolysis test is based on the detection of a damaging effect upon the phospholipids of the erythrocyte membrane by measuring the quantity of hemoglobin released. The macrophage test measures the cell death of peritoneal macrophages when exposed to chemicals and light in vitro. Ketel[201] reported photosensitization in a subject following administration of amantadine.

G. Local Tolerance Studies

If the drug is to be applied topically (e.g., nasal cavity) or administered parenterally, local tolerance studies should be done with animals (e.g., eye irritation,[202-204] skin and muscle,[205] irritation studies. Peteri[206] reported that (N-1-adamatyl)-2-(2-dimethylamine-erthoxy)-acetamide-hydrochloride had no irritancy on skin or mucous membranes.

H. Blood Compatibility Studies

If a compound is to be injected into the vascular system, it should be tested to ensure that it will not cause hemolysis or coagulation of the blood or separation and precipitation of plasma proteins.

I. Organs of Special Sense

Toxic effects on the special senses are distressing and well documented, e.g., aminoglycoside antibiotics and other substances will, under certain conditions, damage the organs of balance and hearing.[207] These adverse effects may be detected by examining fixed flat preparations of these sensory epithelia with phase or interference microscopy.[208]

Toxic effects on the retina may be detected by the method described by Gregory et al.[209] Gehring[210] has reviewed the literature on drugs causing lens changes.

J. Mutagenicity

For 30 years geneticists have known that chemicals may alter the genome, and it has been subsequently shown that this may follow either gene or chromosome mutation. Gene mutation results from deletion or substitution (transitions and transversions) of one of a base pair, intercalation by crosslinking between bases, or a frameshift of base sequences of the DNA. Chromosome mutation occurs by rearrangement, gain, or loss of portions of the chromosome.

All types of gene mutation frequently occur spontaneously. Approximately 3% of all human zygotes surviving for 3 months contain a chromosome aberration, and more than 20% of spontaneous abortions are associated with an aberration. A number of the affected zygotes that develop and are born show well-defined syndromes associated with a characteristic aberration, e.g., Down's and Klinefelter's syndromes. There are at least 2000 characterized human genetic diseases, some of which also carry a raised risk of cancer.

There is evidence that most individuals have several point mutations originating in the germ cells of their parents, but there is none to date that implicates a medicinal product or any other xenobiotic in increasing the mutation rate, i.e., the number of disadvantageous recessive genes in the human gene pool.

Germ cell mutation with reduced fertility and abnormalities of offspring in the next or subsequent generation is not the only potential hazard. Somatic mutations may result in lowered resistance to disease, decreased life span, and, some workers believe, cancer.

The mode of action of some antiviral compounds could conceivably be through reaction with nucleic acids and, therefore, may pose a potential genetic hazard. The actual degree of hazard is difficult to assess as there are many factors which may modify the response. The disposition of the compound in somatic and germinal cells may be difficult to establish, and the mechanisms and capability to repair genetic damage are largely not yet understood. Although several hundred chemicals have been shown to be mutagenic in laboratory tests, there is no evidence that imputes a chemical as a mutagen in man, although there have been occasional reports of suspected occurrences, e.g., Maurer et al.[211] have imputed a trisomy to cytosine arabinoside. It is a matter for debate whether laboratory tests for mutagenicity are sufficiently developed to reliably predict a mutagenic hazard for man.

DNA is universal in all living organisms, and many species of widely differing phyla have been used as test systems. Initial optimism that the combination of the in vitro lymphocyte test, host-mediated assay, and dominant lethal test would be sufficiently accurate and reproducible for general application has given way to a more circumspect approach.

The discipline is however making rapid advances.[212-214] As there is a theoretical possibility of a disaster of considerable magnitude, it is proper to give considerable thought to the need to apply the existing tests, no matter how imperfect, to new antiviral agents. It has already been shown that spurious results can be obtained in individual test systems, but the intelligent application of a panel of tests should be encouraged with full consideration of all factors, particularly the proposed use of the compound. The state-of-the-art does not allow an inflexible approach at present.

Some of the tests that are available are listed in Table 7. Ideally the test would be a sensitive indicator of mutation in somatic and germ cells following administration of the compound to mammals by the route selected for human dosing. Such a test would, of course, show mutation caused by metabolites. At present there is only one whole animal mammalian test for detecting point mutagens, the mouse-specific locus method.[215] In this test treated and untreated outbred mice are mated with a selected strain that is homozygous for a few recessive mutations expressed as coat color and texture. Following mutation the mutant phenotype is expressed in the offspring. The task becomes daunting when it is realized that the spontaneous mutation rate for the outbred mice is about 1:17,000 offspring; to detect a tenfold increase in mutation, ten mutants would have to be found in this number of progeny. To overcome this type of numerical difficulty, microorganisms and *Drosophila melanogaster* have been used. Tests with bacteria (prokaryotic) and yeast (eukaryotic)[216] are inexpensive for rapidly detecting effects on DNA.

Most bacterial tests[217] are based on a specific nutrient requirement of the mutant auxotroph and independence of the wild-type prototroph. Forward mutations detect a wider range of mutagens but are less sensitive than reverse mutation.

Escherichia coli that are DNA polymerase deficient, i.e., lacking DNA repair capacity, were introduced as a more sensitive test. The chief advantage appears to be that positive results are obtained with some animal carcinogens when negative results were found with Salmonella strains. These tests[218] were done with the *E. coli* W3110 (Pol A+) and P3478 (Pol A−) for forward mutations.

The best-known reverse mutation system is the histidine-requiring mutants of *Salmonella typhimurium.* Ames[217] has developed a system in which the compound is converted to putative metabolites by incubation with liver or other tissue homogenates prior to exposure to the bacterial-indicating system. The entry of the compound or its metabolites into the bacterium is facilitated by the use of "deep rough" strains that are deficient in the lipopolysaccharide components of the cell wall.

Several strains of *S. typhimurium* LT-2 are

TABLE 7

Some Screening Tests for Mutagenic Potential

Mammalian indicator cells		Nonmammalian organisms		
		Mammalian metabolic activation		
In vitro exposure	In vivo exposure	In vitro activation	In vivo activation	No mammalian metabolic activation
Human lymphocytes	Human lymphocytes	Ames Liver microsomes/ *S. typhimurium* test	Host-mediated assay	Virus (bacteriophages)
Azoguanine resistant Chinese hamster cells V-79	Animal bone marrow		Urine/bacterial test	Bacteria
	Micronucleus test			*Escherichia coli*
A.K.R. mouse lymphoma	Sister chromatid exchange			*Salmonella typhimurium*
	Dominant lethal test			Fungus
	Heritable translocation			*Saccharomyces cerevisiae*
	Specific locus			*Drosophila*

being used, as they allow different types of damage to be detected.[219]

1. TA 1535 – Reverted well by base pair substitution, but not frameshift mutagens. It detects aromatic nitroso derivatives and metabolically activated aromatic amine mutagens.
2. TA 1537 – This strain contains a frameshift mutation with guanine/cytosine base pair added to a coden triplet. Deletions are detected.
3. TA 1538 – This strain contains a frameshift mutation detecting deletion of two base pairs. It detects metabolically activated polycyclic hydrocarbons. This strain is repair deficient and is used in conjunction with strain TA 1978, which has normal repair capability.

An early method was aimed at exposing the indicator organisms to drug and metabolites by injecting them into the peritoneal cavity of treated animals. The host-mediated assay[220] has not fulfilled its initial promise and has lost favor. A test that has some practical value is that of Commoner et al.[221] They have detected mutagenic metabolites in urine after enzyme destruction of the conjugates.

Other test systems in vitro include Chinese hamster azoguanine- or oubain-resistant cells and the AKR mouse lymphoma.[222] The former has a high or variable back mutation rate, and the latter is only viable in a few strains of mice. Both tests take several weeks to complete.

The remaining tests to be described detect, or are presumed to detect, chromosomal mutations. The dominant lethal assay in the mouse[223] detects postmeiotic damage to male germ cells (lethal damage to earlier stages is generally eliminated) by producing the death of a zygote by heterozygous possession of a dominant lethal mutation. The assay has the advantage that the compound is administered to the intact mammal, and genetic damage is detected (using the zygotes from mating with untreated females as the indicator system). One disadvantage of the test is that its sensitivity is limited by the background level of embryonic death, which may vary between 1 to 10% between strains. The test is not quantitative because only an extremely small proportion of the male germ cells are involved. Treated male animals are paired with a separate pair of females every week for 8 weeks. Post meiotic effects, i.e., on sperm and spermatids, occur in the first 3 weeks, whereas premeiotic effects, i.e., on spermatogonia and spermatocytes, occur later, during the period when induced nondisjunction would be detected.

Thirteen days after successful mating the females are dissected, and the number of live embryos and late and early fetal deaths is counted. The data may be evaluated in several ways.

1. The percentage of implants that died early in gestation.
2. The number of early deaths per pregnancy.
3. The number of pregnancies with one or more early deaths.
4. The preimplantation loss of ova.

Cytogenetic methods for detecting chromosome damage by direct observation with the optical microscope[224] have made rapid advances in recent years. These techniques have the considerable advantage that they may be used to examine human somatic cells, as well as cells from treated animals. Cells usually used are lymphocytes from patients receiving a medicine or from volunteers with cell exposure in vitro and bone marrow from rodents.

The Chinese hamster has 22 individually characteristic chromosomes, whereas the mouse has 40 acrocentric chromosomes,[225] including the 2 sex chromosomes. The former is an excellent subject for such studies. The animals are dosed for 5 days. Following exposure, they are injected with colcemid to arrest cell division at metaphase. Fifty metaphases are analyzed from each of four slides produced from groups of five treated and control animals. The slides are examined[226] for chromosome gaps, chromatid and isochromatid breaks, fragmentation, pulverization, polyploidy, acentrics, dicentrics, and translocation rings.

In vitro studies with phytohemoglutinin-induced mitosis will show the more usual damage to chromatids if the first cell cycle is used (this may be lethal and the cells would be lost in the culture during a second cycle). However, an additional full cycle permits observation of chromosome damage and occurrences such as dicentrics and translocations.

Drugs that break chromosomes (clastogens) may do so in many ways; bifunctional alkylating agents [i.e., highly reactive agents combining with (pairs of) certain functional groups – e.g., amino, carboxyl, phenolic, and phosphoric groups in proteins, nucleic acids, etc.] crosslink the strands of DNA. Bleomycin acts by direct scission of

DNA, methotrexate inhibits inosine production affecting nucleic acids, and others may alter the chromosome protein matrix. Mitomycin C inhibits mitosis.

A test which is claimed to be technically easier than chromosome analysis is the micronucleus test,[227] in which Howell-Jolly bodies in the red cell precursors of the bone marrow are counted.

A newer cytological technique is that of Perry and Evans[228] – the sister chromatid exchange (SCE) method. In this test symmetrical exchange between sister chromatids at one locus is detected by exposing cells to 5-bromodeoxyuridine for two cycles of replication so that chromosomes possess one chromatid unifilarly substituted and its sister bifilarly substituted. Such chromatids stain differentially with a fluorochrome, and when examined by fluorescence microscopy, chromatid exchanges give a harlequin appearance. It is claimed that this method is much more sensitive than the other cytological methods. These workers suggest that SCE events involve recombination of parent and daughter polynucleotide strands at a site of transient discontinuity in both parent strands. The recombination repair process proceeds during DNA replication, and mutation originates as an error in the repair process following mispairing or base deletion or insertion.

The heritable translocation test,[229] which is being developed in several laboratories, promises to improve the sensitivity of mammalian tests for chromosome abnormalities. In this test treated and untreated male mice are mated with untreated females, and male mice that have infertile matings are rebred with untreated female mice. If the male mice remain infertile, they are considered to have undergone "presumptive mutation" of germ cells and are subjected to chromosome analysis.

A World Health Organization scientific group[230] has recommended that compounds which will be given frequently over a period of years should be tested, particularly those that will have widespread prescription, those that will be used in the young, those with structural similarity to known mutagens, those with cytotoxic effects, those that lower fertility, or those that reduce immunological competence.

As with all previous types of toxicity tests, it is important to try to obtain dose responses with positive results. The relevance of the particular test or panel of tests to the antiviral compound's proposed use should be estimated.[231,232]

An extremely contentious subject is the use of mutagenicity tests as short-term predictive tests for carcinogenic activity. Ames claims that the correlation may be as high as 85%, whereas Legator has found that 27% of a group of known carcinogens was mutagenically inactive. Examples supporting both sides of the debate are published frequently, and the authors conclude that it will be some time before the value of these tests for detecting carcinogenic potential rather than mutagenic potential may be judged.

In summary, tests for mutagenesis hold the exciting prospect that they will be of predictive value for detecting genetic damage and may also in due course be useful for screening for carcinogenic action. Until further research is done, their adoption as a mandatory part of the safety evaluation of medicines would be entirely premature. Bridges[233] has proposed that the tests used for screening should be applied in a tier scheme of increasing complexity to prevent wastage of resources.

IX. PREMARKETING STUDIES

A. Long-term Repeat Dose Toxicity Studies

These are the same as the tests described in Section VIII.A, except they are of longer duration and are done to detect toxic effects on the eye, endocrine system, kidney,[234] and other organs. The duration of these studies is at least 6 months for the United Kingdom, Sweden, and Australia, and the requirements of the FDA are set out in Tables 2 and 3.

B. Carcinogenicity Studies

About one out of every six deaths in man is due to cancer. The etiology of most human cancers remains unknown, although it has been established that certain chemicals and ionizing radiation will initiate purposeless persistent proliferation of tissue. Furthermore, oncogenic viruses[235,236] have now been found to induce tumors in higher animals, including subhuman primates.

The epidemiological evidence suggests that the development of chemically induced cancer in man usually follows a long period of exposure to the substance, and there is generally a long latent period between early exposure and clinical appearance of a tumor. This may be 10 to 40 years[237] in man and 3 to 24 months in rodents, long after the exposure to a chemical has ceased.

With the exception of arsenic, all chemicals that are known to be carcinogenic in man[238,239] have been shown to be carcinogenic in one or more species of laboratory animals,[240] although each species may show its own characteristic tumor(s)[241] type, i.e., the tumor may be benign, malignant, may have a different cell type of origin, or may be present in a different organ in different species.

Earlier in the history of chemical carcinogenesis,[242,243] there were high expectations that there may be a close correlation between chemical structure and carcinogenic activity. Although this was shown to be true for certain classes of chemicals, it has now become more common to seek evidence of certain types of chemical reactivity. The molecular events leading to irreversible changes in the cell, which ultimately end in a neoplasm, are poorly understood. Nevertheless, considerable progress has been made in recent years with studies on the interactions of carcinogens with tissue.[244,245]

It has been found that many chemicals are not themselves carcinogens, but rather it is their derivatives produced by the metabolic action of tissues that are the active species. It is not, however, always the organs that possess the metabolizing enzymes that are the targets for carcinogenic action. With a number of carcinogens, the covalent binding of the active species (compound or metabolite) to nucleic acids and proteins of the target organs themselves has been demonstrated. These observations have led to many proposals for short-term predictive tests for carcinogenesis based on the identification of substances that show covalent binding with nucleic acids or bring about mutation of the genome or degranulation of ribosomes.[246] In spite of the above advances and many of its own shortcomings, the long-term administration of compounds to laboratory animals remains the most reliable method for detecting carcinogens.

At present, regulatory agencies only demand the carcinogenic testing of specific classes of drugs (notably FDA and CSM) for which studies with oral contraceptives and beta adrenoceptor blocking agents are mandatory. Tests on antiviral compounds are required only if they satisfy any of the following three criteria, although this situation could change at any time.

1. Compounds (or their metabolites) with structural similarity to known carcinogens.

2. Compounds that damage rapidly growing tissues (e.g., hemopoietic tissue, intestinal mucosa, testes) or interfere with mitosis.

3. Compounds that will be continually administered to patients for prolonged periods (e.g., periods of greater than 1 month), particularly to patients who may be considered to be at greater than usual risk, such as newborn babies and pregnant or lactating women.

Tests are done in one or more species of rodent (oral contraceptives are the exception, with tests required in dogs and monkeys),[247] and the compound is usually administered in their diet for a substantial part of their life, currently 2 years in rats and hamsters and 18 months for mice.

Some authorities believe that with the increased longevity of rodents kept under modern conditions of husbandry, the dosing period should be longer. This has the disadvantage that the spontaneous tumors in the controls will have a higher incidence in the longer trials, and this makes differences between treated and untreated groups more difficult to demonstrate.

There are numerous guidelines for the conduct of carcinogenicity tests.[248-253] The number of animals demanded varies between 25 and 50 per sex per group.[254] The number should take into account the longevity of the strain and the nature of the test, and a number similar to the lowest quoted above should survive the test period. In the authors' laboratory, 65% per group is generally used to ensure a meaningful statistical evaluation. The control group may be double the size of an individual dose group, but it is not valid to summate the tumor incidence at several dose levels to compare with the controls. Although larger numbers in groups will increase mathematical significance for differences between them, it is unlikely that the biological importance will be emphasized.

The number and selection of doses is a further contentious area. It is argued by some workers that only one, or at most two, doses are required. At least two should be used to allow for the unexpected observation of toxicity at the high dose level. The inclusion of three dose levels allows a dose-response curve for a tumorigenic[255] effect to be drawn. The validity and safety of extrapolating downwards[256] from this to a no-effect dose level has not generally been accepted, and it is questioned whether more than two dose levels are necessary. The advantage of a dose-response curve is to estimate a theoretical safe level, as well as to gain confidence that low-incidence effects are due to the drug and are not a cluster of spontaneous tumors.

It is a basic tenet in toxicology that greater prediction may be obtained with a test of unknown sensitivity by increasing the dose. However, it is particularly important in carcinogenicity tests to avoid a false positive response by activating latent oncogenic viruses, by producing persistent chronic tissue damage, or immunosuppressive effects, or by altering the endocrine balance or other physiological mechanisms. The Commission on the Safety of Medicines[257] interpreted the particular tumors observed in rodents with high doses of the oral contraceptives as having no predictive value for man because it was felt that they resulted from alteration of the endocrine status in a species whose endocrine system shows major differences from that of women.

If an exaggerated pharmacological effect or toxicity is not seen, then the highest dose is selected by finding the dose that will lower the weight gain of weanling rodents by about 10%. The high dose level may also be selected on the basis of comparative blood levels in man and animals. Whatever method is used, it should be compatible with survival of the majority of the animals for the duration of the test. The next dose is usually half of the high dose. If a low dose is used, it is usually a small multiple of the proposed human dose.

The compound should be administered by the intended clinical route. If the choice is oral dosing, then the compound is usually administered in the diet or occasionally in the drinking water or by gavage. If parenteral administration is used, the dosage regimen should be relevant to the proposed use.

It is essential that identity of the compound be known and the purity controlled. A contaminant at apparently quite low levels, e.g. 1 ppm, may have considerable importance when administered for years, e.g., *o*- and *p*-toluenesulfonamide in saccharin. Similarly, the thoroughness of the mixing in the diet, the accuracy of the dose, and the stability in diet must be monitored throughout the trial.

For practical and scientific considerations, rodents are usually tested. These animals may be housed in sufficient numbers in a controlled environment, they have a relatively short life span, and there is an increasing body of experience in using them. The author believes that hysterectomy-derived strains housed during the period of the trial in very good conventional conditions, e.g. access to animal house by responsible staff only, will provide the best subjects for the trial.

Careful consideration should be given to all aspects of the environment. Efforts should be made to ensure that the diet is standardized[258] as far as possible and contains insignificant levels of hormones, pesticides, mycotoxins, etc. Animals of both sexes should be used, but caged separately. It should be remembered that the cages must allow for the growth of animals and increased aggressiveness of mature males. The animals should be allocated to the groups at random.

When using large numbers of animals, it is advisable to use a replicate system, which essentially is a method of dividing the experiment into subgroups that are representative of the whole. Each subgroup (replicate) contains one or more cages of animals for each sex at each dose. The position of each replicate (and each cage in the replicate) in the room is randomized, and for all procedures and investigations the animals are handled by one replicate at a time.

At the inception of the trial, 20% of a group of animals larger than that required is weighed, and the standard deviation is used to specify an acceptable weight range. Animals are weighed again, and those within the specified limit are allocated at random to the cages of a replicate; this is repeated until all the replicates are filled. It has been suggested that randomization should also be by litter mates to ensure genetic heterogeneity within replicates.

There are differences of opinion as to whether the strain used should be inbred or outbred. The advantage of the former is that the background incidence of spontaneous tumors is likely to be less variable, whereas the heterogeneity of the latter could confer a greater spectrum of response. A compromise proposal is to use several inbred strains, but this increases the practical difficulties. Whatever strain is used, it should be known to be susceptible to known carcinogens.

It has recently been suggested that prenatal or perinatal exposure to chemicals would increase the sensitivity of the test, but there is no example reported to date of a carcinogen that has been detected by testing in neonates[259] that has not also been found by the use of weanling animals.

All animals should be given a full necropsy regardless of whether or not they died before the end of the study (and may have been undergoing autolysis). Occasionally, it is prudent to cage separately or kill unwell animals to ensure that they are not cannibalized by their cage mates.

Portions of all major tissues should be preserved for histological examination, although only in the absence of an effect may it be necessary to examine the animals from the high dose group and the controls. The cause of unscheduled deaths should be sought. As some authorities place considerable emphasis on an increase in malignant tumors, a very detailed histopathological evaluation of tumor types is necessary.

Interpretation is done using five major criteria:

1. A clear increase in the total incidence of all tumors.
2. An increase in tumors that have been previously observed for the strain in the control group (or the laboratory).
3. An increase in tumors of an uncommon type.
4. An increase in malignant tumors.
5. A decrease in latency of tumor appearance.

In the authors' view, an increase in one of the above factors in both sexes or in both species (if two species are used, or both sexes of both species) carries somewhat more importance than if it occurred alone.

The interpretation of carcinogenicity is not as easy as the above criteria would suggest. The situation may be complicated by early deaths of animals, and the occurrence of spontaneous tumors (i.e., those that are not related to administration of the compound) may be high and variable. At ICI we have observed, for example, a range of 0 to 15% pulmonary tumors in 19 control groups of mice of the same strain that were examined over a period of a few years.

Statistical analysis is frequently applied using one of several recognized methods. One, recommended by the National Academy of Sciences Food Protection Committee,[260] is a chi-squared test with continuity corrections.

The incidence for each group, when data from both sexes are combined, may be found with the Mantel-Haenzel procedure.[261] The *British Journal of Cancer*[262] has given guidelines on methods that compensate for the unscheduled deaths of animals that could prevent the discovery of early tumors.

Roe has called for sanity in carcinogenicity testing.[263] All predictive tests are imperfect, but some of the difficulties could be overcome by the scientific community agreeing that there are shortcomings that require special treatment. It should be recognized that every species or particular strain of a species shows some tumor types that occur with high incidence [e.g., in many strains of male mice, liver parenchymal tumors are common; in many strains of female rats, pituitary (and mammary) tumors reach a high incidence after 12 months]. With a high incidence of this type, the animal model becomes valueless, and an alternative model should be used. Most of these difficulties could be overcome by using the rat and the mouse (if all other factors allow) and by agreeing from the outset that for the above examples of common tumors and for others known for the strain/species/sex, the pursuit of slight differences between figures of high incidence in both treated and untreated animals will not provide a basis for a scientific judgment. We believe that this approach could markedly increase our confidence in the predictive value of carcinogenicity trials.

C. Reproductive Toxicology

The grim testimony of the effects of thalidomide that were seen in 1962 has made everyone acutely aware of the potential dangers of administration of new medicines to women of child-bearing age.[264] Since that time we have become increasingly cognizant of the additional hazards of mutagenesis, fetal toxicity, and the possibility that functional or behavioral abnormalities as well as morphological changes may occur.

To test an antiviral compound for reproductive toxicity, it is necessary to apply tests that challenge the total process from germatogenesis to the weaned offspring and beyond, to the health of future generations. Chemical teratogenesis began in 1909 with the observation of Lehnert[265] that strontium produced fetal bone abnormalities in rabbits. In 1937 vitamin A deficiency was correlated with abnormalities of newborn pigs. In 1941 the dangers of rubella were recognized, and in 1948 Gillman et al.[266] studied the teratogenic effects of trypan blue.

Following the birth of 8000 deformed children, the relationship between phocomelia and the ingestion of thalidomide during the critical period of organogenesis was recognized. When George Somers[267] showed that rabbit fetuses were also susceptible to thalidomide, the teratogenicity testing of new medicines in animals began. The term "terato" means monster and refers to the interference with the normal differentiation of organs resulting in malformations.

In 1966 Goldenthall[268] of the FDA published guidelines for animal screening tests, recognizing that reproductive function is a long and complex process, and proposing tests under three segments.

Segment	Title and nature of test
I	Fertility and general reproductive performance
II	Teratogenesis
III	Peri- and postnatal studies

It is sensible to divide reproductive testing into separate segments rather than one comprehensive test because each segment answers specific questions (Table 8) and may require different dose levels because of the different maternal tolerated doses in the segments, e.g., 5-iododeoxyuridine administered to newborn animals may result in renal cortical and retinal dysplasia.[269] Cyclocytidine produced cleft palate, splanchnocele, ectrodactyly, and clubbed legs in rabbit fetuses when the dams were dosed on days 8 to 17 of pregnancy.[270]

A great number of animal models and guidelines for their application have been produced.[271-277a] The patterns of development of lower animals are usually so different from those of man that they are only of research use. Chick development has been used, but the absence of metabolic exchange with the dam limits its value.

Rabbits (of the Order Lagomorpha) and rats and mice (of the Order Rodentia) have been used extensively. However, prior to the formation of the chorioallantoic placenta, these three species have a yolk sac placenta, and there is interchange between dam and conceptus by this highly permeable structure. A second fundamental difference is the dependence throughout pregnancy of the rabbits and rodents on the ovary for progesterone and estrogens, whereas the placenta of primates (and the guinea pig) takes over this function during gestation.

Wilson[278] has suggested a tier system of teratogenicity testing in which the compound is first tested in rodents and rabbits. The second tier would be done to confirm or revise the embryotoxic dose with a carnivore, (dog, cat, or ferret) or ungulate (pig or sheep), whichever showed the greatest metabolic and disposition similarities to man. The third tier would be done only with a

TABLE 8

Susceptible Stages During Reproductive Toxicology Testing

Segment	Generation involved	Stage of reproductive function
I	Parents (F_0)	Gameteogenesis in gonads; desire and ability for successful mating; function of reproductive tracts and accessory glands; fertilization
	Offspring (F_1) and F_0	Implantation of the blastocyst; absence of toxic effects of the compound or its metabolites on the dam or directly or indirectly on the fetus; uterine development and function, and the integrity of the feto-placental unit for the maintenance of pregnancy
II	F_1	Organogenesis in the embryo
III	$F_1 + F_0$	Development of the fetus; uncomplicated birth; survival of neonates; nursing and suckling behavior; lactation; maturation (morphological, including special senses), functional (including behavior)

Note: To these three segments, we must now add mutation studies, with absence of effects on the genes of the parents and effects on the chromosomes and genes of the conceptus, both for somatic and germ cells.

nonrodent (an alternative to that already used) if there was some uncertainty concerning the therapeutic index. At our laboratories we have a colony of marmosets (*Callathrix jacchus*) which are used for such tests. This species breeds well in captivity and responds to thalidomide with phocomelia.[279] Wilson considers it advisable to have a fourth tier to test for drugs in primates that would unquestionably be required during pregnancy, that would be administered to women of child-bearing age, or where the drug would be given inadvertently before pregnancy was diagnosed. Whether or not this tier system is used, it is essential, where possible, to study the pharmacokinetics and metabolism of the drug in both the dam and the fetus.

At least two species, one a rodent and the other nonrodent, must always be used for the teratogenicity test because it is known that there are marked species differences in the response to teratogens, e.g., diphenylhydantion is teratogenic for mice but not for monkeys. As with other aspects of toxicology, strain differences may, on occasion, be marked, e.g., the susceptibility of A/Jax mice to corticosteroids. If rabbits are used, the strain should be shown to be responsive to thalidomide.

Regulatory authorities (FDA, CSM, Australian Authority) request that wherever possible the drug should be given by the proposed route of clinical use, with the highest dose level in studies eliciting some minimal maternal toxic response.[172] This may be alteration of body weight and may be a slightly different dose for each segment. If the dose is embryocidal, then additional tests using lower doses must be done. Wilson suggests that tolerance levels should be set in reference to the embryotoxic and biologically effective doses with a ratio between 1:1 to 10:1 (or more).

It has been shown that some teratogenic compounds produce their effect at a very specific time during organogenesis. It has been suggested that tests should be done with a number of single (pulse) doses given to separate groups throughout the critical period.

A summary of regulatory requirements is presented in Tables 9, 10, and 11. The investigations to be made on all three segments are shown in Figure 1. Where appropriate, the following studies and measurements should be done:

TABLE 9

Requirements for Reproductive Toxicology Studies at Various Clinical Phases

	Phase required	
Segment	United Kingdom	United States
I	Where appropriate but necessary for Product Licence	After initiation of Phase 2
II	Application for Clinical Trial certificate	Before Phase 2
III	Where appropriate; necessary for pediatric drugs; necessary for Product Licence	After initiation of Phase 2

1. Metabolism and pharmacokinetic studies with the dam.
2. Abortions.
3. Number of implantation sites.
4. Number of corpora lutea.
5. Number of early deaths.
6. Number of late deaths.
7. Number of live fetuses.
8. The weight of the uterus with contents.
9. The weight of each fetus.
10. Calculate the volume of placental fluids.
11. Metabolism and pharmacokinetics in the fetus.
12. External abnormalities.
13. Internal abnormalities.[277b]
14. Skeletal abnormalities (using radiography or the alizarin technique).[277c]
15. Eyes (histopathology of eye).

On offspring born spontaneously, the following studies and measurements should be done:

1. The number of pups at 0, 7, and 21 days.
2. The weight of the pup at 7 and 21 days.
3. The sex of individual fetuses and the sex ratio.
4. The anogenital distance for each sex.
5. Auditory function (startle reflex).
6. Behavioral tests.

One male and one female from each litter should be reserved for testing reproductive function (avoiding brother/sister matings) by production of one litter.

TABLE 10

Outline of Design Requirements for Reproductive Toxicity Tests

Country	Fertility and reproductive performance		Teratology		Peri- and Post-natal studies	
	No. of species	No. of dosages	No. of species	No. of dosages	No. of species	No. of dosages
Australia	1[a]	2	2	2	1	2
EEC	Only if indicated by other tests		2	Not specified	–	–
Sweden	1	3	2	3	1	3
Japan	1 or more	Several	2	Several	1 or more	Several
United Kingdom	1[b]	3	2	3	1	3
United States	1[b]	2	2	2	1	2
Venezuela	1	2	2	2	1	2
Canada	1	2	2	3	1	2

[a]Two tests should be done (with treated animals of one sex and untreated animals of the opposite sex.)
[b]Tests should be done with treated animals of both sexes.

TABLE 11

Minimum Number of Animals per Group and Dosing Period for Reproductive Toxicology Tests

Country	Fertility and reproductive performance in rodents[a]	Teratology: Rodents[b]	Teratology: Rabbits[c]	Teratology: Nonrodent[d]	Peri- and postnatal studies in rodents[e]
Australia	20	10[g]–20	10[g]–20		10
Canada	Not specified	20	15		Not specified
EEC	Not specified	Not specified	Not specified		–
United Kingdom	12 males,[f] 24 females	24	12	(Adequate for clear conclusion)	12
United States	10 male, 20 female[h]	20[g]	10[g]		20
Japan	20 male, 20 female	30[g]	10[g]		20[g] rodents or 20[g] rabbits
Sweden	Not specified	At least 10	10		10

[a]Males, 80 days before mating. Females, 14 days before mating (U.K. dose throughout pregnancy; Australia dose throughout pregnancy; Sweden optional through lactation but at least first week of pregnancy; Japan until day 7 of pregnancy in the rat and day 6 of pregnancy in the mouse).
[b]Mice, days 6 to 15. Rats, days 6 to 15 (Japan day 17). Guinea pigs, days 7 to 35. Hamsters, days 4 to 14 of pregnancy.
[c]Days 6 to 18 of pregnancy (Canada days 7 to 19).
[d]Marmoset, days 25 to 50. Rhesus monkey, days 9 to 40. Pig, days 7 to 35. Cat, days 7 to 17 of pregnancy.
[e]Rodents, day 15 of pregnancy (Japan day 17) to weaning (day 21). Rabbit, day 18 to weaning (day 30).
[f]For veterinary drugs, evidence of absence of potential effects on spermatogenesis in the bull is currently required.
[g]Number of pregnant animals.
[h]Over 40 days of age.

The assessment of teratogenicity studies is occasionally as difficult as carcinogenicity tests. Assessment must consider the following:

1. Alteration of reproductive parameters-dose relationships.

2. Incidence of terata-dose relationships.

3. Type of abnormality – Is it a fetal deformation or an embryonic malformation? A deformation is due to nonspecific effects of the uterine environment on the fetus, e.g., pressure, whereas a malformation is a defect in the development of the embryo. Sullivan[280] has stressed the importance of pharmacological effects on the uterus which lead indirectly to effects on the fetus, e.g., maternal placental blood flow.

a. Is it a major malformation?

b. Is it a common variant, either visceral or skeletal?[281]

c. Is it associated with one embryonic system or more?

4. The laboratory control experience for variants and malformations.

5. Are the findings litter biased?

6. Is the compound teratogenic for more than one species? If it is not, does it nevertheless alter other reproductive parameters in other species?

7. What are the relationships between the "effect" dose and

a. Toxic dose for the dam.

b. LD_{50} for the dam.

c. ED_{50} for the dam.

d. Embryotoxic dose.

The data are not usually normally distributed, and nonparametric statistics (e.g., Mann-Whitney ranking) should be used if erroneous significance levels, i.e., false positives, are to be avoided.[282]

The method of statistical analysis is of considerable importance when there is a low rate of malformations. Haseman[283] believes that the basic unit for analysis is the litter, as this is affected by the genetic constitution of the dam, the litter number, and the environment.

Fertility and general reproductive studies are done to detect any effects of dosing on the total reproductive function of the parent males and/or the parent females.

Young mature males are usually dosed during one complete cycle of germ cell production (through all stages from primary germ cell to spermatazoa). They may be mated with dosed mature females or with females that have been dosed for 2 weeks (see Figure 8). The number of fertile matings and the number of normal conceptuses and young produced from the matings are determined. In some studies the fertility of the parents at a second mating or of the young (investigated by allowing the offspring to mate when mature) is investigated. Brother/sister matings are, of course, avoided. The authors prefer studies using dosed animals of one sex mated with undosed animals of the other sex (see Figure 8).

Peri- and postnatal studies are done to investigate the effects of the compound on the fetus late in pregnancy and also on the capacity of the dam to produce normal young by natural birth and to nurse them to weaning. The dams are dosed during the latter part of pregnancy (i.e., after embryogenesis) and throughout the neonatal period until weaning (see Figure 8).

D. Pharmacology

The pharmacodynamics of a compound is not usually considered after the pharmacokinetics and toxicology, but with an antiviral compound the effects of the general pharmacological actions on the body are of considerable importance, particularly with respect to side effects; nevertheless, they are secondary to the antiviral action. However, it is very necessary to study the effects of the substance on the central nervous system; peripheral neural transmission; autonomic nervous system; cardiovascular, respiratory, and gastrointestinal systems; and renal and hepatic function. It is also necessary to determine whether there is any possibility of pharmacological interactions[284] with other drugs that are likely to be administered concurrently. This may follow displacement from the plasma binding sites,[285] induction or inhibition of enzymes (e.g., monoamine oxidase inhibition), induction of aryl hydroxylases, etc.

E. Pharmacokinetics

This term will be used to encompass both drug disposition and drug metabolism within the living body. The aim of pharmacokinetic studies is to establish the pattern of absorption, distribution, and excretion of the active principle and the products of its biotransformation.[286-289]

These studies begin in man with the first volunteer studies or the clinical pharmacology and are expanded to provide a rational basis for dosage

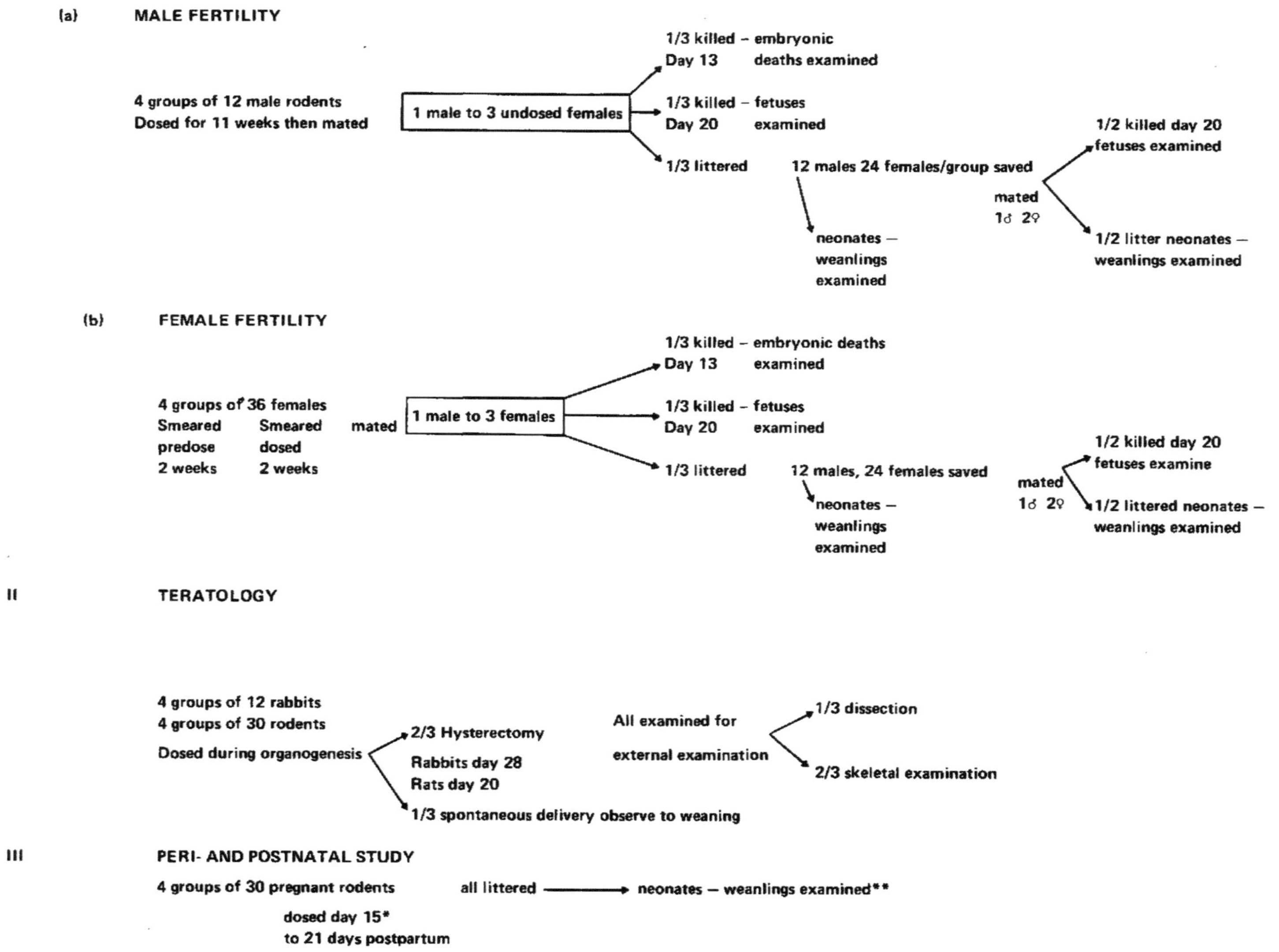

FIGURE 8. Summary of reproductive toxicology test design. (*) The day that sperm is detected in the vagina on day 0 of pregnancy. (**) Under certain circumstances F_1 generation is mated for study of reproductive function and F_1 studied for behavorial abnormalities.

regimens and to contribute to the safety data. They play an essential role in selecting the most appropriate species for toxicity tests[290] and extrapolating the findings from them to man.

Before commencing any studies with the antiviral compound in man, it is advisable to do disposition studies. With other classes of drugs, the biological activity may be measured (e.g., lowering of blood pressure), and the effects of changing the route of administration may be observed (e.g., when the compound is given, say, by the oral route and parenterally); however, with an antiviral compound this may not be possible.

Drug disposition cannot be studied without an extraction and assay method of suitable specificity and sensitivity which may be applied to biological fluids and tissues. Useful techniques include gas chromatography, thin-layer chromatography, high-pressure liquid chromatography, polagraphy, and radioimmune assay. In the absence of methods of suitable specificity and sensitivity, radiolabeling of the compound may be used. The position of the label(s) and the specificity of the labeling must always be reported. The isotopic dilution method is very sensitive.

It is essential to obtain preliminary information from animals before proceeding to man. The extent of absorption may often be conveniently estimated by comparison of area under the curve of plasma concentration/time graphs obtained after administration by parenteral and nonparenteral routes. This method may be used to compare the bioavailability of the active principle in a particular formulation.

Gastrointestinal absorption will be affected by many factors including the pK of the drug, its lipid/water partition, gastrointestinal transit time, metabolism by the gut flora, etc. If absorbed, each drug will be distributed in body fluids in a particular pattern that varies with the drug, dose, and time. Disposition will be affected by binding to plasma proteins or other macromolecules or erythrocytes and entry into extravascular compartments (measured as "volume of distribution"), for example, adipose tissue; passage across blood-brain; blood-placental barriers; metabolism by liver and other tissues; and excretion from the liver in bile, from the kidney in urine, or from the lungs.

The plasma concentration/time graph after a single dose should allow an estimate of peak levels and the half-life; multiple dosing may result in a plateau plasma concentration (for particular dosage regimen).

Some drugs are metabolized during their "first pass" through the liver, whereas others are excreted in the bile and are absorbed again from the bowel, resulting in enterohepatic circulation. The excretion routes and the percentage of the drug excreted during a specified time are of obvious importance and may give an initial indication of accumulation of the drug in the body.

Whole body autoradiography using photographic exposure of frozen sections from animals dosed with radiolabeled compound will give qualitative information on tissue distribution, i.e., high concentrations of drug (or its metabolites) in a particular organ, e.g., binding in pigmented tissues.

The biotransformation of the compound by drug-metabolizing enzymes may increase or decrease the activity and the toxicity of a compound. It may result in products which are more readily eliminated from the body, often after conjugation to make them into more polar derivatives; conversely, the metabolites may bind to tissues in the body. The metabolite may be more toxic, e.g. carcinogenic or mutagenic, in their own right. This is why many in vitro techniques for finding biological effects incorporate a primary stage of metabolic "activation." In early studies it may not be necessary to identify the metabolites, but rather to establish that the separation pattern (chromatographic) is similar for the species being compared.

Metabolism of the drug may depend on the prior possession or artificial stimulation of drug-metabolizing enzymes, e.g., Owen et al.[291] report that the antiviral agent dichlorophenobarbital is a potent self-stimulant of metabolism. The inherent ability to metabolize foreign substances depends on genetic constitution (species and race), sex, age, and physiological state, e.g., pregnancy. If the antiviral agent itself does not possess the capability to stimulate the specific enzymes required for its own metabolism (induction),[292] the prior administration of a compound with that potential may alter the response that the antiviral will give.

In many instances, the marked differences in response to a drug shown between species, sexes, or strains are due to differences in either the route and rate of absorption or the patterns of metabolism. When the unchanged drug is eliminated, the rate frequently depends on the surface area of the animal, whereas the rate of metabolism is independent of surface area and varies with a greater number of extrinsic and intrinsic factors.

When the activity of the drug is directly related

to its plasma level, then metabolic differences are more relevant to the toxicity than the efficacy; if it is not, then greater consideration must be given to the possibility that it is a metabolite that possesses the antiviral activity, or that the drug (or metabolite) is highly bound to its site of action.

If an antiviral compound is used in food-producing animals, then residues of the compound and "major" metabolites in meat, offal, milk, eggs, etc. become extremely important. It is necessary to measure residues and relate these to a drug withdrawal period[293,294] that is enforceable within the normal husbandry procedures for the species. It is also necessary to provide "cold" assay, i.e., one that does not involve the administration of radioisotopes, of sufficient sensitivity to ensure that drug withdrawal policy is respected.

If a drug is used in food animals for limited periods (e.g., estrus synchronization) and is non-carcinogenic, the FDA permits an upper residue limit of 50 ppb ($b = 10^9$) in edible tissues and 5 ppb in milk and eggs. The safety factor of an acceptable daily intake (ADI) by the consumer must exceed 2000 (based on a 90-day toxicity study). If the animals are to be treated for longer periods, e.g., a growth-promoting agent, then the upper limit is 1 ppm in edible tissue and 0.1 ppm in eggs and milk, with the ADI of 1000 (based on a 6-month toxicity study with reproductive toxicology tests). If the compound gives rise to a residue greater than 1 ppm but is not carcinogenic, then a safety factor of at least 100 is required based on lifetime and teratology studies in two rodent species and a reproduction study in one species. In Europe a safety factor of at least 200 times the ADI (based on a 90-day test) is required. In France, Truhaut[295] has proposed that residue problems may be solved by relay toxicity, i.e., feeding the tissues to laboratory animals.

If the drug is to be administered topically (and it can be shown that it is not absorbed from the intact skin, mucous membranes, etc. when they are normal or in conditions similar to the diseased state, e.g., inflamed or abraded), then it is possible that the quantity of systemic toxicity studies could be reduced. As always, there must be a full consideration of all the potential risks in the clinical situation.

X. DRUGS INTENDED FOR ADMINISTRATION TO THE YOUNG

Investigators will want to avoid administering new antiviral compounds to the young until all possible safeguards have been used. It is also important that children are not deprived of good therapeutic substances. Obviously, children are not small adults, and a special approach to safety and efficacy studies must be taken. The FDA has set the following definitions for pediatric groups.[296]

Neonates	0–4 weeks
Infants	0–2 years
Children	2–12 years
Adolescents	12–18 years

There are not mandatory requirements for pediatric drugs, but the following scheme should provide some safeguards. No Phase I studies should be done with the young until a satisfactory outcome has been obtained with

1. Phases I and II in adults.
2. Segments I, II, and III of reproduction toxicity studies.
3. Mutagenesis tests.
4. Additional acute and subacute toxicity studies and pharmacokinetic investigations using young animals.
5. The carcinogenicity potential of the compound.

It is advised that the compound be given to the groups of the young in the reverse order of the above list.

X. CONCLUSION

The discovery and safety evaluation of antiviral medicines is a complex procedure, involving at one extreme the commercial analysis of disease targets and at the other the preparation of statutory submissions for registration authorities. Therefore, it is no longer the province of the isolated virologist, but requires the resources of a multidisciplined organization in order to carry it through. There is, of course, always the possibility of discovery by an individual worker and development by a corporation with the appropriate facilities. Since a commercial organization would only be interested in a discovery if it were commercially viable, it could be to the individual worker's own advantage to be aware of the broad requirements which his new discovery would need to meet.

Much of what has been said in the foregoing

sections must be regarded as guidelines only. Each disease will have its own special virological, commercial, and toxicological considerations, and we hope only to have drawn the reader's attention to some of the more important aspects. The same is true of drug registration requirements. These are by no means rigid, but with the passing of time, a degree of uniformity is emerging which hopefully may lead to a standard of safety testing which will be universally acceptable to registration authorities throughout the world. Meanwhile, in most countries there are government agencies that welcome consultation on the design of toxicological studies at all stages of a drug's development. It must be understood that ultimate responsibility for both the safety and efficacy of a new antiviral drug rests not with government agencies or with registration authorities, but with the drug's sponsor. In the continuing search for antiviral medicines, these twin responsibilities should not be forgotten.

APPENDIX A

Clinical Pathology
PMA/FDA Draft "Preclinical Guidelines"

1. Hematology – Hemoglobin, packed cell volume, total and differential white cell count. Coagulation tests: (e.g., prothrombin time and kaolin activated partial thromboplastic time). Reticulocyte and differential bone marrow count when other findings suggest necessary.

2. Blood chemistry – Glucose, urea, liver enzymes, alanine and aspartate transaminases, and alkaline phosphatase (ALT, APT, ALP).

3. Urinalysis – Include microscopical examination of centrifuged urine sediment.

The author (D.A.R.) suggests that the list should be extended as follows:

1. Hematology – Hemoglobin, packed cell volume, red cell count, platelet count, total and differential white cell count. Prothrombin time and kaolin activated partial thromboplastin time. Reticulocyte and differential bone marrow count when other findings suggest necessary. Methemoglobin for certain compounds when the structure or its biological activity suggests that this substance could be produced.

2. Blood chemistry – Glucose, urea, total protein, albumin, total bilirubin, AST, ALT, ALP, Na^+, K^+.

3. Urinalysis – Volume, color, pH, osmolality, sugar, protein, blood, ketones, bilirubinogen, creatinine. Include microscopical examination of centrifuged urine infranatant. Urinary Na^+ and K^+ may be useful.

APPENDIX B

PMA/FDA Draft "Preclinical Guidelines"

1. Tissues which should be fixed and sectioned routinely – Aorta, adrenal glands, bladder (urinary), bone, bone marrow, brain, epididymides, eyes, heart (including papillary muscle), intestines, (duodenum and ileum), kidneys, liver, lung, lumph node, mammary gland, nerve (peripheral), esophagus, ovary, pancreas, parathyroid, pituitary gland, prostate, spinal cord (cervical attached to brain in the rat), spleen, stomach, striated muscle, testes, thyroid gland, uterus.

2. Tissues to be fixed or sectioned under special circumstances – Bronchi, gall bladder (dog and mouse), inner ear, larynx, salivary gland, skin, spinal cord of dog, thymus, trachea.

3. The author recommends the following additional tissues for routine examination – Cervix, gal bladder, lacrimal gland (rat), intestines (jejunum and colon), salivary glands, skin, spinal cord, thymus, vagina.

REFERENCES

1. **Osdene, T. S.**, Antiviral agents, in *Topics in Medicinal Chemistry,* Rabinowitz, J. L. and Myerson, R. M., Eds., John Wiley & Sons, New York, 1967, 137.
2. **McFazdean, J. A.**, Advances in the chemotherapy of virus diseases, in *Advances in Pharmacology and Chemotherapy,* Vol. 7, Garattini, S., Goldin, A., Hawking, F., and Kopin, I. J., Eds., Academic Press, New York, 1969, 309.
3. **Goz, B. and Prussoff, W. H.**, Pharmacology of viruses, in *Annual Review of Pharmacology,* Vol. 10, Elliot, H. W., Cutting, C. W., and Dreisbach, R. H., Eds., Annual Reviews, Palo Alto, 1970, 143.
4. **Swallow, D. L.**, Antiviral agents, in *Progress in Medicinal Chemistry,* Vol. 8, Ellis, G. P. and West, G. B., Eds., Butterworths, London, 1971, 119.
5. **Bucknall, R. A.**, The continuing search for antiviral drugs, in *Advances in Pharmacology and Chemotherapy,* Vol. 11, Garattini, S., Goldin, A., Hawking, F., and Kopin, I. J., Eds., Academic Press, New York, 1973, 295.
6. **Kapikian, A. Z.**, A collaborative report: Rhinoviruses – extension of the numbering system, *Virology,* 43, 524, 1971.
7. **Fenters, J. D., Gillum, S. S., Holper, J. C., and Marquis, G. S.**, Serotypic relationships among rhinoviruses, *Am. J. Epidemiol.,* 84, 10, 1966.
8. **Yin, F. H., Lonberg-Holm, K., and Chan, S. P.**, Lack of a close relationship between three strains of human rhinoviruses as determined by their RNA sequences, *J. Virol.,* 12, 108, 1973.
9. **Kapikian, A. Z.**, Coronaviruses, in *Diagnostic Procedures for Virus and Rickettsial Infections,* Lennette, E. H., Ed., American Public Health Association, Chicago, 1969, 931.
10. **Fazekas de St. Groth, S. and Hannoun, C.**, Selection par pression immunologique de mutants dominants du virus de la grippe A (Hong Kong), *C. R. Acad. Sci.,* 276, 1917, 1973.
11. **Hoyle, L.**, *The Influenza Viruses,* Springer-Verlag, New York, 1968, 191.
12. **Tyrrell, D. A. J.**, *Common Colds and Related Diseases,* Edward Arnold, London, 1965, 81.
13. **Dolin, R., Blacklow, N. R., Malmgren, R. A., and Chanock, R. M.**, Establishment of human foetal intestinal organ cultures for growth of viruses, *J. Infect. Dis.,* 122, 227, 1970.
14. **Reed, S. E. and Boyd, A.**, Organ cultures of respiratory epithelium infected with rhinovirus or parainfluenza virus studied with the scanning electron microscope, *Infect. Immun.,* 6, 68, 1972.
15. **Wyatt, R. G., Kalika, A. R., and Dolin, R.**, Ultrastructural study of viral replication in human foetal intestinal organ cultures, *Arch. Gesamte Virusforsch.,* 40, 341, 1973.
16. **Dolin, R., Blacklow, N. R., Wyatt, R. G., and Sereno, M. M.**, Detection and localisation of viruses in foetal intestinal organ culture by immunofluorescence, *Infect. Immun.,* 6, 958, 1972.
17. **Wentworth, B. B. and Alexander, E. R.**, Seroepidemiology of infections due to members of the herpesvirus group, *Am. J. Epidemiol.,* 94, 496, 1971.
18. **Douglas, R. G. and Couch, R. B.**, A prospective study of chronic herpes simplex virus infection and recurrent herpes labialis in humans, *J. Immunol.,* 104, 289, 1970.
19. **Wilton, J. M. A., Ivanyi, L., and Leckner, T.**, Cell mediated immunity in herpesvirus hominis infections, *Br. Med. J.,* 1, 723, 1972.
20. **Dowdle, W. R., Coleman, T. M., and Gregg, M. B.**, Natural history of influenza type A in the United States, 1957–1972, in *Progress in Medical Virology,* Vol. 17, Melnick, J. L., Ed., Karger, Basel, 1974, 91.
21. **Chanock, R. M., Parrott, R. H., and Kapikian, A. Z.**, Virus-induced immunopathology, in *Perspectives in Virology,* Vol. 6, Pollard, M., Ed., Academic Press, New York, 1968, 125.
22. **Kim, H. W., Canchola, J. G., and Brandt, P. G.**, Respiratory syncytial virus disease of infants despite prior administration of antigenic inactivated vaccine, *Am. J. Epidemiol.,* 89, 422, 1969.
23. **Kapikian, A. Z., Mitchell, R. H., and Chanock, R. M.**, An epidemiological study of altered clinical reactivity to respiratory syncytial (R.S.) virus infection in children previously vaccinated with an inactivated R.S. virus vaccine, *Am. J. Epidemiol.,* 89, 405, 1969.
24. **Chanock, R. M., Kapikian, A. Z., and Mills, J.**, Influence of immunological factors in respiratory syncytial virus disease, *Arch. Environ. Health,* 21, 347, 1970.
25. **Tyrrell, D. A. J.**, The use of vaccines in respiratory virus diseases, *Trans. Med. Soc. London,* 89, 263, 1973.
26. **Witte, J. J.**, Current status of vaccine-preventable diseases, *Postgrad. Med.,* 56, 55, 1974.
27. **Kapikian, A. Z., Wyatt, R. G., Dolin, R., Thornhill, T. S., Kalica, A. R., and Chanock, R. M.**, Visualization by immune electron microscopy of a 27-nm particle associated with acute infectious nonbacterial gastroenteritis, *J. Virol.,* 10, 1075, 1972.
28. **Kapikian, A. Z., Gerin, J. L., Wyatt, R. G., Thornhill, T. S., and Chanock, R. M.**, Density in cesium chloride of the 27-nm "8F11a" particle associated with acute infectious nonbacterial gastroenteritis: determination by ultracentrifugation and immune electron microscopy, *Proc. Soc. Exp. Biol. Med.,* 142, 874, 1973.
29. **Wyatt, R. G., Dolin, R., Blacklow, N. R., DuPont, H. L., Buscho, R. F., Thornhill, T. S., Kapikian, A. Z., and Chanock, R. M.**, Comparison of three agents of acute infectious nonbacterial gastroenteritis by crosschallenge in volunteers, *J. Infect. Dis.,* 129, 709, 1974.

30. **Kapikian, A. Z., Kim, H. W., Wyatt, R. G., Rodriguez, W. J., Ross, S., Cline, W. L., Parrott, R. H., and Chanock, R. M.**, Reoviruslike agent in stools: association with infantile diarrhea and development of serologic tests, *Science,* 185, 1049, 1974.
31. **Wyatt, R. G., Kapikian, A. Z., Thornhill, T. S., Sereno, M. M., Kim, H. W., and Chanock, R. M.**, In vitro cultivation in human foetal intestinal organ culture of a reovirus-like agent associated with nonbacterial gastroenteritis in infants and children, *J. Infect. Dis.,* 130, 523, 1974.
32. **Flewett, T. H., Bryden, A. S., Davies, H., Woode, G. N., Bridger, J. C., and Derrick, J. M.**, Relation between viruses from acute gastroenteritis of children and newborn calves, *Lancet,* 2, 7872, 1974.
33. **Kapikian, A. Z.**, Acute viral gastroenteritis, *Prev. Med.,* 3, 535, 1974.
34. **Thornhill, T. S., Kalica, A. R., Wyatt, R. G., Kapikian, A. Z., and Chanock, R. M.**, Pattern of shedding of the Norwalk particle in stools during experimentally induced gastroenteritis in volunteers as determined by immune electron microscopy, *J. Infect. Dis.,* 132, 28, 1975.
35. **Holmes, I. H., Ruck, B. J., Bishop, R. F., and Davidson, G. P.**, Infantile enteritis viruses: morphogenesis and morphology, *J. Virol.,* 16, 937, 1975.
36. **Kapikian, A. Z., Cline, W. L., Mebus, C. A., Wyatt, R. G., Kalica, A. R., James, H. D., Jr., VanKirk, D., Chanock, R. M., and Kim, H. W.**, New complement-fixation test for the human reovirus-like agent of infantile gastroenteritis, *Lancet,* 1, 1056, 1975.
37. **Juel-Jensen, B. E.**, Results of the treatment of Zoster with idoxuridine, *Ann. N.Y. Acad. Sci.,* 173, 74, 1970.
38. **Juel-Jensen, B. E., MacCallum, F. O., Mackenzie, A. M. R., and Pike, M. C.**, Treatment of Zoster with idoxuridine in dimethylsulphoxide: results of two double blind controlled trials, *Br. Med. J.,* 4, 776, 1970.
39. **Juel-Jensen, B. E. and MacCallum, F. O.**, *Herpes simplex, Varicella and Zoster,* William Heinmann, London, 1972, 163.
40. **Patel, N., Buthala, D. A., and Walker, J. S.**, Controlled studies of Coxsackie A-21 (COE) virus in volunteers, *J. Infect. Dis.,* 114, 87, 1964.
41. **Douglas, R. G., Cate, T. R., Gerone, P. J., and Couch, R. B.**, Quantitative rhinovirus shedding patterns in volunteers, *Am. Rev. Respir. Dis.,* 94, 159, 1966.
42. **Dawkins, A. T., Gallager, L. R., Togo, Y., Horwick, R. B., and Harris, B. A.**, Studies on induced influenza in man. II. Double blind study designed to assess the prophylactic efficacy of an analogue of amantadine hydrochloride, *JAMA,* 203, 1095, 1968.
43. **Wingfield, W. L., Pollack, D., and Grunert, R. R.**, Therapeutic efficacy of adamantadine HCl and rimantadine HCl in naturally occurring influenza A_2 respiratory illness in man, *N. Engl. J. Med.,* 281, 579, 1969.
44. **Watson, G. I.**, Use of adamantaneamine in an epidemic of Hong Kong influenza type A_2 in a family practice, *Practitioner,* 205, 351, 1970.
45. **Hornick, R. B., Togo, Y., Mahler, S., and Iezzoni, D.**, Evaluation of amantidine hydrochloride in the treatment of A_2 influenzal disease, *Ann. N.Y. Acad. Sci.,* 173, 10, 1970.
46. **Whiffle, H. E.**, Antiviral substances, *Ann. N.Y. Acad. Sci.,* 130, 30, 1965.
47. **Herrmann, E. C. and Stinebring, W. R.**, Second conference on antiviral substances, *Ann. N.Y. Acad. Sci.,* 173, 40, 1970.
48. Third conference on antiviral substances, *Ann. N.Y. Acad. Sci.,* in press.
49. **Erlich, J., Sloan, B., Miller, P. A., and Mamacher, H. E.**, Searching for antiviral materials from microbial fermentations, *Ann. N.Y. Acad. Sci.,* 130, 5, 1965.
50. **Johnson, I. S.**, Observations on antiviral screening, *Ann. N.Y. Acad. Sci.,* 130, 52, 1965.
51. **Hoyle, L.**, *The Influenza Viruses,* Springer-Verlag, New York, 1968, chap. 33.
52. **Kilbourne, E. D.**, The molecular epidemiology of influenza, *J. Infect. Dis.,* 127, 478, 1973.
53. **Bucknall, R. A., Swallow, D. L., Moores, H., and Harrad, J.**, A novel substituted guanidine with high activity against rhinoviruses, *Nature,* 246, 114, 1973.
54. **Swallow, D. L., Stanier, W. E., Hutchinson, A., and Gaskin, H.**, A potent antirhinovirus compound – ICI 73,602, in press.
55. **Buthala, D. A.**, Experience gained in screening for antiviral compounds, *Ann. N.Y. Acad. Sci.,* 130, 17, 1965.
56. **Bauer, J. D.**, in *Modern Trends in Medical Virology,* Heath, R. B. and Waterson, A. F., Eds., Butterworths, London, 1967, 49.
57. **Hurst, E. W.**, Approaches to the chemotherapy of virus diseases, *J. Pharm. Pharmacol.,* 9, 273, 1957.
58. **Hamre, D., Bernstein, J., and Donovick, R.**, Activity of p-aminobenzaldehyde-3-thiosemicarbazone in the chick embryo and the mouse, *Proc. Soc. Exp. Biol. Med.,* 73, 275, 1950.
59. **Fazekas de St. Groth, S. and White, D. O.**, An improved assay for the infectivity of influenza viruses, *J. Hyg.,* 56, 151, 1958.
60. **Fazekas de St. Groth, S. and White, D. O.**, The dose-response relationship between influenza viruses and the surviving allantois, *J. Hyg.,* 56, 523, 1958.
61. **Tamm, I., Folkers, K., Shunk, C. H., and Horsfall, F. L.**, Inhibition of influenza virus multiplication by N-glycosides of benzimidazoles, *J. Exp. Med.,* 99, 227, 1954.
62. **Tamm, I.**, Selective chemical inhibition of influenza B virus multiplication, *J. Bacteriol.,* 72, 42, 1956.
63. **Tamm, I. and Overman, J. R.**, Relationship between the structure of benzimidazole derivatives and inhibitory activity on vaccinia virus multiplication, *Virology,* 3, 185, 1957.

64. **Fulton, F. and Armitage, P.**, Surviving tissue suspensions for influenza virus titrations, *J. Hyg.*, 49, 247, 1951.
65. **Allfrey, V. G., Mirsky, A. E., and Osawa, S.**, Protein synthesis in isolated cell nuclei, *J. Gen. Physiol.*, 40, 451, 1957.
66. **Tamm, I., Nemes, M. M., and Osterhout, S.**, On the role of RNA, *J. Exp. Med.*, 111, 339, 1960.
67. **Tamm, I. and Eggers, H. J.**, in *Viral and Rickettsial Diseases of Man*, Horsfall, F. L. and Tamm, I., Eds., Pitman, London, 1965.
68. **Bucknall, R. A.**, The effects of substituted benzimidazoles on the growth of viruses and the nucleic acid metabolism of the host cell, *J. Gen. Virol.*, 1, 89, 1967.
69. **Rightsel, W. A., Schultz, P., Muething, D., and McLean, I. W.**, Use of vinyl plastic containers in tissue cultures for virus assays, *J. Immunol.*, 74, 464, 1956.
70. **Pauker, K.**, The serological specificity of interferon, *J. Immunol.*, 94, 371, 1965.
71. **DeSomer, P. and Prinzie, A.**, Poliomyelitis virus neutralising antibodies determination by filter paper discs on solidified bottle cultures, *Virology*, 4, 387, 1957.
72. **Porterfield, J. S.**, A simple plaque inhibition test for antiviral agents: application to assay of interferon, *Lancet*, 7, 326, 1959.
73. **Herrmann, E. C., Gabliks, J., Engle, C., and Perlman, P. L.**, Agar diffusion method for detection and bioassay of antiviral antibiotics, *Proc. Soc. Exp. Biol. Med.*, 103, 625, 1960.
74. **Walker, W. E., Waisbren, B. A., Martins, R. R., and Batayias, G. E.**, A method for detecting the sensitivity of antiviral drugs *in vitro* for possible use as clinical consultation, *Am. J. Clin. Pathol.*, 56, 687, 1971.
75. **Bucknall, R. A., Moores, H., Simms, R., and Hesp, B.**, Antiviral effects of Aphidicolin, a new antibiotic produced by *Cephalosporium aphidicola*, *Antimicrob. Agents Chemother.*, 4, 294, 1973.
76. **Finter, N. B.**, Quantitative haemadsorption, a new assay technique, *Virology*, 24, 589, 1964.
77. **Finter, N. B.**, Interferon in mice: protection against small doses of virus, *J. Gen. Virol.*, 1, 395, 1967.
78. **Finter, N. B.**, Methods for screening *in vitro* and *in vivo* for agents active against myxoviruses, *Ann. N.Y. Acad. Sci.*, 173, 131, 1970.
79. **Swallow, D. L., Edwards, P. N., and Finter, N. B.**, Decalins – a new class of compounds active against influenza A viruses, *Ann. N.Y. Acad. Sci.*, 173, 292, 1970.
80. **Hahon, N., Booth, J. A., and Eckert, H. J.**, Quantitative assessment of haemadsorption by myxoviruses: virus haemadsorption assay, *Appl. Microbiol.*, 25, 595, 1973.
81. **Boyd, J. E. and Somerville, R. G.**, A rapid, simple and reliable method for the screening of potential antiviral compounds *in vitro*, *Arch. Gesamte Virusforsch.*, 45, 254, 1974.
82. **Koerting, H. J., Hindersin, P., and Voss, G. R.**, Fluorimetric determination of the relative quantity of virus in cells, *Acta Histochem.*, 50, 84, 1974.
83. **Miller, P. A., Lindsay, N. L., Cormier, M., Mayberry, B. R., and Trown, P. W.**, Rapid semiautomated procedures for assaying antiviral activity, *Ann. N.Y. Acad. Sci.*, 173, 151, 1976.
84. **Barry, R. D., Ives, D. R., and Cruikshank, J. G.**, Participation of deoxyribonucleic acid in the multiplication of influenza virus, *Nature*, 194, 1139, 1962.
85. **Rott, R. and Scholtissek, C.**, Influence of actinomycin D on the growth of myxoviruses, *Z. Naturforsch. Teil B*, 19, 316, 1964.
86. **Bukrinskaya, A. G.**, Resistance of synthesis of influenza virus ribonucleic acid to the action of actinomycin D, *Nature*, 204, 205, 1964.
87. **Hoyle, L.**, *The Influenza Viruses*, Springer-Verlag, New York, 1968, 118.
88. **Seto, J. T. and Rott, R.**, Functional significance of Sialidase during influenza virus multiplication, *Virology*, 30, 731, 1966.
89. **Webster, R. G. and Laver, W. G.**, Preparation and properties of antibody directed specifically against the neuraminidase of influenza virus, *J. Immunol.*, 99, 49, 1967.
90. **Kilbourne, E. D., Laver, W. G., Schulman, J. L., and Webster, R. G.**, Antiviral activity of antiserum specific for an influenza virus neuraminidase, *J. Virol.*, 2, 281, 1968.
91. **Klenk, H. D., Compans, R. W., and Choppin, P. W.**, An electron microscope study of the presence or absence of neuraminic acid in enveloped viruses, *Virology*, 42, 11158, 1970.
92. **Becht, J., Hammerling, V., and Rott, R.**, Undisturbed release of influenza virus in the presence of univalent antineuraminidase antibodies, *Virology*, 46, 337, 1971.
93. **Haskell, T. H., Peterson, F. E., and Watson, D.**, Neuraminidase inhibition and viral chemotherapy, *J. Med. Chem.*, 13, 697, 1970.
94. **Chow, N. L. and Simpson, R. W.**, RNA-dependent RNA polymerase activity associated with virions and subviral particles of myxoviruses, *Proc. Natl. Acad. Sci. U.S.A.*, 68, 752, 1971.
95. **Shatkin, A. J. and Sipe, J. D.**, RNA polymerase activity in purified reoviruses, *Proc. Natl. Acad. Sci. U.S.A.*, 61, 1462, 1968.
96. **Baltimore, D., Huang, A. S., and Stampfer, M.**, Ribonucleic acid synthesis of vesicular stomatitis virus. II. An RNA polymerase in the virion, *Proc. Natl. Acad. Sci. U.S.A.*, 66, 572, 1970.
97. **Huang, A. S., Baltimore, D., and Bratt, M. A.**, Ribonucleic acid polymerase in virions of Newcastle disease virus: comparison with the vesicular stomatitis virus polymerase, *J. Virol.*, 7, 389, 1971.
98. **Marx, P. A., Portner, A., and Kingsbury, D. W.**, Sendai virion transcriptase complex: polypeptide composition and inhibition by virion envelope proteins, *J. Virol.*, 13, 298, 1974.

99. Tem' ., H. and Mizutani, S., RNA dependent DNA polymerase in virions of Rous sarcoma virus, *Nature*, 226, 1211, 1970.
100. Baltimore, D., Viral RNA-dependent DNA polymerase, *Nature*, 226, 1209, 1970.
101. Bader, J. P., Reproduction of RNA tumour viruses, in *Comprehensive Virology*, Fraenkel-Conrat, H. and Wagner, R. R., Eds., Plenum Press, New York, 1975, 4.
102. Strand, M. and August, J. T., Protein kinase and phosphate acceptor proteins in Rauscher murine leukaemia virus, *Nat. New Biol.*, 233, 137, 1971.
103. Hatanaka, M., Twiddy, E., and Gilden, R. V., Protein kinase associated with RNA tumour viruses and other budding RNA viruses, *Virology*, 47, 536, 1972.
104. Kates, J. R. and McAuslan, B. R., Poxvirus DNA-dependent RNA polymerase, *Proc. Natl. Acad. Sci. U.S.A.*, 58, 134, 1967.
105. Brown, M., Dorson, J. W., and Bollum, F. J., Terminal riboadenylate transferase: a poly A polymerase in purified vaccinia virus, *J. Virol.*, 12, 203, 1973.
106. Moss, B., Rosenblum, E. N., and Paoletti, E., Polyadenylate polymerase from vaccinia virions, *Nat. New Biol.*, 245, 59, 1973.
107. Gold, P. and Dales, S., Localisation of nucleotide phosphohydrolase activity within vaccinia, *Proc. Natl. Acad. Sci. U.S.A.*, 60, 845, 1968.
108. Paoletti, E. and Moss, B., Deoxyribonucleic acid-dependent nucleotide phosphohydrolase activity in purified vaccinia virus, *J. Virol.*, 10, 866, 1972.
109. Pogo, B. G. T. and Dales, S., Two deoxyribonuclease activities within purified vaccinia virus, *Proc. Natl. Acad. Sci. U.S.A.*, 63, 820, 1969.
110. Aubertin, A. and McAuslan, B. R., Virus-associated nucleases: evidence for endonuclease and exonuclease activity in rabbitpox and vaccinia viruses, *J. Virol.*, 9, 554, 1972.
111. Kang, H. S. and McAuslan, B. R., Virus-associated nucleases: location and properties of deoxyribonuclease and ribonuclease in purified frog virus 3, *J. Virol.*, 10, 202, 1972.
112. Paoletti, E. and Moss, B., Protein kinase and specific phosphate acceptor proteins associated with vaccinia virus cores, *J. Virol.*, 10, 417, 1972.
113. Kleiman, J. and Moss, B., Protein kinase activity from vaccinia virions: solubilisation and separation into heat-labile and heat-stable components, *J. Virol.*, 12, 684, 1973.
114. Kendal, A. P. and Madeley, C. R., A comparative study of influenza virus neuraminidases using automated techniques, *Biochim. Biophys. Acta*, 185, 163, 1969.
115. Hoorn, B. and Tyrrell, D. A. J., On the growth of certain newer respiratory viruses in organ cultures, *Br. J. Exp. Pathol.*, 46, 109, 1965.
116. Tyrrell, D. A. J. and Hoorn, B., The growth of some myxoviruses in organ cultures, *Br. J. Exp. Pathol.*, 46, 514, 1965.
117. Tyrrell, D. A. J. and Bynoe, M. L., Cultivation of a novel type of common-cold virus in organ cultures, *Br. Med. J.*, 1, 1467, 1965.
118. Hoorn, B. and Tyrrell, D. A. J., Effects of some viruses on ciliated cells, *Am. Rev. Respir. Dis.*, 93, 156, 1966.
119. McIntosh, K., Dees, J. H., Becker, W. B., Kapikian, A. Z., and Chanock, R. M., Recovery in tracheal organ cultures of novel viruses from patients with respiratory disease, *Proc. Natl. Acad. Sci. U.S.A.*, 57, 933, 1967.
120. Almeida, J. D. and Tyrrell, D. A. J., The morphology of three previously uncharacterized human respiratory viruses that grow in organ culture, *J. Gen. Virol.*, 1, 175, 1967.
121. Craighead, J. E. and Brennan, B. J., Cytopathic effects of parainfluenza virus type 3 in organ cultures of human respiratory tract tissue, *Am. J. Pathol.*, 52, 287, 1968.
122. Hoorn, B., Respiratory viruses in model experiments, *Acta Oto Laryngol. Suppl.*, 188, 138, 1963.
123. Hoorn, B. and Tyrrell, D. A. J., A new virus cultivated only in organ cultures of human ciliated epithelium, *Arch. Gesamte Virusforsch.*, 18, 210, 1966.
124. Tyrrell, D. A. J., Bynoe, M. L., and Hoorn, B., Cultivation of "difficult" viruses from patients with common colds, *Br. Med. J.*, 1, 606, 1968.
125. Herbst-Laier, R., Organ cultures in evaluation of antiviral drugs, *Arch. Gesamte Virusforsch.*, 30, 379, 1970.
126. Roome, A. P. C. H., Dickinson, V., and Cail, E. O., Simplified organ cultures of human embryo trachea in the diagnosis of viral respiratory disease of children, *J. Clin. Pathol.*, 24, 487, 1971.
127. Higgins, P. G. and Ellis, E. M., Further observations on the use of organ cultures in the study of acute respiratory tract infections, *J. Med. Microbiol.*, 6, 177, 1973.
128. Clarke, S. K. R., Cook, G. T., Egglestone, S. I., Hall, T. S., Miller, D. L., Reed, S. E., Rubenstein, D., Smith, A. J., and Tyrrell, D. A. J., A virus from epidemic vomiting disease, *Br. Med. J.*, 3, 86, 1972.
129. Albright, D. J., Whalen, R. A., and Blacklow, N. R., Sensitivity of human foetal intestine to interferon, *Nature*, 247, 218, 1974.
130. Baines, P. M., Taylor, P. E., and Zuckerman, A. J., Establishment of organ cultures of human embryonic liver for growth of viruses, *Arch. Gesamte Virusforsch.*, 37, 230, 1972.
131. Itselis, F. G., Ananiev, V. A., Vorozhbieva, T. E., and Safronskaya, N. V., The use of organ cultures for propagation of polio and infectious canine hepatitis viruses, *Acta Virol.* (Prague), 15, 479, 1971.

132. **McCracken, A. W., Smith, K. O., and Armour, V. A.**, Propagation of some common viruses in human embryonic brain cells, *Am. J. Clin. Pathol.*, 61, 482, 1974.
133. **Typskin, L. B., Goryunova, L. G., and Voroshilova, M. K.**, Nervous tissue in monologues and organ cultures, *Trans. Inst. Polio Virus Entsef.*, 22, 118, 1974.
134. **Rubenstein, C. J., Votava, D., and Tyrrell, D. A. J.**, Organ cultures of human fallopian tubes for the propagation of viruses, *Arch. Gesamte Virusforsch.*, 31, 3, 1970.
135. **Weissenbacher, M., Chowchurech, E., Schmunis, G., Sawicki, L., and Galin, M.**, Sensitivity of human conjunctival tissue to adenovirus 8, *Proc. Soc. Exp. Biol. Med.*, 142, 997, 1973.
136. **Finkelstein, M. S., McWilliams, M., and Huizenga, C. G.**, The use of chicken tracheal organ cultures for studying the resistance of respiratory epithelium to viral infection, *J. Immunol.*, 108, 183, 1972.
137. **Masse, E. and Van der Veen, J.**, Growth of mouse hepatitis virus and other indigenous mouse viruses in tracheal organ cultures, *Arch. Gesamte Virusforsch.*, 44, 386, 1974.
138. **Butler, M., Ellaway, W. J., and Hall, T.**, Comparative studies on the infectivity of avian respiratory viruses for eggs, cell cultures and tracheal explants, *J. Comp. Pathol.*, 82, 327, 1972.
139. **Smith, H.**, Mechanisms of virus pathogenicity, *Bacteriol. Rev.*, 36, 291, 1972.
140. **Reed, S. E.**, Viral enhancement of mycoplasma growth in tracheal organ cultures, *J. Comp. Pathol.*, 82, 267, 1972.
141. **James, Z. H. and Povey, R. C.**, Organ cultures of bovine teat skin and its application to study of herpes mammilitis and pseudocowpox infections, *Res. Vet. Sci.*, 15, 40, 1973.
142. **Tyrrell, D. A. J. and Blamire, C. J.**, Improvements in a method of growing respiratory viruses in organ cultures, *Br. J. Exp. Pathol.*, 48, 217, 1967.
143. **Gabridge, M. G.**, Parabiotic chamber for organ culture – improved model, *Appl. Microbiol.*, 28, 774, 1974.
144. **Morris, G. J., Buckland, R., Rubenstein, D., and Farrant, J.**, Preservation of human tracheal organ cultures at −196°C, *J. Med. Microbiol.*, 6, 123, 1973.
145. **Coombes, J. D., Brammer, K. W., Herbst-Laier, R. H., Larin, N. M., McDonald, C. R., Tute, M. S., Wickham, E. A., and Williamson, G. M.**, Laboratory studies on two synthetic dihydroisoquinolines (Pfizer UK 2054 and UK 2371), *Ann. N.Y. Acad. Sci.*, 173, 462, 1970.
146. **Larin, N. M., Herbst-Laier, R. H., and Wenham, R. B. M.**, New test systems for evaluation *in vivo* of potential antiviral drugs against influenza-parainfluenza infections, in *Proc. 3rd Int. Congr. Chemotherapy*, 1964, 775.
147. **Herbst-Laier, R. H.**, A novel technique for *in situ* demonstration of cell viability and cell injury on the respiratory surface of trachea explants, *Arch. Gesamte Virusforsch.*, 30, 248, 1970.
148. **Smorodintsev, A. A.**, The production and effect of interferon in organ cultures of calf trachea, *Br. J. Exp. Pathol.*, 49, 511, 1968.
149. **Bucknall, R. A., Swallow, D. L., Moores, H., and Harrad, J.**, A novel substituted guanidine with high activity *in vitro* against rhinoviruses, *Nature*, 246, 144, 1973.
150. **Hucker, H. B.**, Species differences in drug metabolism, *Annu. Rev. Pharmacol.*, 10, 99, 1970.
151. **Bauer, J. D.**, Introduction to antiviral chemotherapy, in *Chemotherapy of Virus Diseases*, Bauer, J. D., Ed., Pergamon Press, New York, 1972, 1.
152. **Sidwell, R. W., Huffman, J. H., Khare, G. P., Allen, L. B., Witkowski, J. T., and Kobins, R. K.**, Broad-spectrum antiviral activity of Virazole: 1-β-D-Ribofuranosyl-1,2,4-triazole-3-carboxamide, *Science*, 177, 705, 1972.
153. **Huffman, J. H., Sidwell, R. W., Khare, G. P., Witkowski, J. T., Allen, L. B., and Robins, R. K.**, *In vitro* effect of 1-β-D-ribofuranosyl-1,2,4-triazole-3-carboxamide (Virazole, ICN 1229) on deoxyribonucleic acid and ribonucleic acid viruses, *Antimicrob. Agents Chemother.*, 3, 235, 1973.
154. **Witkowski, J. T., Robins, R. K., Sidwell, R. W., and Simon, L. N.**, The design, synthesis and broad spectrum antiviral activity of 1-β-D-ribofuranosyl-1,2,4-triazole-3-carboxamide and related nucleosides, *J. Med. Chem.*, 15, 1150, 1972.
155. **Streeter, D. G., Watkowski, J. T., Khare, G. P., Sidwell, R. W., Bauer, R. J., Robins, R. K., and Simon, L. N.**, Mechanism of action of 1-β-D-ribofuranosyl-1,2,4-triazole-3-carboxamide (Virazole), a new broad-spectrum antiviral agent, *Proc. Natl. Acad. Sci. U.S.A.*, 70, 1174, 1973.
156. **Smith, C. M., Fontanelle, L. J., Muzik, H., Paterson, A. R. P., Unger, H., Brox, L. W., and Henderson, F. J.**, Inhibitors of inosinate dehydrogenase activity in Erlich ascites tumour cells in vitro, *Biochem. Pharmacol.*, 23, 2727, 1974.
157. **Holland, J. J.**, Depression of host controlled RNA synthesis in human cells during poliovirus infection, *Proc. Natl. Acad. Sci. U.S.A.*, 49, 23, 1963.
158. **Brown, F., Martin, S. J., and Underwood, B.**, A study of the kinetics of protein and RNA synthesis induced by foot-and-mouth disease virus, *Biochim. Biophys. Acta*, 129, 166, 1966.
159. **Eggers, H. J. and Tamm, I.**, Spectrum and virus-inhibitory characteristics of 2-(α-hydroxybenzyl)benzimidazole, *J. Exp. Med.*, 113, 657, 1961.
160. **Eggers, H. J. and Tamm, I.**, On the mechanism of selective inhibition of enterovirus multiplication by 2-(α-hydroxybenzyl)benzimidazole, *Virology*, 18, 426, 1961.
161. Guidelines for evaluation of drugs for use in man, *WHO Tech. Rep. Ser.*, p. 563, 1975.
162. First Council Directive 65/65/EEC, *Off. J. Eur. Econ. Commun.*, 22, 369, 1965.
163. Second Council Directive 75/319/EEC, *Off. J. Eur. Econ. Commun.*, 147, 1, 1975.

164. U.S. Food and Drug Administration, Institution committee review of clinical investigations of new drugs in human beings, *Fed. Reg.*, 36, 5037, 1971.
165. The Medicines, Applications for Product Licenses and Clinical Trial and Animal Test Certificates Regulations 1971, No. 973, SI 1971.
166. U.S. Food and Drug Administration, NDA guidelines, *Fed. Reg.*, 32, 108, 1967.
167. **Golberg, L.**, Safety evaluation concepts, *J. Assoc. Off. Anal. Chem.*, 58, 635, 1975.
168. U.K. Committee on Safety of Medicines, Notes on Applications for Clinical Trial Certificates, MAL 4, 1971.
169. U.K. Committee on Safety of Medicines, Notes on Applications for Product Licenses, MAL 2, 1971.
170. U.K. Veterinary Products Committee/Committee on Safety of Medicines, Notes on Applications for Animal Test Certificates, MAL 11, 1972.
171. Association of the British Pharmaceutical Industry, Scientific and Technical Guidelines, The Hennessey Report, Research and Development Committee, 1968.
172. Australian Department of Health, Guidelines for preparing applications for the general marketing or clinical investigation use of a therapeutic substance, New Drug Form 4, 1974.
173. Swedish National Board of Health and Welfare, Submission of Pharmacological and Toxicological Data – Section 4, 06-28, 1973.
174. Principles for pre-clinical testing of drug safety, *WHO Tech. Rep. Ser.*, p. 341, 1966.
175. Appraisal of the Safety of Chemicals in Foods, Drugs and Cosmetics, Association of Food and Drug Officials of the U.S., Texas State Department of Health, Austin, 1959.
176. U.S. Food and Drug Administration, New animal drugs. Definitions and procedural regulations, *Fed. Reg.*, 178, 36, 1971.
177. Pharmaceutical Manufacturers Association, Drug Safety Subsection – U.S. Food and Drug Administration Draft "Preclinical Guidelines," 1973.
178. Principles and Procedure for Evaluating the Safety of Food Additives, Publ. No. 750, National Academy of Sciences U.S.A./National Research Council, Food Protection Committee, Washington, D.C., 1959.
179. U.S. Environmental Protection Agency, Pesticide chemical safety. Proposed toxicology guidelines, *Fed. Reg.*, 37, 19383, 1972; 40, 26801, 1975.
180. **Burns, J. J.**, Committee on problems of drug safety, National Academy of Sciences/National Research Council, *Clin. Pharmacol. Ther.*, 10, 607, 1969.
181. **Morrison, J. K., Quinton, R. M., and Reinert, H.**, The purpose and value of LD50 determinations, *Modern Trends in Toxicology*, Boyland, E. and Goulding, R., Eds., Butterworths, London, 1968.
182. **Anon.**, Rolicin, a new antibiotic active on gram negative bacteria, *J. S. Afr. Vet. Assoc.*, 44, 452, 1973.
183. **Rightsel, W. A.**, Antiviral activity of gliotoxin and gliotoxin acetate, *Nature*, 204, 1333, 1964.
184. **Litchfield, J. T. and Wilcoxon, F.**, A simplified method of evaluating dose-effect experiments, *J. Pharmacol. Exp. Ther.*, 96, 99, 1949.
185. **Mello, N. K.**, Behavioural toxicology, *Fed. Proc.*, 34, 1832, 1975.
186. **De Gruchy, G. C.**, *Drug-induced Blood Disorders*, Blackwell, Oxford, 1975.
187. **Sharratt, M.**, Renal function tests in laboratory animals, in *Metabolic Aspects of Food Safety*, Roe, F. J. C., Ed., Blackwell, Oxford, 1970.
188. **Carrol, S. W.**, Significance of organ weight changes in food safety evaluation, in *Metabolic Aspects of Food Safety*, Roe, F. J. C., Ed., Blackwell, Oxford, 1970.
189. **Schulte-Herman, R.**, Induction of liver growth by xenobiotic compounds and other stimuli, *CRC Crit. Rev. Toxicol.*, 3, 97, 1974.
190. **Goldenthal, E. I.**, Current views on the safety evaluation of drugs, *FDA Pap.*, 2, 1 and 13, 1968.
191. **Vernier, V. G.**, The toxicologic and pharmacologic properties of amantadine hydrochloride, *Toxicol. Appl. Pharmacol.*, 15, 642, 1969.
192. **Phillips, B. M.**, Systemic toxicity of polyinosinic acid: Polycytidylic acid in rodents and dogs, *Toxicol. Appl. Pharmacol.*, 18, 220, 1971.
193. **Horun, E. R., Zendzian, R. P., Schott, L. D., Levy, H. B., and Adamson, R. H.**, Studies on poly 1:C toxicity in experimental animals, *Toxicol. Appl. Pharmacol.*, 23, 579, 1972.
194. **Leonard, B. J., Eccleston, E., and Jones, D.**, Toxicity of interferon inducers of the double stranded RNA type, *Nature*, 224, 1023, 1969.
195. **Sanyer, J. L. and Lee, C. C.**, Preclinical toxicity of cyclocytidine in dogs and monkeys, *Toxicol. Appl. Pharmacol.*, 29 (Abstr. 50), 95, 1974.
196. **Dayan, A. D. and Lewis, P. D.**, Idoxuridine and jaundice, *Lancet*, 2, 1073, 1969.
197. **Fregert, S.**, *Manual of Contact Dermatitis*, Scandinavian University Book, Munksgaard, 1974.
198. **Magnusson, B. and Kligman, A.**, Identification of contact allergens by animal assay – The guniea pig maximisation test, *J. Invest. Dermatol.*, 52, 268, 1969.
199. **Maurer, T. H., Thomann, P., Weirich, E. G., and Hess, R.**, Optimisation test in the guinea pig, *Agents Actions*, 5, 174, 1975.
200. **Carstairs, K. C., Beckenridge, A., Dollery, C. T., and Worlledge, S. M.**, Incidence of a positive direct coomb's test in patients on alpha methyldopa, *Lancet*, 2, 133, 1966.

201. van Ketel, W. G., Fotosensitisation by amantadine (Symmetrel), *Dermatologica*, 148, 124, 1974.
202. U.S. Food and Drug Administration, Illustrated Guide of Grading Eye Irritation by Hazardous Substances, U.S. Government Printing Office, Washington, D.C., 1965.
203. Draize, J. H., Dermal toxicity, *Food Drug Cosmet. Law J.*, 10, 722, 1955.
204. Draize, J. H., Woodward, G., and Calvery, H. L., Methods for the study of irritation and toxicity of substances applied topically to the skin and mucous membranes, *J. Pharmacol. Exp. Ther.*, 82, 377, 1944.
205. Kienel, G., Eine rationelle Methode zur Tierexperirmentallen Prufung der lakalen Vetreglichkeil van intramuskularen Injektionen, *Arzneim. Forsch.*, 23, 263, 1973.
206. Peteri, D. and Sterner, W., Chemie und Toxikologie einer neuen antiviralen Substanz: N-(1-Adamantyl)-2-(2-dimethylaminoäthoxy)-acetamid HCl, *Arzneim. Forsch.*, 23, 577, 1973.
207. Ylikoski, J., Wersäll, J., and Björkroth, B., Hearing loss and cochlear pathology in gentamicin intoxicated guinea pigs, *Acta Pathol. Microbiol. Scand.*, 81 (Suppl. 241), 30, 1973.
208. Engstrom, H., Ades, H. W., and Anderson, A., *Structural Patterns of the Organ of Corti*, Alquist and Wiksell, Stockholm, 1966.
209. Gregory, M., Rutty, D. A., and Wood, D., Differences in the retinotoxic action of chloroquine and phenothiazine derivatives, *J. Pathol.*, 102, 139, 1970.
210. Gehring, P. J., The cataractogenic activity of chemical agents, *CRC Crit. Rev. Toxicol.*, 1, 93, 1971.
211. Maurer, H., Forcier, J., and McIntyre, R., Fetal group C trisomy after cytosine arabinoside and thioguanine, *Ann. Intern. Med.*, 75, 809, 1971.
212. Vogel, F. and Rohrborn, G., *Chemical Mutagenesis in Man*, Springer-Verlag, Berlin, 1970.
213. Fishbein, L., Flamm, W. G., and Falk, H. L., *Chemical Mutagens*, Academic Press, New York, 1970.
214. Hollaender, A., *Chemical Mutagens*, Vol. 1 and 23, Plenum Press, New York, 1973.
215. Searle, A. G., The specific locus test in the mouse, *Mutat. Res.*, 31, 277, 1975.
216. Mitchell, I. de G., A comparison of the sensitivity and specificity of microbial systems for assessing genetic damage, *Agents Actions*, 4, 286, 1974.
217. Ames, B. N., Detection of chemical mutagens with enteric bacteria, in *Chemical Mutagenesis*, Vol. 1, Hollaender, A., Ed., Plenum Press, New York, 1971, 267.
218. Slater, E., Anderson, M. D., and Rosenkranz, S., Rapid detection of mutagens and carcinogens, *Cancer Res.*, 31, 970, 1971.
219. Ames, B. N., Lee, F. D., and Durston, W. E., An improved bacterial test system for detection and classification of mutagens and carcinogens, *Proc. Natl. Acad. Sci. U.S.A.*, 70, 782, 1973.
220. Gabridge, M. G. and Legator, M. S., A host mediated microbiological assay for the detection of mutagenic compounds, *Proc. Soc. Exp. Biol. Med.*, 130, 831, 1969.
221. Commoner, B., Vithayathel, A. J., and Henry, I. J., Detection of metabolic carcinogen intermediates in urine of carcinogen fed rats by means of bacterial mutagenesis, *Nature*, 249, 852, 1974.
222. Fox, M. and Anderson, D., Characteristics of spontaneous and induced thymidine and 5-ido-2-deoxyuridine resistant clones of mouse lymphoma cells, *Mutat. Res.*, 25, 89, 1974.
223. Bateman, A. J. and Epstein, S. S., Dominant lethal mutations in mammals, *Chem. Mutagens*, 2, 541, 1973.
224. Balson, P. J., Damage to chromosomes by drugs, *Drug Reaction Bull.*, 37, 116, 1972.
225. Dean, B. J., Chemical induced chromosome damage, *Lab. Anim.*, 3, 157, 1969.
226. Schleiermacher, E., Chromosome aberrations in mitoses and meioses *in vivo*, *Arch. Toxicol.*, 28, 105, 1971.
227. Schmid, W., Chemical mutagen testing on *in vivo* somatic mammalian cells, *Agents Actions*, 3/2, 77, 1973.
228. Perry, P. and Evans, H. J., Cytological detection of mutagenic-carcinogenic exposure by sister chromatid exchange, *Nature*, 258, 121, 1975.
229. Leonard, A., Tests for heritable translocations in male mammals, *Mutat. Res.*, 31, 291, 1975.
230. Evaluation and testing of drugs for mutagenicity: principles and problems, *WHO Tech. Rep. Ser.*, p. 482, 1971.
231. Provisional Standards for the Examination of Hereditary Safety of Food Additives and Contaminants, Japanese Ministry of Health and Welfare, Division of Food Chemistry, Tokyo, 1975.
232. Testing and Evaluation of Drugs, Food Additives, Poisons, and Other Substances of Genetic Toxicity, Swedish National Board of Health and Welfare, February 13, 1974.
233. Bridges, B. A., The three-tier approach to mutagenicity screening and the concept of radiation-equivalent dose, *Mutat. Res.*, 26, 335, 1974.
234. Wiseman, E. H. and Reinert, H., Anti-inflammatory drugs and renal papillary necrosis, *Agents Actions*, 5, 322, 1975.
235. Dalton, A., Heine, U., Ursula, I., Albert, J., and Melnick, M. D., Characterization of Oncornaviruses and related viruses, *J. Natl. Cancer Inst.*, 55, 941, 1975.
236. Kobayashi, S. and Mukai, N., Retinoblastoma-like tumours induced by human adeno virus type 12 in rats, *Cancer Res.*, 34, 1641, 1974.
237. Herbts, A. L., Ulfelder, H., and Poskanzer, D. C., Adenocarcinoma of the vagina: association of maternal stilboestrol therapy with tumour appearance in young women, *N. Engl. J. Med.*, 284, 878, 1971.
238. Hoover, R. and Fraumeni, J., Drugs in clinical use which cause cancer, *J. Clin. Pharmacol.*, 15, 16, 1975.
239. Carter, R. L. and Roe, F. J., Chemical carcinogens in industry, *J. Soc. Occup. Med.*, 25, 86, 1975.

240. **Hartwell, J. L.**, Survey of Compounds which have been Tested for Carcinogenic Activity, Publ. No. 149, U.S. Public Health Service, Washington, D.C., 1947; Shubik, P. and Hartwell, J. L., Suppl. 1, 1948–53, Suppl. 2, 1954–60; Vol. 1961–67, Vol. 1968–69, Vol. 1970–71, Carcinogenesis Area, Division of Cancer Cause and Prevention, National Cancer Institute, Bethesda, Md.
241. **Tomatis, L., Partensky, C., and Mantesana, R.**, The predictive value of mouse live tumour induction in carcinogenicity testing – A literature survey, *Int. J. Cancer*, 12, 1, 1973.
242. **Clayson, D. B.**, *Chemical Carcinogenesis*, Churchill, London, 1962.
243. **Arcos, J. and Argus, M.**, *Chemical Induction of Cancer*, Vol. I (1968), Vol. IIA (1974), Vol. IIB (1974), Academic Press, New York.
244. **Miller, J. A. and Miller, E. C.**, Some current threshold of research in chemical carcinogenesis, in *Chemical Carcinogenesis, Part A*, Ts'O, P. O. and Dipaolo, J. A., Eds., Marcel Dekker, New York, 1974.
245. **Berenblum, I.**, Carcinogenesis as a biological problem, in *Frontiers of Biology*, Neuberger, A. and Tatum, E. L., Eds., North-Holland Amsterdam, 1974, 34.
246. **Williams, D. J. and Robin, B. R.**, Disruption by carcinogens of the hormone dependent association of membranes with polysomes, *Nature*, 232, 102, 1971.
247. **Berliner, V. R.**, U.S. F.D.A.: Requirements for toxicity testing of contraceptive products, in *Pharmacological Models in Contraceptive Development*, WHO Symp., Briggs, M. M. and Diczfalusy, E., Eds., WHO Stockholm, 1974.
248. Principles of the testing and evaluation of drugs for carcinogenicity, *WHO Tech. Rep. Ser.*, p. 426, 1969.
249. Assessment of the carcinogenicity and mutagenicity of chemicals, *WHO Tech. Rep. Ser.*, p. 546, 1974.
250. **Grice, H. C.**, The Testing of Chemicals for Carcinogenicity, Mutagenicity and Teratology, Health and Welfare Department, Canada, 1973.
251. U.S. Food and Drug Administration, Advisory committee on protocols for safety evaluation, Panel on carcinogenesis report on cancer testing in the safety evaluation of food additives and pesticides, *Toxicol. Appl. Pharmacol.*, 20, 419, 1971.
252. Problems in Evaluation of Carcinogenic Hazard from Use of Food Additives, Publ. No. 749, National Academy of Sciences U.S.A./National Research Council, Food Protection Committee, Washington, D.C., 1959.
253. **Berenblum, I.**, Carcinogenicity testing, *UICC Tech. Rep. Ser.*, p. 2, 1969.
254. **Golberg, L.**, *Carcinogenesis Testing of Chemicals*, CRC Press, Cleveland, 1974.
255. **Druckery, H.**, Quantitative aspects in chemical carcinogenesis, in *Potential Carcinogenic Hazards from Drugs*, UICC Monogr. 7, Truhaut, R., Ed., Springer-Verlag, Berlin, 1967.
256. **Mantel, N.**, The concept of threshold in carcinogenesis, *Clin. Pharmacol. Ther.*, 4, 104, 1963.
257. U.K. Committee on Safety of Medicines, Carcinogenicity Tests of Oral Contraceptives, HMSO, London, 1972.
258. **Clayson, D. B.**, Nutrition and experimental carcinogenesis – A review, *Cancer Res.*, 35, 3292, 1975.
259. **Toth, B.**, A critical review of experiments in chemical carcinogenesis, using newborn animals, *Cancer Res.*, 28, 727, 1968.
260. National Academy of Sciences Food Protection Committee, Natl. Acad. Sci. Publ., p. 749, 1959.
261. **Mantel, N. and Haenszel, W.**, Statistical aspects of the analysis of data from retrospective studies of disease, *J. Natl. Cancer Inst.*, 22, 719, 1959.
262. **Peto, R. and Pike, M. C.**, Conservatism of the approximation (O-E) 2/E in logrank test for survival data on tumour incidence data, *Biometrics*, 29, 579, 1973.
263. **Roe, F. J. C.**, Carcinogenesis and sanity, *Food Cosmet. Toxicol.*, 6, 485, 1968.
264. **Wilson, J. G.**, Present status of drugs as teratogens in man, *Teratology*, 7, 3, 1973.
265. **Lehnerdt, F.**, Zur Frage der Substitution des Calciums im Knochensystem durch Strontium, *Beitr. Pathol. Anat. Allg. Pathol.*, 46, 468, 1909.
266. **Gillman, J., Gilbert, C., and Gillman, T.**, A preliminary report on hydrocephalus, spina bifida and other congenital anomalies in the rat produced by trypan blue, *S. Afr. J. Med. Sci.*, 13, 47, 1948.
267. **Somers, G. F.**, Thalidomide and congenital abnormalities, *Lancet*, 1, 912, 1962.
268. **Goldenthal, E. I.**, Guidelines for Reproduction Studies for Safety Evaluation of Drugs for Human Use, Food and Drug Administration Communication, 1966.
269. **Percy, D. H. and Albert, D. M.**, Developmental defects in rats treated postnatally with 5-iododeotyuridine (IUDR), *Teratology*, 9, 275, 1974.
270. **Ohkuma, H., Tsuyama, S., and Hirayania, H.**, Teratogenicity of cyclocytidine on rabbit or rat fetuses, *Teratology*, 8, 102, 1973.
271. Principles for the testing of drugs for teratogenicity, *WHO Tech. Rep. Ser.*, p. 364, 1967
272. U.K. Veterinary Products Committee/Committee on Safety of Medicines, Notes for Guidance on Reproductive Studies, MAL 38, 1975.
273. U.K. Committee on Safety of Medicines, Notes for Guidance on Reproductive Studies. II. Implemented January 1975, MAL 36, 1975.
274. Animal Test Procedures Regarding Drug Effects on Reproduction, Notification 529, Japanese Ministry of Health and Welfare, Pharmaceutical Affairs Bureau, Tokyo, 1975.
275. U.S. Food and Drug Administration, Advisory Committee on Protocols for Safety Evaluation, Panel on reproduction report on reproduction studies in the safety evaluation of food additives and pesticide residues, *Toxicol. Appl. Pharmacol.*, 16, 264, 1970.

276. **Tuchman-Duplessis, H.,** Teratogenic drug screening. Present procedures and requirements, *Teratology,* 5, 271, 1972.
277. **Wilson, J. G. and Warkany, J.,** *Teratology Principles and Techniques,* University of Chicago Press, Chicago, 1965, 262, 162.
278. **Wilson, J. G.,** Factors determining the teratogenicity of drugs, *Annu. Rev. Pharmacol.,* 205, 1974.
279. **Posthwillo, D. E., Hamilton, W. J., and Sopher, D.,** The marmoset as an animal model for teratological research, *Nature,* 239, 460, 1972.
280. **Robson, J. M. and Sullivan, F. M.,** *Modern Trends in Toxicology, Teratology,* Boyland, E. and Goulding R., Eds., Butterworths, London, 1968, 1.
281. **Palmer, A. K.,** Sporadic malformations in laboratory animals and their influence on drug testing, in *Drugs and Foetal Development,* Klingberg, M. A., Ed., Plenum Press, New York, 1972.
282. **Palmer, A. K.,** Some thoughts on reproductive studies for safety evaluation, *Proc. Eur. Soc. Study Drug Toxicol.,* 14, 79, 1973.
283. **Haseman, J. K. and Hogan, M. D.,** Selection of the experimental unit in teratology studies, *Teratology,* 12, 165, 1975.
284. **Griffin, J. P. and D'Arcy, P. F.,** *A Manual of Adverse Drug Interactions,* John Wright, Bristol, 1975.
285. **Thorp, J. M.,** The influence of plasma proteins on the action of drugs, in *Absorption and Distribution of Drugs,* Binns, T. B. and Dodds, C., Eds., Williams and Wilkins, Baltimore, 1964, 64.
286. **La Du, B. N., Mandel, H. G., and Way, E. L.,** *Fundamentals of Drug Metabolism and Drug Disposition,* Williams and Wilkins, Baltimore, 1971.
287. **Currey, S. H.,** *Drug Disposition and Pharmacokinetics,* Blackwell, Oxford, 1974.
288. **Williams, R. T.,** *Detoxification Mechanisms,* Chapman Hall, London, 1959.
289. **Parke, D. V.,** *The Biochemistry of Foreign Compounds,* Pergamon Press, Oxford, 1968.
290. **Gillette, J. R.,** Comparative metabolism and the choice of experimental animals, in *The Laboratory Animal in Drug Testing,* Vol. 1, 5th Symp. Int. Committee on Laboratory Animals, Fischer-Verlag, Stuttgart, 1973.
291. **Owen, N. V., Griffing, W. J., Hoffman, D. G., Gibson, W. R., and Anderson, R. C.,** Effects of dietary administration of 5-(3,4-dichlorophenyl) 5-ethylbarbituric acid (dichlorophenobarbital) to rats: Emphasis on hepatic drug-metabolising enzymes and morphology, *Toxicol. Appl. Pharmacol.,* 18, 720, 1971.
292. **Comley, A. H.,** Pharmacological implications of microsomal enzyme induction, *Pharmacol. Rev.,* 19, 317, 1967.
293. U.S. Food and Drug Administration, Compounds used in food producing animal procedures for determining acceptability of assay methods used for assuring the absence of residues in edible products of such animals, *Fed. Reg.,* 138, 38, 1973.
294. U.K. Veterinary Products Committee/Committee on Safety of Medicines, Notes on Applications for Veterinary Product Licences for Substances or Articles to be used for Incorporation in Animal Feeding Stuffs for Medicinal Products, MAL 27, 1974.
295. **Truhaut, R. and Ferrando, R.,** La toxicité de relais, *Toxicology,* 3, 361, 1975.
296. **Lockhart, J. D.,** The information gap in pediatric drug therapy, *FDA Pap.,* 5, 6, 1971.

Chapter 7

THE POTENTIAL OF INTERFERON AND INTERFERON INDUCERS

Michael J. Clemens
D. A. J. Tyrrell

TABLE OF CONTENTS

SECTION A
THE MOLECULAR BIOLOGY OF INTERFERON ACTION AT THE MOLECULAR LEVEL

Michael J. Clemens

I. INTRODUCTION

Since the discovery of interferon by Isaacs and Lindenmann in 1957,[1] a considerable amount of time and effort has been devoted to the analysis of how this agent inhibits viral replication in animal cells. The pace of the work has increased significantly in recent years, largely as a result of our improved understanding of the molecular biology of virus replication and as a consequence of better methods for the preparation and purification of interferon. This section concentrates on the most recent aspects of knowledge about the action of interferon; for a wider coverage of the literature, the reader is referred to several review articles available.[2-6] I will describe the evidence for the current view that interferon interacts with its target cell to bring about the production of an inhibitor or inhibitors of viral RNA and/or protein synthesis in infected cells. I will also consider the evidence which suggests that the interferon-treated cell can discriminate under some circumstances between the synthesis of viral and cellular macromolecules. Finally, I will deal briefly with the evidence that interferon has some effects on nonviral functions within animal cells.

II. BIOLOGICAL REQUIREMENTS FOR THE ANTIVIRAL ACTIVITY OF INTERFERON

Interferon is a glycoprotein that is produced by cells in response to viral infections or, experimentally, in response to a number of other nonviral-inducing agents. However, before it can render cells resistant to subsequent infection, it probably has to interact with the external cell surface. Vengris et al.[7] have shown that anti-interferon serum will inhibit development of the antiviral state in cells already induced to synthesize interferon, suggesting that the interferon has to appear outside the cell before it can act. Interferon which is covalently linked to Sepharose is active in inducing an antiviral state in cells,[8] a fact which indicates that the interferon does not have to penetrate into cells to be active. However, it is likely that some interferon may be detached from its inert support on contact with the cells in this type of experiment. It may be at the level of binding to cell surface receptors that the species specificity of interferon is manifested.[9]

Following binding of interferon molecules to their surface membrane receptors, unknown intracellular events must happen which functionally link the surface sites to the cell nucleus. Virtually nothing is known about this stage in the mechanism of interferon action, which may take several hours to occur. However, we do know that the nucleus is essential for the development of resistance to virus infection during interferon treatment, although subsequent enucleation of cells after exposure to interferon still leaves them resistant to later infection. Thus, the antiviral state seems quite stable and is no longer dependent on nuclear functions once it has been established. Experiments on whole cells with inhibitors of transcription suggest that DNA-dependent RNA synthesis is required for induction of the antiviral state by interferon,[12,13] thus explaining the nuclear requirement. Recently it has been possible to assign a gene required for a response to interferon in human cells to chromosome 21. Thus, fibroblasts which are trisomic for this chromosome show a three- to sevenfold increased sensitivity to interferon compared to normal diploid cells.[14] It has also been proposed that there may be an additional regulatory gene concerned with the interferon response in human cells.[15]

In addition to the requirement for RNA synthesis, protein synthesis appears to be necessary for the development of resistance to virus infection during treatment with interferon.[16] This conclusion is again based on the use of metabolic inhibitors and is therefore indirect. It has led to the search for an "antiviral protein" in interferon-treated cells. As will be described later, there is a recent claim for the identification of such a protein but so far it has not been purified.

In view of the above requirements for macromolecular synthesis in interferon-treated cells leading to induction of the antiviral state (summarized in Figure 1), it is not surprising to find that cells usually have to be exposed to interferon for several hours before inoculation with a virus in order to become resistant to infection. However, a recent report claims that the survival level of mice infected with vesicular stomatitis virus is increased by interferon even when the latter is given as long as 4 days after inoculation of the virus.[17] Such a finding could be of clinical importance for therapy of acute virus diseases late in the course of infection.

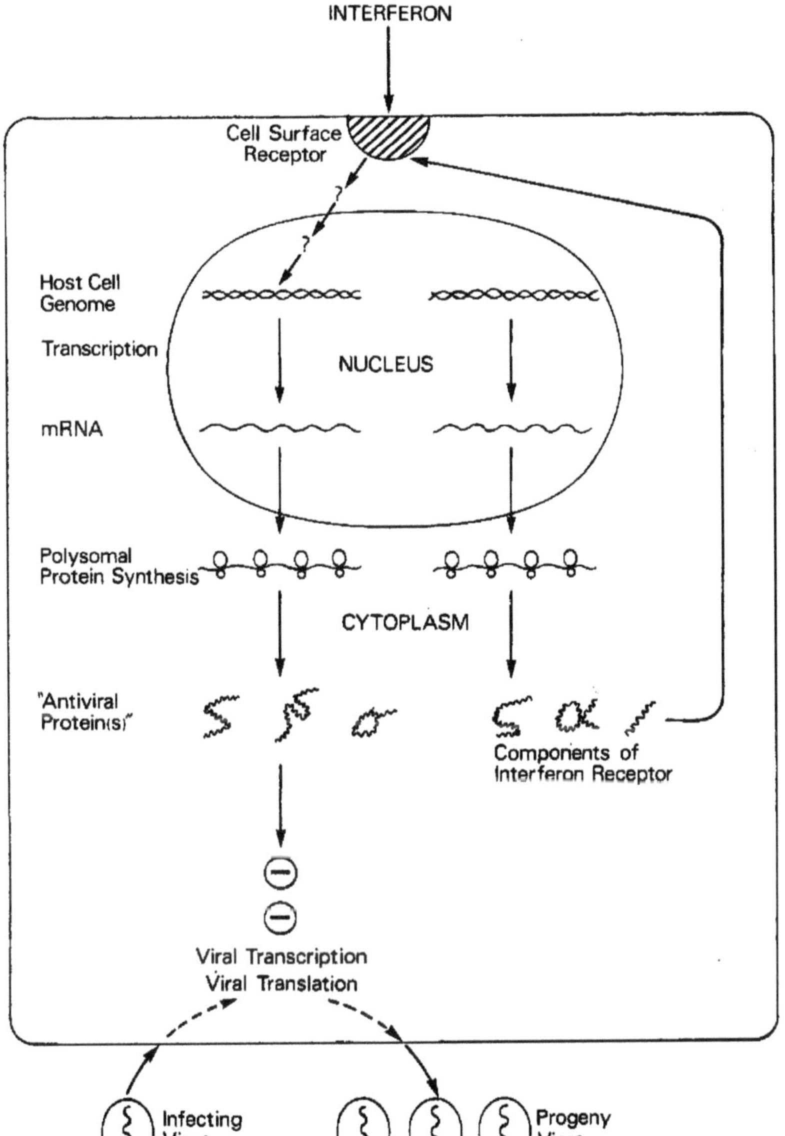

FIGURE 1. Summary of the biological requirements for induction of the antiviral state by interferon. Interferon is thought to bind to a receptor on the external cell surface, initiating unknown event(s) whereby one or more "antiviral proteins" are induced in a process involving the cell nucleus. The antiviral protein(s) inhibit virus replication by interfering with one or more steps in virus macromolecular synthesis. The host genome codes for both the interferon receptor and the antiviral protein(s), and different chromosomes may be involved in these two functions.

III. DISCRIMINATION BETWEEN VIRUS AND HOST CELL FUNCTIONS

A very significant property of interferon, which needs to be understood before its mechanism of action can be fully explained, is its ability in some situations to distinguish between viral and host cell functions. It is important to remember that, although interferon-treated infected cells will not support virus growth, interferon-treated *uninfected* cells usually continue to produce new protein and grow at almost the normal rate;[18] in other words, interferon is not a toxic agent. It has also been shown that induction of new cellular functions such as enzyme synthesis in uninfected cells is far less sensitive to inhibition by interferon than is the replication of viruses in the same cell type.[19,20] For example, synthesis of tyrosine aminotransferase in rat hepatoma cells in response to glucocorticoid hormones is partially inhibited only by a several-hundred-fold greater concentration of interferon than is required for comparable inhibition of vesicular stomatitis or reovirus multiplication.[19] Discrimination between virus and host functions can be observed within a single cell population if infection with the virus does not itself immediately affect activities in the host cell. Gupta et al.[21] showed that interferon treatment of mouse L cells inhibited the accumulation of proteins of the reovirus virion at early times after infection when there was no effect of either the virus or the interferon on total (host cell) protein synthesis. However, later in the infection period they found that pretreatment with interferon resulted in a generalized inhibition of total protein synthesis.

Cells that have been infected or transformed in their growth characteristics by tumor viruses have also been used in experiments to test the discriminatory ability of interferon. A particularly useful system is the Friend leukemia virus-transformed erythroblast cell.[22] Certain lines of these cells continue to release biological active virus, but can also be induced to express a host cell function (viz; hemoglobin synthesis) when exposed to dimethyl sulfoxide. Swetly and Ostertag[23] showed that interferon treatment of such cells inhibited Friend virus release or replication but had no effect on the DMSO-induced erythroid differentiation culminating in synthesis of hemoglobin, as illustrated in Figure 2. Another example of discrimination between viral and host functions by interferon is the synthesis of RNA and protein in cells infected by SV40 virus. Interferon treatment inhibits production of viral RNA sequences and synthesis of virus-specific T antigen while having no effect on total cellular transcription or translation.[24,25] However, it must be noted that SV40-specific RNA and T antigen synthesis in transformed nonpermissive cells (rather than infected permissive cells) is no longer sensitive to interferon but behaves like a cellular process.[26]

There have been relatively few reports of apparent discrimination between virus and host cell activities in cell-free extracts prepared from interferon-treated cells. One of the most con-

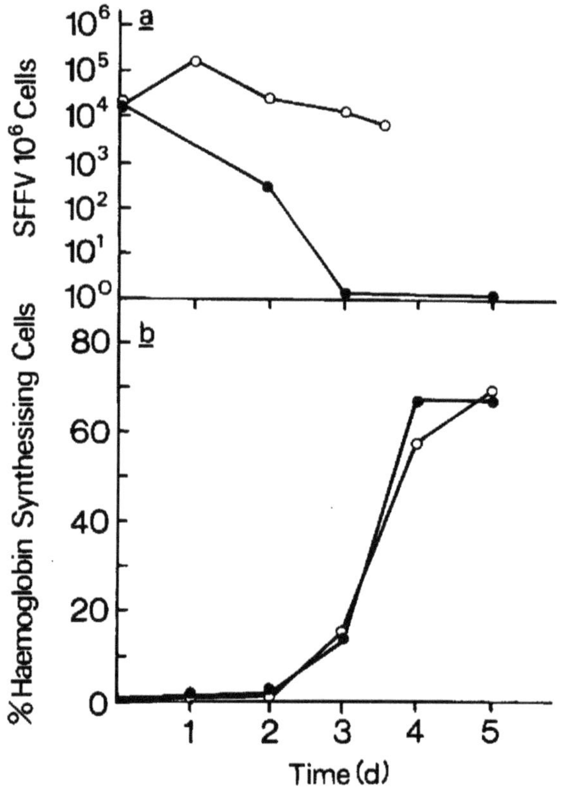

FIGURE 2. Selectivity of interferon action. Effect of exogenous interferon on the production of spleen focus forming virus (SFFV) and on hemoglobin synthesis during dimethyl sulfoxide-induced differentiation of Friend leukemia virus transformed cells. Cells were transferred daily to fresh medium with (•—•) or without (0–0) interferon (900 units/ml). (a) Virus production shown on a logarithmic scale. (b) Erythroid differentiation, estimated by staining cells for hemoglobin with benzidine, shown on a linear scale as a function of time of exposure to 1.5% DMSO. (From Swetly, P. and Ostertag, W., *Nature*, 251, 642, 1974. With permission.)

spicuous has been the claim by Samuel and Joklik[27] that the translation of mRNAs coding for reovirus or vaccinia virus proteins is severely inhibited in preincubated cell-free systems derived from interferon-treated Krebs ascites cells, under conditions where cellular mRNAs are utilized for protein synthesis equally as efficiently as in extracts from untreated cells. This important observation will be discussed further in a later section of this chapter.

IV. EFFECTS OF INTERFERON ON VIRAL PROTEIN SYNTHESIS

It has been known for over 10 years that viral protein synthesis is a major target for the inhibitory action of interferon in infected cells. The classic study by Joklik and Merigan[28] showed that vaccinia virus messenger RNA is transcribed from DNA of the infecting virions in both control and interferon-treated cells (indeed at an enhanced rate in the latter), but it does not give rise to virus-specific polysomes in interferon-treated cells. This indicates that there is a block between synthesis of the mRNA and its utilization for protein synthesis. Similar results have been obtained by Metz and Esteban.[29] In the case of reovirus infection, it has also been shown that adsorption of virus to cells and conversion of parental virions to subviral particles are completely resistant to interferon and that transcription of early viral mRNA is only slightly sensitive.[30] However, the translation of this early mRNA is very strongly inhibited in interferon-treated cells. This is particularly true in the case of the mRNA coding for one particular polypeptide which plays a key role in the later events of virus replication. Thus, viral protein synthesis is the limiting step for the formation of infectious reovirus progeny in interferon-treated cells.

In cases where the infecting virus inhibits host cell protein synthesis, pretreatment of the cells with interferon seems to enhance this inhibitory effect.[23,28,31] Indeed, it has been shown that in the case of interferon-treated L cells infected with vaccinia virus, cell death occurs within 5 to 6 hr of infection whereas control cells survive for as long as 24 hr after infection.[28] Clearly, there is no selectivity of interferon action between host and virus functions in this case.

Much of the information concerning the molecular mechanism of interferon action comes from the use of cell-free protein-synthesizing systems prepared from control and treated cells. Cell extracts can be obtained which reflect reasonably well the translational activity of the original intact cells, in spite of the fact that the rate of protein synthesis in such extracts is always less than in vivo. A number of different groups have reported an effect on the ability of in vitro systems prepared from interferon-treated cells to translate added mRNAs.[32-37] Of particular interest is the finding by Friedman et al.[33,34] that both interferon treatment and virus infection of mouse L cells were necessary in order to see any inhibition of EMC viral RNA translation in extracts subsequently prepared from these cells. Since vaccinia was the infecting virus used in this case, this observation agrees with the results described earlier concerning the inhibition of protein synthesis in vaccinia-infected control and interferon-treated whole cells. In many other reports, however, virus infection of the cells has not been necessary in order to see an interferon-induced inhibition of mRNA translation in cell-free extracts prepared subsequently.[35-41]

There is good correlation between the antiviral effect of interferon (as expressed in whole cells) and the effect on the subsequent ability of cell-free systems to translate exogenous mRNAs. For example, the interferon dose response of both activities is similar[34,35] (see Figure 3), and the inhibitor of RNA synthesis, actinomycin D, prevents the development of both responses in cells exposed to interferon.[36]

There has been a great deal of work in attempts to analyze the mechanism by which interferon-treated cell extracts are impaired in their ability to translate added mRNAs. Certain aspects of the activities of the extracts, such as their ability to translate endogenous polysomal mRNA or poly (U), are virtually normal.[32-37] However, with one exception, there appears to be relatively little selectivity with respect to inhibition of translation of naturally occurring exogenous mRNAs. For example, translation of rabbit globin mRNA or of mixed L-cell mRNAs is impaired as much as translation of Mengo or EMC virus RNAs in interferon-treated L-cell extracts.[35,41] Fractionation of the crude post-mitochondrial supernatant systems which have been generally used in these experiments has revealed that the interferon-induced inhibition of activity is associated with both the ribosome and

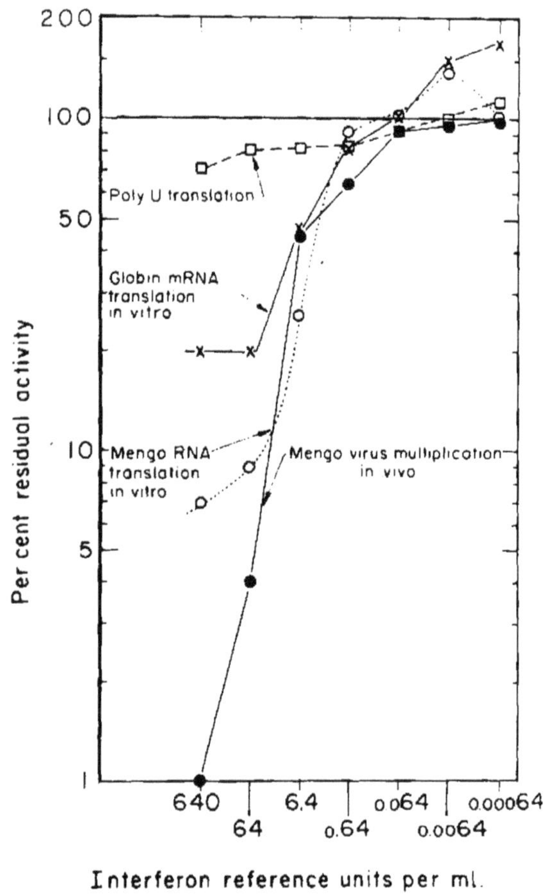

FIGURE 3. Inhibition of virus multiplication in vivo and inhibition of messenger RNA translation in vitro following treatment of L cells with various concentrations of interferon. Mouse L cells were treated in suspension culture for 18 hr, with the concentrations of mouse interferon shown. A portion of each culture was then used to assay growth of Mengo virus, and the remainder was used to prepare cell-free extracts for measurement of mRNA translation, using Mengo virus RNA, globin mRNA, or poly (U) as template. The results are shown as percentages of control (untreated cell) values on a logarithmic scale. (From Falcoff, E., Folcoff, R., Lebleu, B., and Revel, M., *J. Virol.*, 12, 421, 1973. With permission.)

soluble cell sap fractions,[32,33,35-37] and evidence has been presented for the presence of one or more inhibitors in these fractions.[37-39,41] Whether such inhibitors correspond to the putative antiviral protein remains to be established. Samuel and Joklik[27] have described the presence of an inhibitory protein which is associated with ribosomes from interferon-treated cells and which can be washed off the ribosomes under conditions of moderately high ionic strength. The inhibitory activity was associated with the presence of an extra polypeptide of molecular weight 48,000 in the ribosomal salt washes from interferon-treated cells, as shown in Figure 4. Clearly, it will be important to see whether such a component can be reproducibly detected in extracts from interferon-treated cells of various types prepared under a variety of conditions and to determine its mechanism of action on protein synthesis.

It is not clear from present work which precise step or steps in protein synthesis constitute the sites at which interferon treatment inhibits translation. Analysis of virus-specific polysome sizes in vaccinia-infected L cells[42] has shown that interferon treatment causes a reduction in the average number of ribosomes per mRNA molecule (see Figure 5).[43] This suggests a major defect of the protein synthetic machinery at the level of polypeptide chain initiation, although there also appears to be an increase in average ribosome transit time along each mRNA molecule.[43] However, the conclusion from certain studies using cell-free extracts from interferon-treated cells with exogenous mRNAs has been that polypeptide chain elongation is the major site of inhibition in vitro. In particular, it has been shown that the inhibition of picornavirus RNA translation can be overcome by the addition of purified mammalian transfer RNA to preincubated cell-free systems prepared from interferon-treated L cells.[38-41] Gupta et al.[38] reported that EMC viral RNA accumulated in large polysomes when its cell-free translation ceased, but these polysomes could be reactivated by the addition of tRNA even when new polypeptide initiation was blocked by the antibiotic pactamycin. Furthermore, they described an impairment of the endogenous acceptor capacity of tRNA for leucine charging in interferon-treated L-cell systems.[38] Thus, the implication is that interferon treatment of cells induces or activates one or more inhibitors which in turn inactivate one or more tRNA species essential for the normal translation of viral mRNAs during the course of an infection. Content et al.[40] have extended the hypothesis to include the fact that, in vitro, polypeptide chain initiation is also impaired. These authors have proposed that the latter effect is secondary to a block in chain elongation since both processes are reactivated by the restoring tRNA. However, in spite of the confidence with which the "defective tRNA" theory has been proposed by certain groups, it must be pointed out that several observations

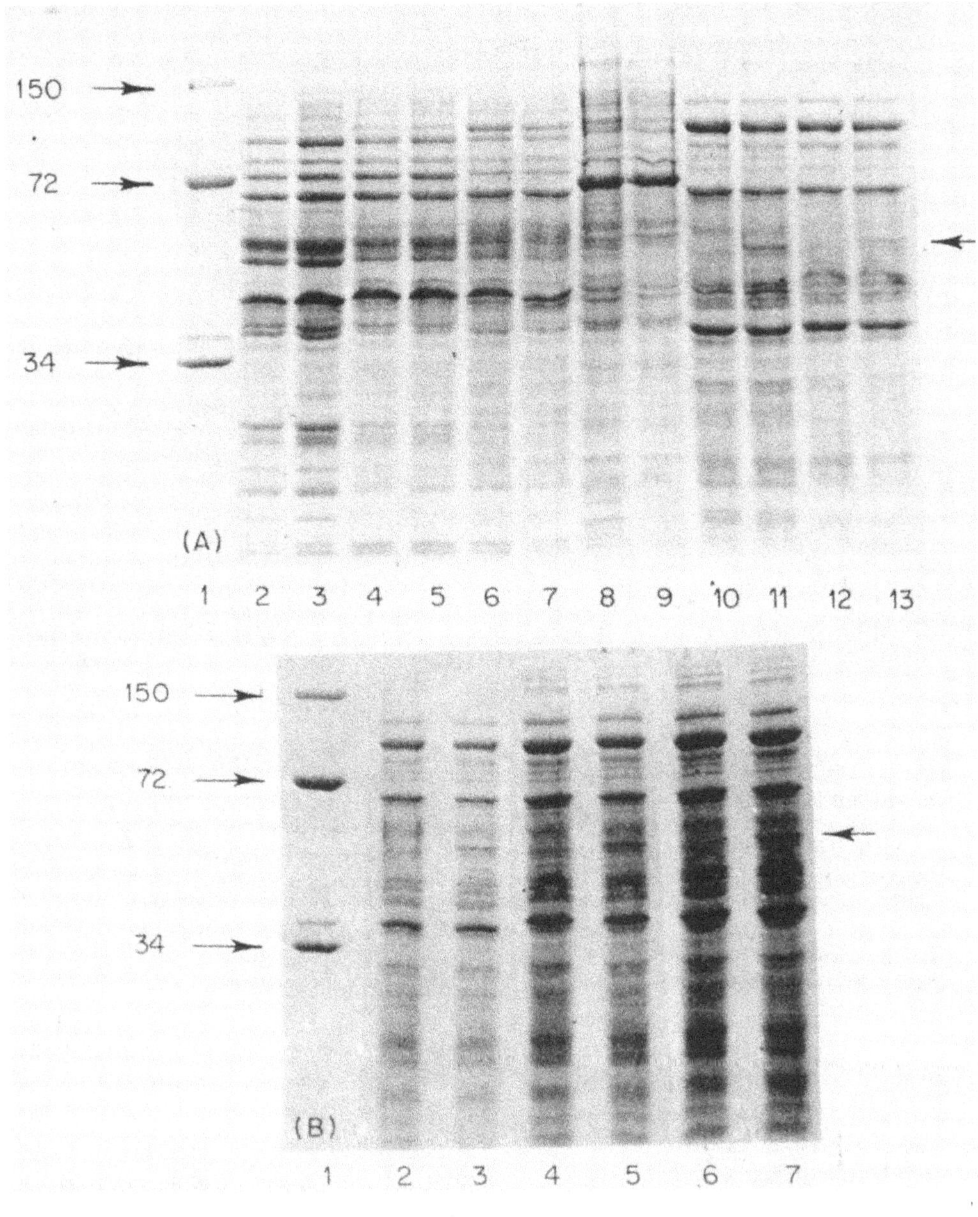

FIGURE 4. Evidence for appearance of a new polypeptide component in the salt wash of ribosomes from interferon-treated cells. Krebs ascites cells growing in mice were either untreated or treated in vivo with two doses of 10^4 units of interferon at 18 and 26 hr before collection. Cells were then harvested and ribosomes prepared from them. The ribosomes were washed with increasing concentrations of KCl, and the washes were recovered after high-speed centrifugation and subjected to sodium dodecyl sulfate-polyacrylamide gel electrophoresis (from top to bottom). Gel A – (1) Reovirus polypeptide molecular weight markers with molecular weights (in thousands) indicated by the arrows; (2 to 13) ribosomal salt-wash fractions from untreated (even numbers) and interferon-treated (odd numbers) cells: (2 and 3), 0.05 *M* KCl wash; (4 and 5), 0.05 to 0.07 *M* KCl wash; (6 and 7), 0.07 to 0.17 *M* KCl wash; (8 and 9), 0.17 to 0.30 *M* KCl wash; (10 and 11), 0.30 to 0.60 *M* KCl wash; (12 and 13), 0.60 to 1.0 *M* KCl wash. Gel B – (1) as A1; (2, 4, and 6) increasing amounts of 0.3 to 0.6 *M* KCl wash of ribosomes from untreated cells; (3, 5, and 7) increasing amounts of 0.3 to 0.6 *M* KCl wash of ribosomes from interferon-treated cells. The 48,000-dalton polypeptide, which is found only in interferon-treated cells, is indicated by arrows. (From Samuel, C. E. and Joklik, W K., *Virology*, 58, 476, 1974. With permission.)

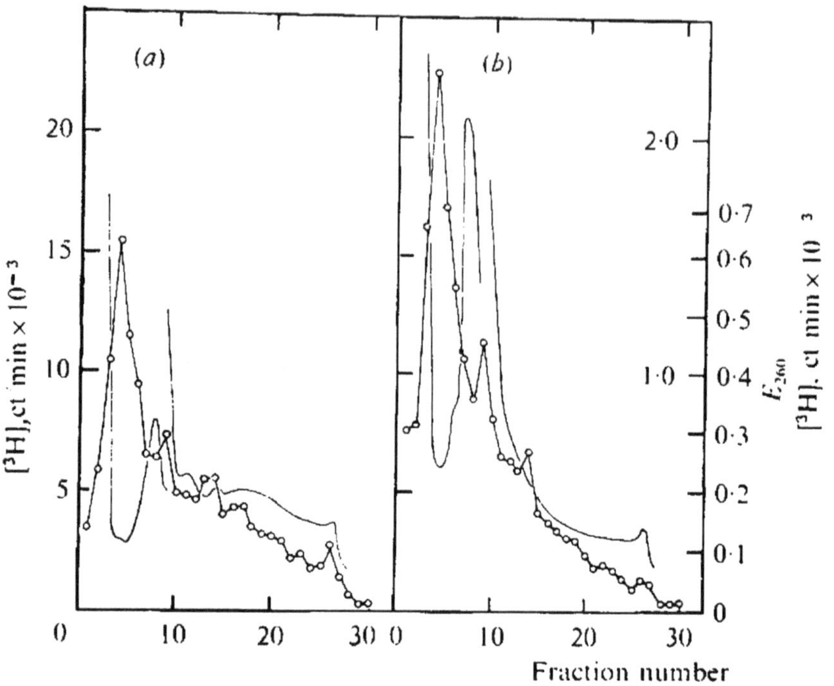

FIGURE 5. Polyribosomes from control and interferon-treated vaccinia virus-infected cells. Mouse L cells were (a) untreated or (b) treated with interferon (40 units/ml) for 20 hr before infection with vaccinia virus. Newly synthesized viral RNA was labeled with [^{3}H]-uridine (2 μCi/ml) in the presence of actinomycin D (1 μg/ml) to inhibit host RNA synthesis. After 30 min cell extracts were prepared and subjected to sucrose density gradient centrifugation. Fractions of the gradients were collected for measurement of cold acid-insoluble radioactivity in RNA. (0–0) Counts per minute; (–), A_{260}. In (b), the left-hand absorbance scale applies to the left of fraction 9 and the right-hand scale applies to the remainder. (From Metz, D. H., Esteban, M., and Danielescu, G., *J. Gen. Virol.*, 27, 197, 1975. With permission.)

cannot yet be explained by it. First, preparations of tRNA from interferon-treated cells are just as capable of restoring translation in vitro as are preparations of tRNA from control cells,[39] suggesting that any inactivation of tRNA function by interferon must at least be reversible. Second, the hypothesis does not provide a satisfactory basis for understanding the cases of discrimination between viral and host cell protein synthesis in interferon-treated cells (see previous section), unless it is assumed that all viral protein synthesis requires a tRNA species which is not important for the synthesis of the majority of cellular proteins. Finally, the phenomenon of tRNA reversal of interferon-induced inhibition of cell-free protein synthesis has not been observed in some systems. In particular, the enhanced inhibition of EMC-RNA translation brought about by double-stranded RNA in the presence of extracts from interferon-treated cells (see below) is not reversed by tRNA from a variety of eukaryotic sources.[101]

An additional possible mechanism by which the translation of some newly synthesized viral mRNAs may be inhibited in interferon-treated cells involves the recently discovered process of methylation of modified 5′ end groups of mRNA.[44-47] Sen et al.[48] have recently reported that there is inhibition of the methylation reaction in extracts from interferon-treated Ehrlich ascites cells, using unmethylated reovirus mRNA as substrate. Since methylation appears to be required for translation of most species of mRNA,[49] such an effect may clearly constitute part of the antiviral action of interferon. However, it seems that the picornavirus RNAs are not methylated at the 5′ end[50] and yet their translation is equally sensitive to the inhibitory effects of interferon, so that methylation of mRNA cannot constitute the sole site of action on protein synthesis.

V. INTERFERON AND DOUBLE-STRANDED RNA

A significant feature of the biology of interferon is its relationship to the effects of double-stranded RNA (dsRNA) on animal cells.[51] It has been known for some time that interferon production can be triggered not only by virus infection, but also in response to double-stranded polyribonucleotides (natural or synthetic).[52-54] Furthermore, dsRNA may also play a role in the mechanism of action of interferon as well as its induction. This can be concluded from the observations of Kerr et al.[55] that viral RNA-directed protein synthesis in extracts from interferon-treated L cells is highly sensitive to inhibition by very low concentrations of dsRNA. Control-cell extracts show considerably less, if any, sensitivity to inhibition by dsRNA. More recently it has been established that there is a potent activity in the ribosome-free supernatant of interferon-treated cell extracts which confers enhanced sensitivity towards dsRNA when added to control cell-free system,[101,102] as shown in Table 1. The presence of dsRNA in these systems thus reveals a latent effect of interferon treatment on uninfected cells. This result is particularly interesting in view of the fact that both interferon treatment and virus infection of cells were previously needed to inhibit subsequent mRNA translation under the conditions used by Kerr and co-workers.[33,34] Therefore, it is possible that small amounts of dsRNA produced during the infection cycle of a virus may act as the signal by which the interferon-treated cell recognizes that it has been infected and shuts down its translational machinery. It is not certain whether all viruses produce dsRNA during infection, but the presence of such material has been demonstrated in the case of polio-infected HeLa cells[56,57] and in chick cells infected with vaccinia (a DNA virus).[58] The molecular mechanism by which dsRNA inhibits protein synthesis in

TABLE 1

Enhanced Sensitivity of EMC-RNA Translation to Inhibition by Double-stranded RNA in the Presence of Cell Sap from Interferon-treated Cells

A preincubated postmitochondrial supernatant system from control L cells was used to translate exogenous EMC-RNA (40 μg/ml) under the conditions previously described.[34] Where indicated, dSRNA (1 μg/ml) and ribosome-free cell sap from L cells previously incubated with or without interferon for 18 hr were added to the incubations. Aliquots of each incubation mixture were removed at the times indicated for assay of [^{35}S]-methionine incorporation into protein. Incorporation of radioactivity in the absence of added EMC-RNA amounted to approximately 1,000 cpm/5 μl of incubation mixture.

	Incorporation of methionine (cpm/5 μl assay)			
Time of incubation (min)	Plus control cell sap	Plus dsRNA and control cell sap	Plus interferon cell sap	Plus dsRNA and interferon cell sap
5	1,791	1,147	1,139	1,108
15	1,805	1,744	1,622	1,462
30	4,152	2,986	3,827	1,953
45	8,785	4,555	6,701	2,292
60	11,282	6,936	9,947	2,678
90	15,714	9,673	14,683	2,817
120	18,670	12,815	18,807	2,732
150	19,548	14,357	21,842	2,638

interferon-treated cells or their extracts has not yet been clearly established, so it is not known whether there could be any basis for discrimination between viral and host cell translation. One possibility is that the dsRNA-induced inhibition of protein synthesis is nonselective and that prolonged exposure of interferon-treated cells to dsRNA could result in "cell suicide," thereby limiting the production of progeny virus and the spread of infection. Indeed, the phenomenon of enhanced toxicity of the synthetic dsRNA poly (I):poly (C) towards cells previously exposed to interferon[59] may reflect this effect. On a practical level, the enhanced inhibitory effects of dsRNA in interferon-treated cell extracts mean that care must be taken in interpreting comparisons of the translation of different mRNA species. A number of viral mRNAs, in particular, could be contaminated with small quantities of viral dsRNA, giving rise to increased inhibition of the translation of these messengers. Whether such an effect contributed to the apparent specificity of the cell-free system described by Samuel and Joklik,[27] which showed inhibition of reovirus or vaccinia virus mRNA translation but no inhibition of cellular mRNA translation, remains to be established. Clearly it would be useful to know the structural requirements of dsRNA for it to be inhibitory in interferon-treated cells or their extracts; such information might help to define the precise conditions under which inhibition could occur and could indicate whether the possible presence of double-stranded regions in the secondary structures of various mRNAs could ever influence their relative rates of translation.

VI. EFFECTS OF INTERFERON ON VIRAL RNA SYNTHESIS

There are a number of well-documented cases of interferon-induced inhibition of viral RNA synthesis,[60–63] and in the past much speculation has centered on the question of whether transcription or translation is the primary target for interferon action. It is probably fair to summarize the current view by saying that in many of the systems studied there is evidence for a block in transcription consequent upon an earlier inhibition of viral protein synthesis. However, there are some notable exceptions to this generalization, in which inhibition of translation cannot account for the effects of interferon on RNA synthesis.

A good example of the difficulties that can arise in attempting to analyze the relative interdependence of transcription and translation in virus-infected cells is the infection of mouse L cells or chick embryo fibroblasts with vesicular stomatitis virus (VSV). Synthesis of viral RNA in these cells occurs in two phases following infection. Primary transcription of parental RNA by the endogenous virion polymerase occurs in the absence of protein synthesis (as shown by the use of inhibitors), and earlier work appeared to suggest that interferon treatment of cells could inhibit this process.[61,62] Thus, there would be less viral mRNA produced, and consequently less virus-specific protein synthesis would occur. This in turn would lead to reduced amounts of new viral RNA polymerase and inhibition of secondary transcription. More recently, however, Repik et al.[64] have demonstrated by RNA-RNA hybridization that primary transcription does in fact occur normally in interferon pretreated cells. They have suggested that a step between the two phases of RNA synthesis is the main target for interferon action. This step could be the synthesis of viral proteins. A rather similar conclusion arises from the work of Wiebe and Joklik[30] on reovirus infection of L cells. In this case the transcription of early viral mRNA species was only slightly inhibited by interferon pretreatment, and the RNA synthetic activity of isolated subviral particles was normal, regardless of previous exposure of the cells to interferon. However, formation of double-stranded RNA containing progeny and transcription of late mRNA were severely inhibited in interferon-treated cells. This is thought to be due to inhibition of the synthesis of the λ_1 virus protein, an essential constituent of the dsRNA containing progeny structures in which late mRNA synthesis occurs.

In spite of the implications of the above examples, there is no *a priori* reason to assume that interferon treatment of cells should necessarily directly affect only transcription *or* translation. It is quite possible that under some circumstances interferon does induce an inhibitor of transcription; in cases of the kind just described, primary transcription processes may appear particularly resistant to interferon because they occur within functional compartments constituted by the nature of the infecting subviral particles. There is an additional possibility that an apparently dual effect of interferon on RNA metabolism and protein synthesis may be provided through induction of cytoplasmic ribonuclease activity. Marcus et al.[65] have recently described a tenfold enhancement of membrane-bound alkaline ribonuclease activity in chick embryo cells exposed to interferon. As these authors point out, enhanced degradation of newly synthesized viral RNA could be mistakenly interpreted as an inhibition of transcription in vivo.

Perhaps the most clear-cut example of an effect of interferon on viral RNA synthesis which does *not* appear to be a consequence of inhibition of translation concerns infection of monkey cells with the small DNA tumor virus SV40. In this system it has been shown that interferon treatment causes a decreased amount of early viral mRNA to appear in the cytoplasm of infected cells,[25] and it has been possible to eliminate changes in virus penetration or in cytoplasmic degradation of RNA as explanations for this result.[66] Moreover, inhibition of overall protein synthesis with cycloheximide does not mimic the effect of interferon on production of viral RNA sequences. Action of interferon at the nuclear level seems a probable mechanism for its effects here. Once again, however, the possibility of multiple sites of action must be considered, for SV40 T-antigen synthesis brought about by microinjecting SV40 RNA into monkey cells is also inhibited when the cells have been pretreated with interferon.[67]

VII. NONANTIVIRAL EFFECTS OF INTERFERON

It is now established that interferon preparations can exert a number of effects on target cells which are not directly connected with inhibition of virus replication. As already mentioned, one such effect is enhancement of the toxicity of the synthetic double-stranded RNA poly (I):poly (C) towards cells in culture.[59] Mouse L cells, secondary mouse embryo cells, and human skin fibroblasts each show a cytotoxic response to poly (I):poly (C) after exposure to the homologous species of interferon, the degree of the effect being dependent on the interferon concentration. The toxicity enhancement effect of interferon exhibits species specificity towards the target cell and is destroyed by heating or by trypsin,[59] properties which are all characteristic of the antiviral activity of interferon. Another effect of interferon appears to be the sensitization of cells to the effects of deprivation of nutrients such as essential amino acids. Arginine starvation of L cells treated with interferon leads to cell death in a dose-dependent manner, whereas control cells merely show an inhibition of growth when deprived of this amino acid.[68]

An intriguing nonantiviral function of interferon is its ability to "prime" cells to produce more interferon when subsequently challenged with an infecting virus or with a synthetic inducer.[69] Cells which are normally nonproducers of interferon can thus be converted into producers by previous exposure to interferon. This priming effect has been shown to copurify with the antiviral effect of interferon over a one-million-fold range of interferon-specific activities,[70] as does the dsRNA toxicity-enhancing property of interferon.[70] Such evidence supports the view that all these activities of interferon preparations are brought about by the same component in them, but this cannot be rigorously established until the active constituent has been purified to molecular homogeneity. Almost nothing is known about the biochemical mechanisms that are responsible for the nonantiviral effects of interferon; indeed, very little work has been done in this area. It would clearly be of interest to know the relationship of these mechanisms to those concerned with the inhibition of virus replication in infected cells.

A potentially exciting aspect of current research concerns the inhibition of tumor cell growth by interferon, both in vivo and in vitro (see review by Gresser[71]). The multiplication of several cell lines is inhibited by preparations of homologous interferon, and, in particular, a number of transformed lymphoblastoid cell types have been studied. Gresser et al.[72] showed that the degree of inhibition of mouse leukemia cell division by interferon in vitro correlated with the antiviral titers of the various interferon preparations used. Moreover, various lines of Epstein-Barr virus-transformed human lymphoid cells showed similar variations in their susceptibilities to both the growth-inhibitory and the antiviral effects of human interferon.[73] As with the other nonantiviral effects of interferon, relatively little is known about the biochemistry of the inhibition of tumor cell growth. Recently, Tovey et al.[74] described a very rapid effect of interferon on thymidine incorporation into DNA of mouse leukemia cells growing under steady-state conditions, but there were minimal effects on precursor incorporation into total RNA or protein. The ability of different types of tumor cells to respond to interferon in vitro is extremely variable (see Figure 6), and it would be of interest to know why this is so. In vivo there is even evidence that interferon pretreatment can enhance the growth of certain tumors in mice,[75] but this effect may be a consequence of the depressive effect of interferon on cell-mediated immune responses under some circumstances.[76]

VIII. SUMMARY

This section describes the current state of knowledge concerning the biochemical mechanisms by which interferon inhibits viral replication in animal cells. The initial events involved in development of the antiviral state in cells exposed to interferon are summarized, and it is concluded that interferon interacts with the cell surface to bring about intracellular changes which result in

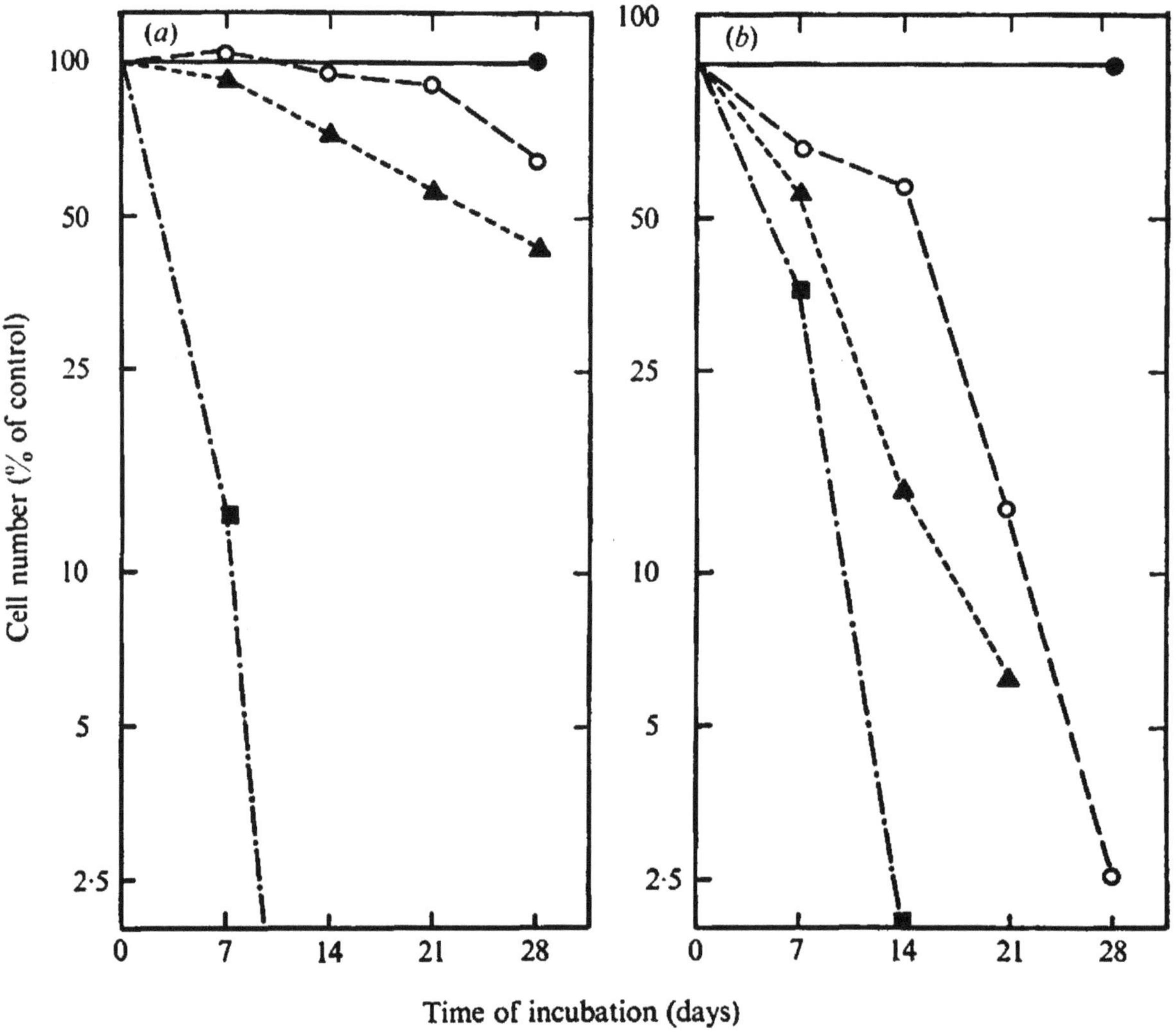

FIGURE 6. Effect of interferon on the growth of human lymphoid cell lines. (a) Interferon-sensitive Daudi cells grown in cultures containing various concentrations of human interferon: (0– –0) 0.1 units/ml; (▲- -▲) 1 unit/ml; (■-·■) 10 units/ml. Interferon-resistant Raji cells grown in the presence of a large excess of interferon: (●–●) 10,000 units/ml. (b) Interferon-sensitive P3Hr-1 cells grown in cultures containing various concentrations of human interferon: (0– –0) 10 units/ml; (▲- -▲) 100 units/ml; (■-·-■) 1,000 units/ml. Interferon-resistant Maku cells grown in the presence of a large excess of interferon: (●–●) 10,000 units/ml. At weekly intervals the number of living cells in cultures was determined and the cell density adjusted to 10^5 living cells per milliter. (From Adams, A., Strander, H., and Cantell, K., *J. Gen. Virol.*, 28, 207, 1975. With permission.)

induction of one or more inhibitors ("antiviral proteins"). Evidence is presented which indicates that cells treated with interferon can discriminate between viral and cellular functions in some situations, although the possibility of "cell suicide" occurring as a result of generalized inhibition of virus and host metabolism is also discussed.

The molecular mechanisms through which interferon treatment impairs virus macromolecular synthesis are considered, and the reasons for concluding that viral protein synthesis is a major target for the action of interferon are discussed. Dominant inhibitors of translation have been identified in interferon-treated cell extracts, and there is evidence that both the initiation and elongation phases of protein synthesis are affected by these inhibitors. The relationships between double-stranded RNA and the biological effects of interferon are described, with particular reference to protein synthesis. The current views on the influence of interferon treatment on viral RNA synthesis are presented, including the evidence that interferon may have primary effects on transcription, independent of translational impairment, in some systems.

Finally, the nonantiviral effects of interferon are summarized, although the relationship between the mechanisms responsible for these effects and the mechanisms causing inhibition of virus multiplication remains uncertain.

ACKNOWLEDGMENT

I wish to express my thanks to those authors and publishers whose data are reproduced in several of the figures in this section. I am also grateful to Dr. I. M. Kerr for helpful discussions and to the Royal Society for financial support in the form of a Mr. and Mrs. John Jaffé Research Fellowship.

SECTION B
BIOLOGICAL AND CLINICAL EFFECTS OF INTERFERONS AND INTERFERON INDUCERS

D. A. J. Tyrrell

I. INTRODUCTION

These substances have given rise to a great deal of scientific literature, much of it concerning research on the most fundamental aspects of virus replication, biochemical mechanisms concerned with the control of protein synthesis, and the experimental therapy of malignant disease. Nevertheless, it is fair to say that during most of the time since the original paper of Isaacs and Lindenmann[1] in 1957, the driving force behind much of the research on interferons (and subsequently on interferon inducers) was the hope that they might provide useful antiviral agents. Since respiratory viruses probably cause more infections of man than any other group of viruses (and since many of these viruses turned out to be sensitive to interferon), it is not surprising that much of the research was on the prevention of respiratory virus disease. Enthusiasm for interferon or inducers as antiviral medicaments has waxed and waned with the years, but in the latter part of this section we shall summarize the results of a significant number of successful experiments in which these substances were tested for their ability to prevent experimental infections in man. However, it is not possible at present to exploit the results or any other aspect of the interferon phenomenon in everyday clinical medicine. The general prospects are, however, sufficiently promising, and the field of research is sufficiently active to warrant a chapter in a book of this type. This section must necessarily be very condensed and highly selective. There are many sources of more detailed reading for those who are interested. The most recent and comprehensive review is a monograph edited by Finter,[77] and subsequent publications are listed by abstracting services. For those who require less detail and a clinical emphasis, I have written a short book on the subject.[78]

II. THE BIOLOGY OF INTERFERON FORMATION

The discovery of interferon arose out of systematic studies by Isaacs of the phenomenon of virus interference, in particular the reasons why cells which had been exposed to noninfectious influenza virus were unable to support the growth of infectious virus. He showed that these cells, which were in a state of "interference," were releasing into the medium a substance which could be added to other normal cells and which then induced the same state in them, i.e., they became resistant to infection with viruses. He showed that the interferon, as he named the substance, was not virus particle and was not neutralized by antiviral serum, but was a protein which resisted treatment at pH 6. It protected cells against a wide range of viruses but was "species specific" in that interferons were usually active only in cells from closely related species and not in those from distant species: for example, chicken cell interferon does not protect human cells and human interferon does not protect chicken cells, while human and monkey interferons cross react.

Since those early days, a great deal more information has been obtained about interferon. The exact trigger for the production of virus-induced interferon is not yet certain; it may well be the formation in the cell of double-stranded ribonucleic acid (dsRNA) (see Section A). Whatever the trigger is, it apparently derepresses some genetic function which leads to the formation of a new species of messenger RNA (mRNA) which can be extracted from the cell experimentally. Normally this mRNA then attaches to ribosomes which synthesize the biologically active protein interferon. It can be inferred from experiments with inhibitors that the presence of interferon in cells may then trigger a mechanism which cuts off the formation of interferon so that the cell, if it survives, returns to a resting state in which interferon is no longer being synthesized.

It has been shown on numerous occasions that when a virus infection occurs in an animal, including man, interferon is often produced. This is most easily detected in the organ in which virus infection is actively proceeding. The interferon may, however, enter the circulation and be able to make distant cells resistant to virus infection. There is still dispute about whether the production of interferon plays an important part in bringing a virus infection to a conclusion. Some of the difficulty has arisen from unjustified assumptions that all virus infections would behave in the same way and from wide generalizations from limited data. It was thought at one time that interferon reached its peak concentration and virus infection fell before antibody appeared.[79] I would think that, in relation to infections of the respiratory tract at least, the amount of interferon produced mainly reflects the extent of virus infection[80] and that this is mainly terminated by the production of antibody. However, in a matter as important as resistance against viruses there is probably a series of biological mechanisms for dealing with them, and interferon production is probably one of these. It may well be more important in subjects who cannot make normal antibody response, for example patients with hypogammaglobulinemia who nevertheless successfully combat a respiratory virus infection.

III. THE EFFECT OF INTERFERON

There is now evidence that interferon attaches to the surface of susceptible cells[81] and not to the surfaces of cells which are interferon resistant. The present hypothesis is that the molecule interacts with the cell surface and transmits a "message," perhaps analogous to that of certain peptide hormones. As a result it is postulated that an antiviral protein is produced. However, this has never been detected. Nevertheless, it is possible to detect biochemical differences between treated and untreated cells. The ribosomes can be shown in vitro to fail to translate a viral RNA message, although they still translate other messages such as that of cellular mRNA. Biochemical details remain to be worked out, but the general biological consequences have been reproduced repeatedly. Cells treated with hundreds of times more interferon than is required to inhibit virus synthesis are normal structurally and continue to metabolize and grow; however, when exposed to fully infectious virus of various sorts (DNA or RNA viruses with single- or double-stranded nucleic acid), those with cubic or helical symmetry, with or without an envelope, are resistant to infection. In some cases it has been shown that infection by virus nucleic acid is also blocked. In some circumstances, e.g., vaccinia virus infection, when large doses of virus are given, virus replication is blocked but cell damage is not. In all cases virus protein synthesis is blocked, but the synthesis of virus nucleic acid is more complicated. In some cases no nucleic acid is formed, but in others, such as the synthesis of messenger RNA by vaccinia virus, there is no inhibition – which is not surprising since the DNA and enzyme required are present in the infecting particle. Interferon also inhibits cell growth; the growth of other intracellular parasites such as bacteria, protozoa, and chlamydia; and it may affect immune response. These and other biochemical aspects have been reviewed in more detail by Metz.[82]

IV. THE PRODUCTION OF INTERFERON

Interferon was first detected as a response of the cell to inactivated virus, but it also occurs after virus infection. The study of mutant viruses as interferon inducers suggests that their effect may be due to the occurrence of double-stranded RNA within the cell.[83] Whatever the exact means, many viruses are interferon inducers, but certainly are much more effective and convenient than others. Therefore, parainfluenza viruses, such as Sendai virus and Newcastle disease virus (NDV), have been widely used both in research and in the production of interferon for clinical trials. The amount of interferon found also depends on the cell used, and it is clear that transformed cells are often poor producers. Different strains of one cell type may vary greatly, and it has proved necessary to screen a number of strains of human fibroblasts and lymphoblastoid cells to detect those which give high yields. The reasons are not clearly understood, but experiments with hybrid cell lines have shown that only if certain chromosomes[84] are present can interferon be synthesized.

It was found that extracts of filtrates of penicillium cultures (named Statolon or Helenine) protected animals against lethal infections with neurotropic viruses. The protection was associated with the fact that these materials induced interferon in the circulation. This effect was found to be due to the presence of double-stranded RNA which was contained within previously unsuspected virus-like particles which were derived from the mycelium of the fungus.[85,86] As an elegant part of the proof, it was shown that polyribose inosinic:polyribose cytidylic acid (poly I:C) induced interferon both in animals and in tissue cultures.[52] Although many other nonviral inducers have been found, poly I:C has been the most widely studied member of this group (Table 2). It has also been used to stimulate interferon production by fibroblasts for use in clinical trials.

TABLE 2

Nonviral Substances which Induce the Formation of Interferon

Microorganisms	Other substances
Bacteria, e.g., *Brucella*	Endotoxins
Chlamydia	Double-stranded RNA, e.g., from phage
Rickettsiae	Synthetic nucleic acid-like polymers, e.g., poly I:C
Mycoplasmas	Other synthetic polymers, e.g., COAM
Protozoa	Antibiotics
Mycophage, e.g., virus from *Penicillium chrysogenum*	Antigens – on immune cells
	Mitogens

V. EFFECT OF INTERFERON AND INDUCERS IN ANIMAL VIRUS RESPIRATORY INFECTIONS

Before interferon and interferon inducers were used to treat respiratory infections of man, preliminary experiments were done in animals. For example, mouse interferon was produced by inoculating the mouse brain with interferon. This was then delivered by aerosol into the respiratory tract of groups of mice; these were then inoculated by aerosol with influenza A virus. The virus content of the mouse lung was titrated at intervals and was found to be significantly lower in treated mice than in controls.[77] This method was used because no significant effect was seen in previous experiments using intranasal drops and in evaluating the results by other methods, such as the death of the mice. Interferon inducers have also been tested in the mouse lung. It was shown that the effect of poly I:C depended very much on the challenge virus used, and PVM pneumonia seemed to be much more readily affected than other models tested;[13] in addition, statolon when given parenterally had little effect, whereas if it was given intranasally, it protected. This was presumably because in the first instance the interferon or the inducer failed to get into those parts of the respiratory tract where significant virus multiplication was taking place.

These results are significantly different from those obtained when other types of virus infections were tested, such as neurotropic infections with viruses such as Semliki Forest virus or local infections of the skin with vaccinia. In these cases if enough interferon was given early enough, the lesions were prevented almost completely. In general, it was found that interferon must be given either before or shortly after the virus if virus multiplication and disease are to be prevented.

Experiments such as these suggest that it would be worthwhile to try to prevent respiratory infection in human volunteers with interferon or inducers, and work on these topics will constitute most of the rest of this chapter. However, it is first necessary to summarize work on preparing suitable interferon for clinical trials.

VI. THE PREPARATION OF INTERFERON FOR HUMAN USE

In the early days it was impractical to prepare much satisfactory interferon for this purpose from human cells, although various human cells were screened, and some batches were made from cultures of human amnion. In the United Kingdom it was decided to use interferon in monkey kidney cells, using the basic technology developed for the production of cell substrates for poliovirus vaccine. The cultures were inoculated with an influenza A virus as inducer, and the interferon was freeze dried and reconstituted just prior to use.[88]

Although this interferon prevented infection of the skin with vaccinia virus, it had no effect on respiratory infections. In later studies more potent interferon was produced by Cantell from cultured human buffy-coat leucocytes which were stimulated with a parainfluenza virus[89] and concentrated and purified prior to use. Apparently, rather similar material was used by Russian workers without concentration.[90] In order to make the results of assays of human interferon in different laboratories and by different methods as comparable as possible, a research reference preparation was established for human interferon and interferon from some animals. An arbitary unitage was assigned to each which corresponded roughly to the sort of titer usually obtained when it was assayed. The activity of all human interferon since then has been described in terms of units of activity of this reference material, although most laboratories perform their tests using a local secondary reference standard. The tests used are various,[77] but many are based on the reduction of plaques produced by an interferon-sensitive virus such as vesicular stomatitis virus (VSV); such a test is illustrated in Figure 7. Other test methods are available, and for greater precision, dose-response curves can be obtained and analyzed statistically.

It is important to concentrate and purify the interferon. Serum or human albumin was needed in the medium in which leucocytes were suspended if good yields were to be obtained. Therefore, the culture medium which contained interferon also contained much cell debris, particularly proteins, which were difficult to separate from interferon. Although the titer sounded impressive, the biological activity of interferon is so great that culture supernatants consist almost entirely of contaminating material. Nevertheless, using modern methods of protein separation it has been possible to substantially increase the activity and to make purified interferon with an activity of 10^6 units per milliliter and per milligram of protein.[91,92] It may soon be possible to improve on this by affinity chromotography by binding the interferon to antibody columns or to albumin columns and then eluting it.

The purified and concentrated interferon could not be used until it had been safety tested. Safety was partly achieved by selecting the blood donors carefully from an area of the world with little hepatitis and also checking them individually. The processing was done in a suitable closed area in a blood transfusion department in a commercial laboratory and not in an ordinary research laboratory. After processing (which would in itself have been harmful to many microbes), the material was tested much as a virus vaccine is tested before use (for the presence of bacteria, fungi, mycoplasmas, and contaminating viruses, as well as for interferon potency and for innocuity) by injecting it into mice and guinea pigs. Early trials showed that human interferon produced no harmful effects when given parenterally, so no harmful effects

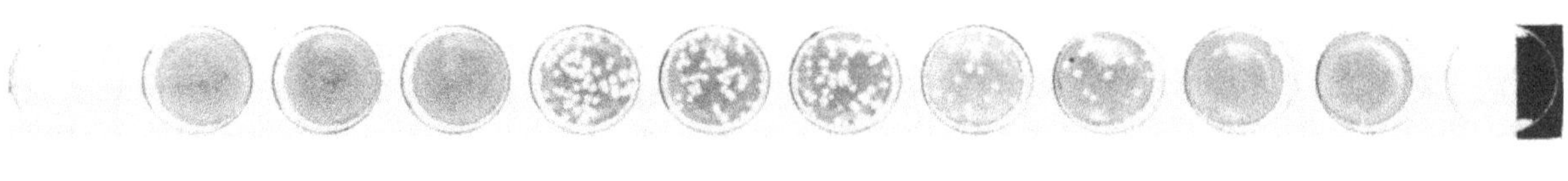

FIGURE 7. A simple assay for human interferon. Tissue cultures of a line of monkey kidney cells which are sensitive to human interferon (V3) are put up in wells in a plastic tray. These are then overlaid with dilutions of interferon overnight. Next day each well receives a uniform dose of interferon-sensitive virus, in this case vesicular stomatitis virus (VSV), and then an overlay medium. The cultures are reincubated, and when plaques have appeared they are fixed, stained, and inspected. The plate shows the reduction in the number of plaques which can also be represented as a dose-response curve for quantitative purposes (Courtesy of Dr. T. H. J. Mathews.)

were expected when it was used locally in the respiratory tract; in fact, none was observed.

Methods are now being developed to produce human fibroblast interferon to similar standards. The work has not yet reached the clinical stage.[93,94]

Poly I:C for use in man was prepared by carefully polymerizing the nucleotides with polymerases and then annealing the strands thus formed to produce double-stranded complexes which were standardized as far as possible by physical methods such as estimates of its molecular weight and the degree of double strandedness of the molecule. They were also assayed for biological activity in inducing interferon in vitro and protecting experimentally infected animals.[95]

VII. CLINICAL TRIALS

A series of trials of interferon was conducted at the Common Cold Unit, Salisbury (a branch of the Clinical Research Centre and supported jointly by the Department of Health and Social Security), which maintains the facilities and staff for the direct support of the volunteers, and the Medical Research Council, which maintains the laboratories and the scientific and technical staff. The results are summarized in Table 3. The first results were disappointing, but it was thought possible that interferon might appear to be ineffective because the concentration was too low or because it was removed too rapidly by the mucociliary blanket. Therefore, when larger amounts of more potent interferon became available, the experiment was repeated using slightly different viruses and methods of spraying. As there was still no effect, it was confirmed that the interferon used was active in human tracheal epithelial cells in organ culture and that it was as active against the viruses used in the volunteers (rhinoviruses and influenza B) as it was against the viruses used in the assay.[96] Finally, significant protection was obtained; although there is no certain proof, we think it likely that this was largely because we finally used a large enough amount.

Interferon inducers of several types have been used, for example, other viruses. Particularly because other workers have used them, we employed an enterovirus vaccine. This was a living type 1 echovirus which was given orally and induced an alimentary tract infection and an immune response. However, it failed to induce measurable amounts of interferon in the circulation; further, it had no detectable effect on the shedding of a rhinovirus infection induced in the nose or on the accompanying clinical symptoms. On the other hand, attenuated influenza virus was given intranasally and seemed to "take" less often in volunteers who had been given a rhinovirus than in those given a coronavirus. It is unclear whether or not this was due to interferon induction.

As can be seen from Table 3, poly I:C which was given to volunteers in the U.S. induced small amounts of interferon in the respiratory secretions and produced protection against infection and symptoms. However, the effect was less marked than that produced by exogenous interferon, but this must be regarded as only a preliminary judgment on the relative merits of the two measures; only a few arbitrarily chosen dosages and schedules have been tested so far. Low molecular weight inducers have also been tested in the upper respiratory tract of man. Quite marked protection has been reported,[97] but only a slight effect on virus shedding was seen by other workers.[98,99]

VIII. CONCLUSION

Interferons are naturally occurring antiviral substances which have been shown to prevent experimental infections of the respiratory tract of man and animals. Double-stranded ribonucleic acids and the synthetic polymer poly I:C, although toxic, induce interferon when given locally into the respiratory tract, and poly I:C protects to a limited extent. Further work is needed to develop dosage schedules which will protect more completely and to find out whether a useful effect can be obtained when the antiviral is given after symptoms have developed.

ABBREVIATIONS

DMSO	Dimethyl sulfoxide
dsRNA	Double-stranded RNA
EMC virus	Encephalomyocarditis virus
mRNA	Messenger ribonucleic acid
poly (I):poly (C)	Polyriboinosinic acid:polyribocytidylic acid
poly (U)	Polyribouridylic acid
SV40	Simian virus 40
tRNA	Transfer ribonucleic acid
VSV	Vesicular stomatitis virus

TABLE 3

Summary of Selected Experiments in which Respiratory Infections in Volunteers were Successfully Modified by Previous Treatment with Interferon or an Interferon Inducer

					Proportion of volunteers			
					Treated		Untreated	
Treatment	Amount given	Time of administration	Method	Virus given[a]	Infected	Ill	Infected	Ill
Human leucocyte interferon[c]	14 × 10^6 units	30 doses from −24 to +72 hr	Nasal spray	Rhinovirus 4	1/16	0/16	6/16	5/16
Poly I:C[d]	7 mg	14 doses −1 day	Nasal spray	Influenza A	6/9	4/9[b]	8/10	7/10
	17.5 mg	7 daily doses 0–5 days	Nasal spray	Rhinovirus 13	13/18	6/18[b]	15/18	12/18
Rhinovirus[e]	–	−30 hr	Intranasal drops	Influenza A	1/8	–	17/27	–

[a]All viruses were given intranasally by drops or spray.
[b]There was a statistically significant reduction in the symptoms score.
[c]From Merigan et al.[96]
[d]From Hill et al.[95]
[e]From Tyrrell and Reed.[100]

REFERENCES

1. **Isaacs, A. and Lindenmann, J.**, Virus interference. I. The interferon, *Proc. R. Soc. London Ser. B*, 147, 258, 1957.
2. **Ng, M. H. and Vilcek, J.**, Interferons: physicochemical properties and control of cellular synthesis, *Adv. Protein Chem.*, 26, 173, 1972.
3. **Finter, N. B., Ed.**, *Interferons and Interferon Inducers*, 2nd ed., North-Holland, Amsterdam, 1973.
4. **Baron, S., Finter, N. B., Galasso, G. J., Glasgow, L. A., Levy, H. B., and Youngner, J. S.**, Interferon, *Science*, 180, 779, 1973.
5. **Metz, D. H.**, Discrimination between viral and cellular macromolecular synthesis, in *25th Symp. of the Society for General Microbiology*, Burke, D. C. and Russell, W. C., Eds., Cambridge University Press, Cambridge, 1975.
6. **Metz, D. H.**, The mechanism of action of interferon, *Cell*, 6, 429, 1975.
7. **Vengris, V. E., Stollar, B. D., and Pitha, P. M.**, Interferon externalization by producing cell before induction of antiviral state, *Virology*, 65, 410, 1975.
8. **Ankel, H., Chany, C., Galliot, B., Chevalier, M. J., and Robert, M.**, Antiviral effect of interferon covalently bound to Sepharose, *Proc. Natl. Acad. Sci. U.S.A.*, 70, 2360, 1973.
9. **Chany, C., Ankel, H., and Bourgeade, M.-F.**, Interferon cell receptor interactions, *Perspect. Virol.*, 9, 269, 1975.
10. **Radke, K. L., Colby, C., Kates, J. R., Krider, N. M., and Prescott, D. M.**, Establishment and maintenance of the interferon-induced antiviral state: studies in enucleated cells, *J. Virol.*, 13, 623, 1974.
11. **Young, C. S. H., Pringle, C. R., and Follett, E. A. C.**, Action of interferon in enucleated cells, *J. Virol.*, 15, 428, 1975.
12. **Levine, S.**, Effect of actinomycin D and puromycin dihydrochloride on action of interferon, *Virology*, 24, 586, 1964.
13. **Taylor, J.**, Inhibition of interferon action by actinomycin, *Biochem. Biophys. Res. Commun.*, 14, 447, 1964.
14. **Tan, Y. H., Schneider, E. L., Tischfield, J., Epstein, C. J., and Ruddle, F. H.**, Human chromosome 21 dosage: effect on the expression of the interferon induced antiviral state, *Science*, 186, 61, 1974.
15. **Tan, Y. H.**, Chromosome-21-dosage effect on inducibility of antiviral gene(s), *Nature*, 253, 280, 1975.
16. **Friedman, R. M. and Sonnabend, J. A.**, Inhibition of interferon action by p-fluorophenylalanine, *Nature*, 203, 366, 1964.
17. **Gresser, I., Tovey, M. G., and Bourali-Maury, C.**, Efficacy of exogenous interferon treatment initiated after onset of multiplication of vesicular stomatitis virus in mouse brains, *J. Gen. Virol.*, 27, 395, 1975.
18. **Sonnabend, J. A. and Friedman, R. M.**, Mechanisms of interferon action, in *Interferons and Interferon Inducers*, 2nd ed., Finter, N. B., Ed., North-Holland, Amsterdam, 1973.
19. **Vassef, A., Spencer, C., Gelehrter, T. D., and Lengyel, P.**, Selectivity of interferon action: hormonal induction of tyrosine aminotransferase in rat hepatoma cells is much less sensitive to interferon than the replication of vesicular stomatitis virus or reovirus, *Biochim. Biophys. Acta*, 353, 115, 1974.
20. **Beck, G., Poindron, P., Illinger, D., Beck, J.-P., Ebel, J.-P., and Falcoff, R.**, Inhibition of steroid inducible tyrosine aminotransferase by mouse and rat interferon in hepatoma tissue culture cells, *FEBS Lett.*, 48, 297, 1974.
21. **Gupta, S. L., Graziadei, W. D., Weideli, H., Sopori, M. L., and Lengyel, P.**, Selective inhibition of viral protein accumulation in interferon-treated cells; nondiscriminate inhibition of the translation of added viral and cellular messenger RNAs in their extracts, *Virology*, 57, 49, 1974.
22. **Friend, C., Scher, W., Holland, J. G., and Sato, T.**, Haemoglobin synthesis in murine virus-induced leukemic cells *in vitro*: stimulation of erythroid differentiation by dimethyl sulfoxide, *Proc. Natl. Acad. Sci. U.S.A.*, 68, 378, 1971.
23. **Swetly, P. and Ostertag, W.**, Friend virus release and induction of haemoglobin synthesis in erythroleukaemic cells respond differently to interferon, *Nature*, 251, 642, 1974.
24. **Oxman, M. N., Rowe, W. P., and Black, P. H.**, Studies on adenovirus-SV40 hybrid viruses. VI. Differential effects of interferon on SV40 and adenovirus T antigen formation in cells infected with SV40 virus, adenoviruses and adenovirus-SV40 virus hybrids, *Proc. Natl. Acad. Sci. U.S.A.*, 57, 941, 1967.
25. **Oxman, M. N. and Levin, K. J.**, Interferon and transcription of early virus-specific RNA in cells infected with Simian virus 40, *Proc. Natl. Acad. Sci. U.S.A.*, 68, 299, 1971.
26. **Oxman, M. N., Baron, S., Black, P. H., Takemoto, K. K., Habel, K., and Rowe, W. P.**, The effect of interferon on SV-40 T antigen production in SV-40 transformed cells, *Virology*, 32, 122, 1967.
27. **Samuel, C. E. and Joklik, W. K.**, A protein synthesising system from interferon-treated cells that discriminates between cellular and viral messenger RNAs, *Virology*, 58, 476, 1974.
28. **Joklik, W. K. and Merigan, T. C.**, Concerning the mechanism of action of interferon, *Proc. Natl. Acad. Sci. U.S.A.*, 56, 558, 1966.
29. **Metz, D. H. and Esteban, M.**, Interferon inhibits viral protein synthesis in L cells infected with vaccinia virus, *Nature*, 238, 385, 1972.
30. **Wiebe, M. E. and Joklik, W. K.**, The mechanism of inhibition of reovirus replication by interferon, *Virology*, 66, 229, 1975.
31. **Suh, M., Bodo, G., Wolf, W., Viehhauser, G., and Jungwirth, C.**, Interferon effect on cellular functions: enhancement of virus induced inhibition of host cell protein synthesis in interferon-treated cells, *Z. Naturforsch.*, 29c, 623, 1974.
32. **Kerr, I. M.**, Protein synthesis in cell-free systems: an effect of interferon, *J. Virol.*, 7, 448, 1971.

33. **Friedman, R. M., Esteban, M., Metz, D. H., Tovell, D. R., Kerr, I. M., and Williamson, R.**, Translation of RNA by L cell extracts: effect of interferon, *FEBS Lett.*, 24, 273, 1972.
34. **Friedman, R. M., Metz, D. H., Esteban, M., Tovell, D. R., Ball, L. A., and Kerr, I. M.**, Mechanism of interferon action: inhibition of viral messenger ribonucleic acid translation in L-cell extracts, *J. Virol.*, 10, 1184, 1972.
35. **Falcoff, E., Falcoff, R., Lebleu, B., and Revel, M.**, Correlation between the antiviral effect of interferon treatment and the inhibition of *in vitro* mRNA translation in noninfected L cells, *J. Virol.*, 12, 421, 1973.
36. **Falcoff, E., Falcoff, R., Lebleu, B., and Revel, M.**, Interferon treatment inhibits Mengo RNA and haemoglobin mRNA translation in cell-free extracts of L cells, *Nat. New Biol.*, 240, 145, 1972.
37. **Gupta, S. L., Sopori, M. L., and Lengyel, P.**, Inhibition of protein synthesis directed by added viral and cellular messenger RNAs in extracts of interferon treated Ehrlich ascites tumor cells: location and dominance of inhibitors, *Biochem. Biophys. Res. Commun.*, 54, 777, 1973.
38. **Gupta, S. L., Sopori, M. L., and Lengyel, P.**, Release of the inhibition of messenger RNA translation in extracts of interferon treated Ehrlich ascites tumor cells by added transfer RNA, *Biochem. Biophys. Res. Commun.*, 57, 763, 1974.
39. **Content, J., Lebleu, B., Zilberstein, A., Berissi, H., and Revel, M.**, Mechanism of the interferon-induced block of mRNA translation in mouse L cells: reversal of the block by transfer RNA, *FEBS Lett.*, 41, 125, 1974.
40. **Content, J., Lebleu, B., Nudel, U., Zilberstein, A., Berissi, H., and Revel M.**, Blocks in elongation and initiation of protein synthesis induced by interferon treatment in mouse L cells, *Eur. J. Biochem.*, 54, 1, 1975.
41. **Revel, M., Content, J., Zilberstein, A., Lebleu, B., Falcoff, E., and Falcoff, R.**, Mechanism of interferon-induced block in mRNA translation, *Perspect. Virol.*, 9, 233, 1975.
42. **Metz, D. H., Esteban, M., and Danielescu, G.**, Formation of virus polyribosomes in L cells infected with vaccinia virus, *J. Gen. Virol.*, 27, 181, 1975.
43. **Metz, D. H., Esteban, M., and Danielescu, G.**, The effect of interferon on the formation of viral polyribosomes in L cells infected with vaccinia virus, *J. Gen. Virol.*, 27, 197, 1975.
44. **Furuichi, Y., Morgan, M., Muthukrishnan, S., and Shatkin, A. J.**, Reovirus messenger RNA contains a methylated, blocked 5'-terminal structure: $m^7G(5')ppp(5')G^mpCp$-, *Proc. Natl. Acad. Sci. U.S.A.*, 72, 362, 1975.
45. **Wei, C. M. and Moss, B.**, Methylated nucleotides block 5'-terminus of vaccinia virus messenger RNA, *Proc. Natl. Acad. Sci. U.S.A.*, 72, 318, 1975.
46. **Furuichi, Y., Muthukrishnan, S., and Shatkin, A. J.**, 5'-terminal $m^7G(5')ppp(5')G^mp$ *in vivo:* identification in reovirus genome RNA, *Proc. Natl. Acad. Sci. U.S.A.*, 72, 742, 1975.
47. **Rottman, F., Shatkin, A. J., and Perry, R. P.**, Sequences containing methylated nucleotides at the 5' termini of messenger RNAs: possible implications for processing, *Cell*, 3, 197, 1974.
48. **Sen, G. C., Lebleu, B., Brown, G. E., Rebello, M. A., Furuichi, Y., Morgan, M., Shatkin, A. J., and Lengyel P.**, Inhibition of reovirus messenger RNA methylation in extracts of interferon treated Ehrlich ascites tumor cells, *Biochem. Biophys. Res. Commun.*, 65, 427, 1975.
49. **Both, G. W., Banerjee, A. K., and Shatkin, A. J.**, Methylation-dependent translation of viral messenger RNAs *in vitro*, *Proc. Natl. Acad. Sci. U.S.A.*, 72, 1189, 1975.
50. **Hewlett, M. J., Rose, J. K., and Baltimore, D.**, The 5' terminal structure of poliovirus polyribosomal RNA is pUp, *Proc. Natl. Acad. Sci. U.S.A.*, 73, 327, 1976.
51. **Carter, W. A. and de Clercq, E.**, Viral infection and host defense, *Science*, 186, 1172, 1974.
52. **Field, A. K., Tytell, A. A., Lampson, G. P., and Hilleman, M. R.**, Inducers of interferon and host resistance. II. Multi-stranded synthetic polynucleotide complexes, *Proc. Natl. Acad. Sci. U.S.A.*, 58, 1004, 1967.
53. **Colby, C. and Chamberlin, M. J.**, The specificity of interferon induction in chick embryo cells by helical RNA, *Proc. Natl. Acad. Sci. U.S.A.*, 63, 160, 1969.
54. **De Clercq, E. and Merigan, T. C.**, Requirement of a stable secondary structure for the antiviral activity of polynucleotides, *Nature*, 222, 1148, 1969.
55. **Kerr, I. M., Brown, R. E., and Ball, L. A.**, Increased sensitivity of cell-free protein synthesis to double-stranded RNA after interferon treatment, *Nature*, 250, 57, 1974.
56. **Hunt, T. and Ehrenfeld, E.**, Cytoplasm from poliovirus-infected HeLa cells inhibits cell-free haemoglobin synthesis, *Nat. New Biol.*, 230, 91, 1971.
57. **Ehrenfeld, E. and Hunt, T.**, Double-stranded poliovirus RNA inhibits initiation of protein synthesis by reticulocyte lysates, *Proc. Natl. Acad. Sci. U.S.A.*, 68, 1075, 1971.
58. **Colby, C. and Duesberg, P. H.**, Double-stranded RNA in vaccinia virus infected cells, *Nature*, 222, 940, 1969.
59. **Stewart, W. E., de Clercq, E., Billiau, A., Desmyter, J., and de Somer, P.**, Increased susceptibility of cells treated with interferon to the toxicity of polyriboinosinic polyribocytidylic acid, *Proc. Natl. Acad. Sci U.S.A.*, 69, 1851, 1972.
60. **Mecs, E., Sonnabend, J. A., Martin, E. M., and Fantes, K. H.**, The effect of interferon on the synthesis of RNA in chick cells infected with Semliki forest virus, *J. Gen. Virol.*, 1, 25, 1967.
61. **Manders, E. K., Tilles, J. G., and Huang, A. S.**, Interferon mediated inhibition of virion-directed transcription, *Virology*, 49, 573, 1972.
62. **Marcus, P. I., Engelhardt, D. L., Hunt, S. M., and Sekellick, M. J.**, Interferon action: inhibition of vesicular stomatitis virus RNA synthesis induced by virion-bound polymerase, *Science*, 174, 593, 1971.

63. **Bean, W. J. and Simpson, R. W.**, Primary transcription of the influenza virus genome in permissive cells, *Virology*, 56, 646, 1973.
64. **Repik, P., Flamand, A., and Bishop, D. H. L.**, Effect of interferon upon the primary and secondary transcription of vesicular stomatitis and influenza viruses, *J. Virol.*, 14, 1169, 1974.
65. **Marcus, P. I., Terry, T. M., and Levine, S.**, Interferon action. II. Membrane-bound alkaline ribonuclease activity in chick embryo cells manifesting interferon-mediated interference, *Proc. Natl. Acad. Sci. U.S.A.*, 72, 182, 1975.
66. **Metz, D. H., Levin, M. J., and Oxman, M. N.**, Mechanism of interferon action. Further evidence for transcription as the primary site of action in Simian virus 40 infection, *J. Gen. Virol.*, in press.
67. **Graessmann, A., Graessmann, M., Hoffmann, H., Niebel, J., Brandner, G., and Mueller, N.**, Inhibition by interferon of SV40 tumour antigen formation in monkey cells injected with SV40 cRNA transcribed *in vitro*, *FEBS Lett.*, 39, 249, 1974.
68. **Lee, S. H. S. and Rozee, K. R.**, Enhanced responsiveness of L-929 cells treated with interferon to arginine starvation, *Exp. Cell Res.*, 93, 143, 1975.
69. **Stewart, W. E., Gosser, L. B., and Lockhart, R. Z.**, Priming: a non-antiviral function of interferon, *J. Virol.*, 7, 792, 1971.
70. **Stewart, W. E., de Clercq, E., de Somer, P., Berg, K., Ogburn, C. A., and Paucker, K.**, Antiviral and non-antiviral activity of highly purified interferon, *Nat. New Biol.*, 246, 141, 1973.
71. **Gresser, I.**, Antitumor effects of interferon, *Adv. Cancer Res.*, 16, 97, 1972.
72. **Gresser, I., Brouty-Boyé, D., Thomas, M. T., and Macieira-Coelho, A.**, Interferon and cell division. I. Inhibition of the multiplication of mouse leukemia L1210 cells *in vitro* by interferon preparations, *Proc. Natl Acad. Sci. U.S.A.* 66, 1052, 1970.
73. **Adams, A., Strander, H., and Cantell, K.**, Sensitivity of the Epstein-Barr virus transformed human lymphoid cell lines to interferon, *J. Gen. Virol.*, 28, 207, 1975.
74. **Tovey, M., Brouty-Boyé, D., and Gresser, I.**, Early effect of interferon on mouse leukemia cells cultivated in a chemostat, *Proc. Natl. Acad. Sci. U.S.A.*, 72, 2265, 1975.
75. **Gazdar, A. F., Sims, H., Spahn, G. F., and Baron, S.**, Interferon mediates enhancement of tumour growth and virus-induced sarcomas in mice, *Nat. New Biol.*, 245, 77, 1973.
76. **de Maeyer, E., de Maeyer-Guignard, J., and Vandeputte, M.**, Inhibition by interferon of delayed-type hypersensitivity in the mouse, *Proc. Natl. Acad. Sci. U.S.A.*, 72, 1753, 1975.
77. **Finter, N. B., Ed.**, *Interferons and Interferon Inducers*, North-Holland, Amsterdam, 1973.
78. **Tyrrell, D. A. J.**, *Interferon and its Clinical Potential*, William Heineman, London, in press.
79. **Isaacs, A. and Hitchcock, G.**, Role of interferon in recovery from virus infections, *Lancet*, 2, 69, 1960.
80. **Smorodintsev, A. A., Beare, A. S., Bynoe, M. L., Head, B., and Tyrrell, D. A. J.**, The formation of interferon during acute respiratory virus infection of volunteers, *Arch. Gesamte Virusforsch.*, 33, 9, 1971.
81. **Berman, B. and Vilček, J.**, Cellular binding characteristics of human interferon, *Virology*, 57, 378, 1974.
82. **Metz, D. H.**, *Advances in Drug Research*, Academic Press, New York, in press.
83. **Lomniczi, N. B. and Burke, D. C.**, Interferon production by temperature-sensitive mutants of Semliki forest virus, *J. Gen. Virol.*, 8, 55, 1970.
84. **Tan, Y. H., Tischfield, J., and Ruddle, F. H.**, The linkage of genes for the human interferon-induced antiviral protein and indophenol oxidase-B traits to chromosome G-21, *J. Exp. Med.*, 137, 317, 1973.
85. **Kleinschmidt, W. J. and Murphy, F. B.**, Investigations on interferon induced by statolon, *Virology*, 27, 484, 1965.
86. **Lampson, G. P., Tytell, A. A., Field, A. K., Nemes, M. M., and Hilleman, M. R.**, Inducers of interferon and host resistance. I. Double-stranded RNA from extracts of *Penicillium funiculosum*, *Proc. Natl. Acad. Sci. U.S.A.*, 58, 782, 1967.
87. **Nemes, M. M., Tytell, A. A., Lampson, G. P., Field, A. K., and Hilleman, M. R.**, Inducers of interferon and host resistance. VI. Antiviral efficacy of poly I:C in animal models, *Proc. Soc. Exp. Biol. Med.*, 132, 776, 1969.
88. Scientific Committee on Interferon, Effect of interferon on vaccination in volunteers, *Lancet*, 1, 873, 1962.
89. **Cantell, K., Strander, H., Hadházy, G. Y., and Nevanlinna, H. R.**, How much interferon can be prepared in human leukocytes, in *The Interferons*, Rita, G., Ed., Academic Press, New York, 1968, 223.
90. **Soloviev, V. D** The results of controlled observations on the prophylaxis of influenza with interferon, *Bull. WHO*, 41, 683, 1969.
91. Scientific Committee on Interferon, Progress towards trials of human interferon in man, *Ann. N.Y. Acad. Sci.*, 173, 770, 1970.
92. **Fantes, K. H.**, Purification and physico-chemical properties of interferons, in *Interferons and Interferon Inducers*, Finter, N. B., Ed., North-Holland, Amsterdam, 1973, 171.
93. **Billiau, A., Joniau, M., and De Somer, P.**, Mass production of human interferon in diploid cells stimulated by poly I:C, *J. Gen. Virol.*, 19, 1, 1973.
94. **Vilček, J. and Havell, E. A.**, Stabilization of interferon messenger RNA activity by the treatment of cells with metabolic inhibitors and lowering of the incubation temperature, *Proc. Natl. Acad. Sci. U.S.A.*, 70, 3909, 1973.
95. **Hill, D. A. et al.**, Evaluation of an interferon inducer in viral respiratory disease, *JAMA*, 219, 1178, 1972.
96. **Merigan, T. C., Reed, S. E., Hall, T. S., and Tyrrell, D. A. J.**, Inhibition of respiratory virus infection by locally applied interferon, *Lancet*, 1, 563, 1973.

97. **Panusarn, C., Stanley, E. D., Dirda, V., Rubenis, M., and Jackson, G. G.,** Prevention of illness from rhinovirus infection by a topical interferon inducer, *N. Engl. J. Med.*, 291, 57, 1974.
98. **Douglas, R. G., Jr. and Betts, R. F.,** Effect of induced interferon in experimental rhinovirus infections in volunteers, *Infect. Immun.*, 9, 506, 1974.
99. **Reed, S. E.,** unpublished observations.
100. **Tyrrell, D. A. J. and Reed, S. E.,** Some possible practical implications of interferon and interference, in *Non-specific Factors Influencing Host Resistance,* Braun, W. and Ungar, J., Eds., Karger, Basel, 1973, 438.
101. **Kerr, I. M., Brown, R. E., Clemens, M. J., and Gilbert, C. S.,** Interferon-mediated inhibition of cell-free protein synthesis in response to double-stranded RNA, *Eur. J. Biochem.*, in press.
102. **Roberts, W. K., Clemens, M. J., and Kerr, I. M.,** Interferon-induced inhibition of protein synthesis in L-cell extracts: an ATP-dependent step in the activation of an inhibitor by double-stranded RNA, *Proc. Natl. Acad. Sci. U.S.A.*, in press.

Chapter 8

CHEMOPROPHYLAXIS AND VIRUSES OF THE RESPIRATORY TRACT: SOME NEW APPROACHES TO ANTIVIRAL CHEMOTHERAPY, METALLOENZYME INHIBITION, AND TRANSPORT BY LIPOSOMES

D. D. Perrin

TABLE OF CONTENTS

I. THE HOST-VIRUS RELATIONSHIP

Many diseases of viral origin (including smallpox, yellow fever, and poliomyelitis) can be controlled adequately by the use of vaccines. On the other hand, in diseases such as influenza or the common cold, vaccines are of only limited use because of variability in antigenicity or multiplicity in the number of serotypes (Chapter 1, Volume I). The syndrome of the common cold can be produced by more than 89 rhinoviruses[1] and by other viruses including coronaviruses, myxoviruses (influenza, parainfluenza, and respiratory syncytial viruses), adenoviruses, herpesviruses, enteroviruses, echoviruses, and coxsackieviruses. Prospects for preparing a general common cold vaccine are very poor. Possible avenues for the chemotherapeutic control of virus infections are also more restricted than for diseases of bacterial, fungal, or protozoal origin. This is because of the very intimate relation between virus and host. The latter not only provides the energy and structural requirements of the virus, but it also carries out, under the control of the virus, replication and assembly of further virus particles, thereby increasing the intensity of the viral infection and facilitating its spread among a susceptible population.

The differences between viruses and other particulate matter have directed attention away from their similarities. For example, tumors and viruses in a host are characteristically foreign nucleic acid material, and it is not surprising that at least some viruses are oncogenic.[2] The rate of replication of virus particles varies widely: acute viral infections, for example, of influenza are fully established within several days. In such cases the average time for doubling virion concentration is probably only 1 or 2 hr. On the other hand, much slower replication times are to be expected for slow viruses which require several years to produce their effects. These include such chronic diseases of the central nervous system as kuru (in the Central Highlands of New Guinea), Creutzfeldt-Jakob disease, and visna and maedi (diseases of sheep in Iceland). There are probably many viruses to which the host cell has become adapted, so that the viruses are "passengers" and do not give rise to symptoms of disease.

Progress in antiviral chemotherapy will depend increasingly on the exploitation of differences in chemical and biochemical properties of viruses and their host organisms: this in turn will reflect growth of basic knowledge in this field. One of the potentially most rewarding areas in which to seek such chemoprophylactic agents is composed of influenza, common cold, and other viral diseases of the human respiratory tract.

II. MOLECULAR BIOLOGY OF RNA VIRUSES

Viruses commonly involved in upper respiratory tract infection include influenza viruses, parainfluenza viruses, respiratory syncytial virus, and the rhinoviruses, all of which contain RNA. They have many similar points.

The free influenza virion is an irregularly shaped particle bounded by a lipid membrane which overlies a lipoprotein envelope (Chapter 5, Volume I). The virion interacts with its environment through two sets of spikes that it carries dispersed over this envelope. The first of these contains hemagglutinin that binds to sialic acid residues on cell membranes. The second type of spike contains the enzyme neuraminidase which removes neuraminic acid side chains from mucoproteins. This process may facilitate penetration of the virion into the host cell or help in the release of newly assembled virus particles.

Thus, when the free virion comes into contact with the outer surface of an appropriate cell, it binds to sialic acid residues at receptor sites, triggering its entry into the cell by viropexis (pinocytosis).[3,4] It is possible that viral adsorption and penetration are accompanied by some structural change in the cell membrane or by some spatial rearrangement of membrane components such as glycoproteins and glycolipids. Inside the cell the virion is stripped of its envelope (possibly by cytoplasmic proteases and lipases), and the core of the virion is exposed. The core comprises a ribonucleoprotein coat or capsid surrounding the viral RNA. It includes a virion transcriptase enzyme which transcribes a copy of the single-stranded RNA (complementary strand mRNA). Penetration of the RNA virus into the cell is followed by a decrease in cell DNA-dependent RNA polymerase activity and the appearance of viral (RNA-dependent) RNA polymerase. This is because removal of the outer envelope of the virion leads to activation of the viral transcriptase enzyme and release of RNA from the nucleocapsid. The complementary strand mRNA in the cytoplasm directs the host cell to synthesize structural proteins and enzymes such as the virus-specific RNA-dependent RNA polymerase enzyme (RNA replicase) which is necessary for the replication of the virus RNA. These components are reassembled, and new virus particles are finally released by budding from the cell membrane.

Viral architecture and pathways of infection and replication vary for different types of viruses. However, the parainfluenza viruses (including Sendai virus) and the respiratory syncytial virus resemble influenza viruses; they are helically symmetrical and surrounded by an outer envelope studded with spikes (Chapter 1, Volume I).

III. STRATEGIES OF CURRENT ANTIVIRAL THERAPY

Albert[5] lists five stages in viral multiplication that should be vulnerable to selective chemotherapeutic attack.

1. Attachment of the virion to the host cell.
2. Transcription of viral nucleic acid in the host cell.
3. Translation of viral mRNA into protein.
4. Replication of viral nucleic acid.
5. Coating of the nucleic acid with protein.

Similarly, Oxford[6] listed various stages in the infection of mammalian cells by influenza virus and endeavored to assign the points of action of antiviral compounds. Except for amantadine, which interferes with the cell penetration and/or uncoating of the virus, most antiviral agents impede one or more of the processes involving nucleic acid. The hope has been expresssed[6] that "knowledge of the details of replication of the [influenza] virus should lead to more logical approaches for the selection of virus inhibitory compounds."

However, much more information will be needed about the molecular biology of viruses before antiviral action can be confidently forecast. The empirical nature of many of the present approaches is illustrated by the fact that amantidine is active against influenza A viruses but not against influenza B or C viruses. Currently antiviral drugs have been discovered almost entirely by chance. For, example, following the observation in 1950[7] that *p*-aminobenzaldehyde 3-thiosemicarbazone was weakly active against vaccinia virus, methisazone was developed as a member in a long series of antiviral thiosemicarbazones. Methisazone has a very wide antiviral spectrum and is currently clinically used against smallpox and a number of other DNA viruses. Similarly, idoxuridine (a thymine antagonist) and cytarabine (a cytosine antagonist) are synthetic nucleosides that have been found empirically to be active against DNA virus infections.

IV. RESPIRATORY TRACT INFECTIONS OF VIRAL ORIGIN

The usual route for viral infection of the respiratory tract is by inhalation of particles of infected material which lodge within the respiratory tree. If particles are less than 5 μm in diameter, they tend to penetrate to the lungs, whereas larger particles are trapped in the nasal turbinates.[8] Most droplets containing infective material are much larger than this, so that airborne infections usually have their primary foci in the upper respiratory tract.

Two main groups of viruses can be distinguished.

1. Those that are not "invasive" and have a pathology that is characteristically associated with the respiratory tract. Typically, these are the "respiratory viruses" such as myxovirus, adenovirus, and rhinovirus which spread outward over the surface of respiratory mucosa.
2. Viruses that invade by the respiratory route but subsequently produce generalized disease. Examples include exanthems, mumps, Newcastle disease, smallpox, and measles.

Only the first group of viruses is discussed in this chapter.

The upper respiratory tract – nasal passages, pharynx, larynx, trachea, and bronchi – is lined predominantly with ciliated "respiratory epithelium" which is interspersed with goblet cells. (There is specialized epithelium in the olfactory area and near the vocal cords.) The respiratory epithelium rests on a basement membrane which separates it from underlying connective tissue. Leucocytes and lymphocytes readily migrate into this epithelium. Mucin, mixed with water, is slowly and continuously secreted by the goblet cells to form the mucus which, as a thin film, keeps the walls of the nasal cavity moist and serves as a barrier against infection. The cilia propel particles lodged on the mucus away from the lungs and hence also have a defensive role. Mucus is rich in glycoproteins, the carbohydrate portions of which include neuraminic acid (from which it has been isolated) and its derivatives. The neuraminidase of myxoviruses, by hydrolyzing protective neuraminic acid molecules, is thus seen as a device by which these viruses can invade respiratory tract tissue.

V. METAL IONS AND METALLOENZYMES

An extensive range of compounds is active against viruses in tissue culture or in the presence of reagents that strip off the viral envelope. These include many chelating agents for metal ions, so that this behavior suggests the involvement of metal ions or metalloenzymes in the "life cycles" of viruses.

Organic thiosemicarbazones are chelating agents by virtue of the thiosemicarbazone moiety, and antiviral activity was demonstrated more than 20 years ago for thiosemicarbazones of *p*-aminobenzaldehyde,[7] nicotinaldehyde,[9] isonicotinaldehyde,[9] 2-thenylaldehyde,[9] 3-thenylaldehyde,[9] and isatin.[9] The greater activity of isatin 3-thiosemicarbazone (against vaccinia virus) correlates with its improved metal-chelating ability; this ligand has three sites through which it can coordinate to a metal ion to form complexes such as those shown in Figures 1 and 2.

Derivatives of isatin thiosemicarbazone with greater antiviral activity have been synthesized.[10] One of these, methisazone (1-methylisatin 3-thiosemicarbazone, see Figure 3), was used successfully in clinical trials as a prophylaxis against smallpox.[11] This was the first time that a

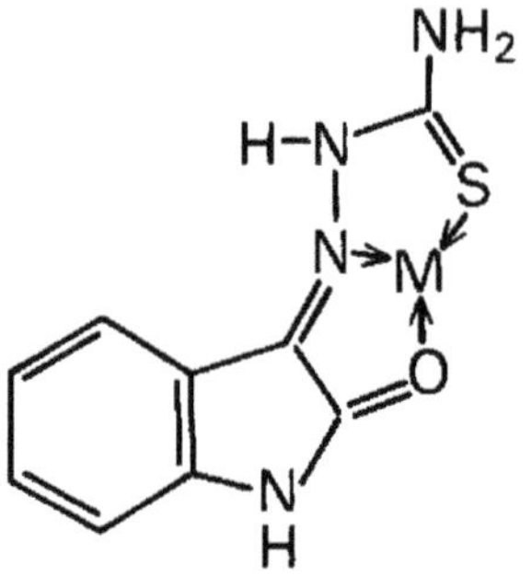

FIGURE 1. Methisazone forming a terdentate chelate with a divalent metal ion (M) by bonding through a sulfur.

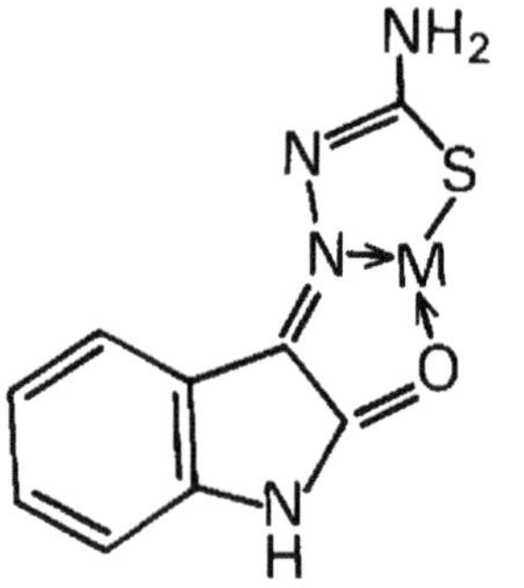

FIGURE 2. Methisazone forming a terdentate chelate through a nitrogen and an oxygen.

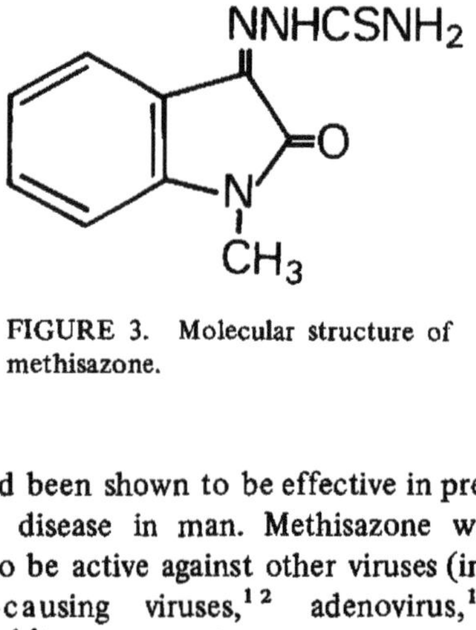

FIGURE 3. Molecular structure of methisazone.

drug had been shown to be effective in preventing a virus disease in man. Methisazone was later found to be active against other viruses (including tumor-causing viruses,[12] adenovirus,[13] and alastrim,[14]) and useful in the treatment of vaccinia infections.[15] Another metal-chelating agent, 4-bromo-3-methylisothiazole-5-carboxaldehyde thiosemicarbazone, also afforded significant protection against smallpox when administered prophylactically, but was ineffective once the disease was established.[16] The effect of methisazone on adenovirus was abolished if the sulfur atom was replaced by oxygen.[13] Similarly, some *S*-alkylated derivatives of N^1-alkylated isatin thiosemicarbazones reversibly inhibit mengovirus replication in vitro.[17]

The foregoing observations are consistent with the antiviral activity due to the ability of methisazone to form a terdentate chelate with an essential divalent heavy metal ion by bonding through a sulfur, a nitrogen, and an oxygen, as shown in Figures 1 and 2. A similar structure-activity relation has been suggested[18] to explain the anticancer effects of a number of related compounds. Implication of nucleic acid in the process catalyzed by the metal ion or metalloenzyme was indicated by the observation that the antiviral activity of methisazone and related thiosemicarbazones affects a late stage of the pox virus cycle[19] by interfering with viral *m*RNA translation.[20] Similarly, methisazone decreases the reverse transcriptase activity and cytopathic effect of the RNA slow viruses, which produce visna and maedi,[21] and inhibits the RNA-dependent DNA polymerase of Rous sarcoma virus and other RNA tumor viruses.[22] Further evidence for the involvement of one or more essential metal ions in virus activity is that the influenza virus RNA-dependent RNA polymerase can be inhibited in vitro by a number of chelating agents, including selenocystamine, bathocuproine (Figure 4) bathophenanthroline, some heterocyclic thiosemicarbazones,[23] and selenocystine.[24] The inhibitory effects of related compounds, such as isatin 3-semicarbazone and 3- and 4-acetylpyridine thiosemicarbazones, fall off with a decrease in chelating ability. Results suggested that influenza RNA polymerase, which is essential for the replication of the virus, is a zinc metalloenzyme in which the zinc is firmly bound to the protein. Inhibition of the enzyme by chelating agents probably involves the formation of an apoenzyme-zinc-ligand complex (Figure 5): the process is reversible, and activity is restored if the solution is dialyzed against distilled water.

Such ternary complexes have been proposed to explain inactivation of other zinc-containing enzymes by chelating agents. For example, it was suggested that the inhibition of carbonic anhydrase by chelating agents was due to the rapid, reversible formation of such a ternary complex, followed by a much slower loss of zinc from the enzyme.[25] Similarly, with yeast phosphoglucomutase, inhibition by EDTA was attributed to the formation of an apoenzyme-zinc-EDTA ternary complex.[26]

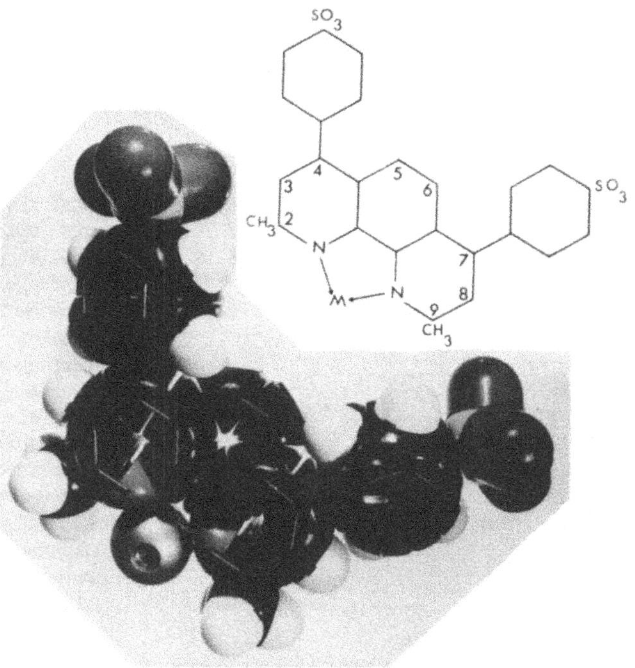

FIGURE 4. Molecular structure of a 1:1 bathocuproine-zinc complex.

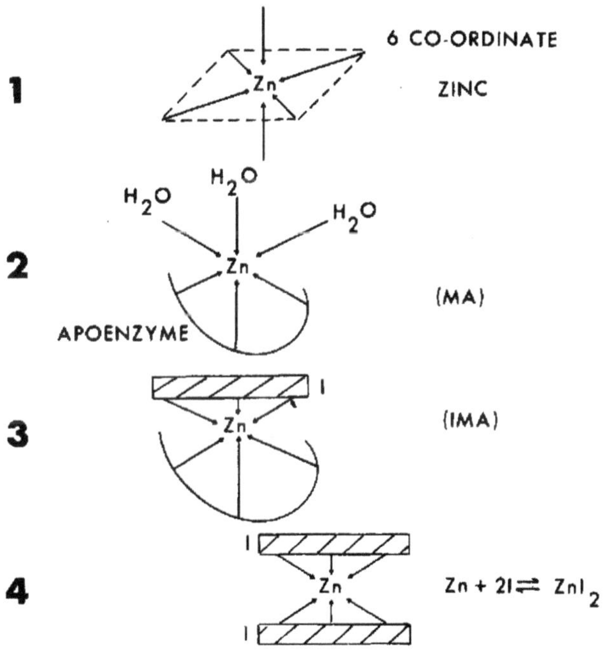

FIGURE 5. Hypothetical modes of inhibition of a zinc metalloenzyme. (MA) Metal-apoenzyme complex. (IMA) Inhibitor-metal-apoenzyme complex. Either one molecule of inhibitor binds to the zinc so causing steric hindrance effects on enzyme functioning or two molecules of the chelating agent remove the zinc from the enzyme causing collapse of enzyme tertiary structure.

Other polymerases have been shown to be zinc metalloenzymes (Table 1), including the RNA polymerase of *Escherichia coli*[27] and the DNA polymerases of *E. coli*,[28] sea urchin,[29] bacteriophage,[30] and avian myeloblastosis virus.[31,32] Removal of zinc from *E. coli* DNA polymerase led to loss of activity; addition of zinc to the apoenzyme fully restored activity.[28] Bathocuproine disulfonate and selenocystamine inhibit the DNA-dependent RNA polymerase of *E. coli*.[6]

Terminal deoxynucleotidyltransferase of calf thymus gland may contain a tightly bound cation, probably zinc.[33] Zinc is also believed to be present in DNA polymerase from chick embryo.[34] These results (taken in conjunction with inhibition studies on DNA-dependent RNA polymerases from rat liver, sea urchin (*Strongylocentrotus franciscanus blastulae*), and *E. coli;* RNA-dependent DNA polymerase (reverse transcriptase) from Rous sarcoma virus; and calf thymus terminal deoxynucleotidyltransferase, using 1,10-phenanthroline[27,29,33,35] suggest that many if not all DNA and RNA polymerases and nucleotidyl transferases are metalloenzymes. It is likely that the tightly bound zinc (or other metal ion) is involved in the interaction between nucleic acid and the enzyme, and the added ligand acts as

TABLE 1

Evidence for the Presence of Zinc in Polymerases and Related Enzymes

Enzyme	Source	No. of Zns	Method	Reference
RNA polymerase	*Escherichia coli*	2	Atomic absorption	35
DNA polymerase'	*E. coli*	~2	Atomic absorption	29
DNA polymerase I	*E. coli*	1	Atomic absorption	28
RNA polymerase	T_7 phage	2–4	Atomic absorption	30
Reverse transcriptase	Avian myeloblastosis virus	2	Microwave emission spectrometry	32
Reverse transcriptase	AMV	~1	Atomic absorption	31
DNA polymerase	Sea urchin nuclei	~4	Atomic absorption	29
RNA polymerase	Influenza virus	?	Atomic absorption	23
RNA polymerase	Sea urchin	?	Inhibited by 1,10-phenanthroline	27
Reverse transcriptase	Rous sarcoma virus	?	Inhibited by 1,10-phenanthroline	22, 27
RNA polymerase	Rat liver	?	Inhibited by 1,10-phenanthroline	27
DNA polymerase	Chicken embryo	?	Inhibited by 1,10-phenanthroline	34
DNA-terminal nucleotidyl transferase	Calf thymus	?	Inhibited by 1,10-phenanthroline, reversed by Zn^{2+}	33

a competitive inhibitor of substrate addition. This would explain the insensitivity of DNA polymerases to inhibition by 1,10-phenanthroline at saturating levels of DNA. The role of the heavy metal such as zinc, which is an integral part of the metalloenzyme, must not be confused with that of magnesium or manganese ions, ordinarily present in large excess and also essential cofactors in the enzymic activity.[36,37]

In many cases, heavy metal ions other than zinc are present in metalloenzymes or as activators of enzymes. Examples include copper in oxidases and iron(II) in ribonucleoside diphosphate reductase.[38] The latter, which catalyzes the reductive conversion of ribonucleotides to deoxyribonucleotides, is possibly a rate-limiting step in DNA biosynthesis.[39] Tumor-derived ribonucleoside diphosphate reductase and herpesviruses are inhibited by α-(*N*)-heterocyclic carboxaldehyde thiosemicarbazones such as 2-formylpyridine thiosemicarbazone, which can form metal chelates analogous to formulae 4 and 5.[40] The more effective of these inhibitors were also active against mouse tumors in vivo.[40,41] Reduction of cytidine diphosphate by rat tumor nucleoside diphosphate reductase was inhibited by 1,10-phenanthroline and α,α′-bipyridyl; activity was fully restored by adding an equimolar amount of ferrous ammonium sulfate.[38] Evidence suggested that inhibition by the thiosemicarbazones was due to the formation of a chelate with iron(II); this illustrates a general principle that any given chelating agent can form chelates with more than one kind of transition or heavy metal ion.

It is also possible that chelating agents might act on more than one component of a biological system. Thus, in a study on vaccinia virus, 5-hydroxypyridine-2-aldehyde thiosemicarbazone and 1-formylisoquinoline thiosemicarbazone may have inhibited DNA synthesis, whereas other thiosemicarbazones were active against RNA synthesis.[42] It may be significant that both of these ligands bond to metal ions through two nitrogens and a sulfur, whereas methisazone bonds through an oxygen, a nitrogen, and a sulfur. It seems probable that in this case two different metalloenzymes are involved; which of these two enzymes is inhibited more by a given chelating agent depends on quantitative differences in the relevant stability constants.

The dependence of influenza virus replication on the action of one or more zinc metalloenzymes opens up a new avenue for control of this disease. In the presence of a suitable chelating agent, the level of free zinc ion in the appropriate tissues may be decreased to such an extent that synthesis of the enzyme is impeded. Alternatively, formation of the apoenzyme-zinc-ligand complex would block the assembly of the RNA needed in the replication of the virions. Either or both of these processes would allow more time for the mobilization of the body's defense mechanisms. This approach is currently being explored.[43]

More generally, the involvement of zinc metalloenzymes in the synthesis of both RNA and DNA raises the possibility of using suitable chelating agents for metal ions as broad-spectrum antiviral agents. The feasibility of this approach is illustrated by isatin 3-thiosemicarbazone and its *N*-methyl and *N*-ethyl derivatives which have shown antiviral activity against DNA and RNA viruses, including members of the following groups: poxvirus, adenovirus, herpesvirus, picornavirus, reovirus, arbovirus, myxovirus, and paramyxovirus.[44] The first three groups are DNA viruses; the remainder are RNA viruses. Unfortunately, most of the testing has been done only on tissue cultures or some other artificial system so that little can be concluded about antiviral activity in the intact animal. Typical examples are the inhibition by isatin thiosemicarbazone or its *N*-methyl derivative of the replication in tissue culture, of infectious bovine rhinotracheitis virus,[45] maedi,[21] and progressive pneumonia virus.[21] (The first of these is a herpesvirus; the other two are RNA slow viruses.)

A comparable suggestion has been made[29] that "a zinc-chelating agent or a radioisotope of zinc could be brought to the site of DNA replication or repair and inhibit those processes .. (This) may be of value in the design of chemotherapeutic agents for malignant disease."

It is likely that in RNA-dependent RNA polymerase the zinc ion shows its usual octahedral stereochemistry and that three or four sites around the zinc are occupied by binding groups on the enzyme. (Two sites would not hold zinc strongly enough to withstand removal of the zinc by EDTA, while more than four sites occupied would leave too few for stable ternary complex formation. Three occupied sites and three free sites would favor terdentate ligands and explain why 2-acetylpyridine thiosemicarbazone is more effective than 1,10-phenanthroline.)

The nature of the groups binding the zinc in the enzyme is also important because these groups will affect the strength with which the other ligand is bound. Probable binding groups in other zinc enzymes are amino and S^- groups in carboxypeptidases and S^-, imidazole, and amino groups in dehydrogenases and carbonic anhydrase.[46] Sulfur groups bind more strongly to zinc than do the corresponding oxygen or NH groups; however, in ligands for potential use in therapy, –SH groups provide difficulties because they tend to be too readily oxidized. One of the advantages of thiosemicarbazones is that they exist in dynamic equilibrium with small amounts of the tautomeric forms (isothiosemicarbazones) in which an –SH group is present. Quantitative information on potentially useful zinc-ligand complexes is very limited.

In 2-acetylpyridine thiosemicarbazone the ring nitrogen, the imino nitrogen, and the sulfur atom can be disposed around a central metal ion to form two five-membered chelate rings. By keto-enol tautomerism, the sulfur can become ionized to $-S^-$, allowing stronger metal binding. Further evidence for terdentate chelate formation by these complexing agents comes from the X-ray structure determination of crystalline bis(isoquinoline-1-carboxaldehyde thiosemicarbazanato)nickel(II).[47] Of the enormous number of thiosemicarbazones tested for antivaccinia, antipolio, or antileukemia activity, only those able to form terdentate chelates would appear to be suitable for the present purpose. This is probably an important consideration in designing further possible therapeutic agents to act in this way.

VI. SOME PHARMACOLOGICAL CONSIDERATIONS

Zinc is essential for the growth and survival of all organisms. It occurs in more than 50 enzymes, including peptidases (e.g., carboxypeptidase), dehydrogenases (e.g., alcohol dehydrogenases, glutamic dehydrogenase, and lactic dehydrogenase), alkaline phosphatase, and carbonic anhydrase.[46] Some of the better known zinc metalloenzymes and the tissues in which they occur are given in Table 2. Zinc-deficient media severely reduce the growth of the flagellate *Euglena gracilis,* decreasing the RNA content, arresting mitosis, and leading to an increase in polyphosphate.[48] Similar results have been found with *Mycobacterium smegmatis;* it was also suggested that energy- and phosphate-using reactions

TABLE 2

Some Zinc Metalloenzymes Present in Mammals

Enzyme	In
Carboxypeptidases A and B	Porcine and bovine pancreas
Dipeptidase	Mammalian tissues
Leucine aminopeptidase	Porcine kidney
Carnosinase	Porcine kidney
Alcohol dehydrogenase	Human and equine liver
Malate dehydrogenase	Porcine liver
Glutamate dehydrogenase	Mammalian liver
Lactate dehydrogenase	Rabbit skeletal muscle
D-Glyceraldehyde 3-phosphate dehydrogenase	Rabbit and bovine muscle
Spermine oxidase	Bovine plasma
Alkaline phosphatase	Mammalian kidney, human leucocytes
Carbonic anhydrase	Mammalian erythrocytes, parietal cells

were inhibited, but that ATP remained high.[49] Further evidence for the involvement of zinc in nucleic acid biosynthesis comes from feeding experiments. Addition of zinc ion stimulated nuclear DNA synthesis in the livers of partially hepatectomized zinc-deficient rats, but did not affect normally fed rats.[50] Perfusion of such rat livers with EDTA solution prevented the normal increase in the rate of liver DNA synthesis; the effect was reversed when Zn^{2+} was added.[51] DNA-dependent RNA polymerase activity in liver nuclei from suckling rats nursed by zinc-deficient dams decreased to about half the level in control animals.[52] On the other hand, addition of zinc ions in vitro to liver nuclei led to strong inhibition of RNA polymerase activity, but it was not known whether this was due to reaction of zinc ions with some chelating sites in the enzyme molecule with the monomer nucleotide triphosphates or with the priming DNA. The intraperitoneal injection of zinc ions into normal but partially hepatectomized rats decreased thymidine incorporation in DNA synthesis.[50]

In human cancer patients serum and cellular zinc levels are less than normal, suggesting that cancer cells bind zinc more strongly than normal tissues; this raises the possibility that decreased levels of zinc might inhibit tumor growth.[53] Thus, survival of rats with implanted carcinosarcomas was significantly increased, and tumor growth was markedly reduced if the rats were maintained on a diet deficient in zinc.[53] In China antimony-EDTA complexes have been developed as anticancer agents; their activity depends on their ability to inhibit the incorporation of zinc into tumor cells.[54] Depletion of zinc in cells of the respiratory tract should similarly decrease the extent to which influenza virus can induce the formation, in the cell cytoplasm, of the RNA-dependent RNA polymerase which is characteristic of virus-infected cells.

A convenient method for the in vitro measurement of RNA-dependent RNA polymerase activity of influenza virus is based on the use of a detergent (such as Nonidet P40 or Triton® X-100) to remove the viral envelope and expose the enzyme to a buffered solution containing Mg^{2+}, Mn^{2+}, ATP, CTP, GTP, UTP, and [^{3}H]-UTP (see Chapter 5, Volume I, Appendix). The extent of nucleic acid formation in a given time is subsequently measured on the acid-precipitated fraction using a scintillation counter.[36] In tris buffer, pH 8.0 and 37°, the chelating agent selenocystamine gave 50% inhibition of polymerase activity of influenza A and B viruses at ligand concentrations of 0.003 to 0.012 m*M*.[55] Further results for a range of other effective chelating agents, tested on a B virus,[23] are given in Table 3. In some cases the dose-response curves of inhibitor concentration flattened out at less than complete inhibition (so that 2-acetylpyridine thiosemicarbazone, for example, gave maximally only 70% inhibition). This may have been due to the solubility of the inhibitor as a limiting factor, as many heterocyclic thiosemicarbazones are only sparingly soluble in water.

However, results from in vitro testing cannot be applied directly to the in vivo situation because of the greater complexity and multiphasic nature of the intact animal and because of kinetic considerations. In all living organisms there is constant dynamic equilibrium between metal ions (such as zinc, copper, iron, calcium, and magnesium) and

TABLE 3

Inhibition of RNA-dependent RNA Polymerase Activity of Influenza B/Lee/40 Virus by Chelating Agents

Chelating agent	Concentration (m*M*) causing 50% inhibition
2-Acetylpyridine thiosemicarbazone	0.002
Isatin 3-thiosemicarbazone	0.02
Isatin 3-selenosemicarbazone	0.03
Isatin 4'-ethylthiosemicarbazone	0.04
Bathophenanthroline disulfonate, disodium salt	0.05
Bathocuproine disulfonate, disodium salt	0.08
Methylisatin 3-thiosemicarbazone	0.10
Isatin 4'-methylthiosemicarbazone	0.10
Solochrome black	0.13
Eriochrome blue	0.15
Zincon	0.19
PAR	0.26
Cystamine	0.3
Dithizone	0.4

From Oxford, J. S. and Perrin, D. D., *J. Gen. Virol.*, 23, 59, 1974. With permission.

their complexes with chelating agents (such as amino acids, peptides, proteins, carboxylic acids, and phosphates). Different kinds of metal ions may be bound at different sites, or competitively at the same site, on a given protein molecule. The effectiveness of an added complexing agent in vitro is likely to be greater than in vivo, where it has to compete with natural chelating agents. Also, the simultaneous involvement of two or more metal ions in living processes and the possible effects of extraneous species further complicates the overall picture. Thus, yeast phosphoglucomutase is inhibited by EDTA, which forms a ternary complex; however, low concentrations of EDTA stimulated enzyme activity, probably by removing traces of undesirable heavy metal ions.[26] Systems containing metal ions and complexing agents (in excess) constitute metal-ion buffer systems in which the free metal ion concentration is controlled by the other components in a manner analogous to pH buffers of weak acids and their salts. This is probably the major reason for the frequently observed failure to reverse the inhibitory effect of a chelating agent by adding excess of a metal ion, especially if there is competition at more than one site. If a metal ion is added to a solution of a metalloenzyme (the activity of which has been inhibited by a chelating agent such as EDTA), restoration of enzyme activity depends on two factors — the relative stabilities of the ternary apoenzyme-metal ligand and the metal-ligand complexes and the availability in the system of alternative binding sites for the added metal ion. If the metal is bound at a site remote from the enzymically active portion, or if it is bound only by the added ligand, activity is likely to be restored; if the metal is bound at a site near the active portion, inhibition by the metal ion would be observed.

Most metal chelates formed by first-row transition metal ions exist in dynamic equilibrium with their components, so that the effect of adding a preformed ligand-metal complex to a solution depends on the natures and metal-binding abilities of other species that are present. Particularly with proteins, more than one type of metal-binding site ($-NH_2$, $-COO^-$; $-S^-$) may be available. Thus, it is possible that the inhibition observed when cadmium-2-acetylpyridine thiosemicarbazone is added to influenza RNA polymerase solution is twofold. The complex is rapidly dissociated, with the cadmium going to a binding site on the enzyme while the thiosemicarbazone is free to bind to the zinc-containing portion: either or both processes might inhibit enzyme activity. This is a likely explanation of the cumulative inhibition of RNA-dependent DNA polymerase by copper(II) and 1-methylisatin 3-thiosemicarbazone.[22]

An improved understanding of multiple equilibria among metal ions and ligands in biological systems is possible using a suitable computer program such as COMICS,[56] provided that relevant stability constants are known. In this way, based on a copper(II), zinc, and 17 amino acid model of blood plasma,[57] we have examined the effectiveness of 2-acetylpyridine thiosemicarbazone as a competitor for zinc.[58] When the model was extended to include copper, zinc, calcium, magnesium, amino acids, albumin, and other ligands, almost all of the exchangeable copper and zinc was bound to albumin. If the system was made 2 m*M* in 2-acetylpyridine thiosemicarbazone (which is inhibitory in vitro to influenza virus), 85% of the copper and 99% of the zinc was transferred from the albumin to the new ligand (Table 4). Thus, it seemed feasible to maintain concentration of this thiosemicarbazone in the body for a protracted time, while at the same time producing a deficiency in readily exchangeable zinc. However, feeding experiments using mice inoculated with influenza virus were only marginally successful; the agent was too toxic.[43]

TABLE 4

Effect on Metal Ion Distribution of Adding 2-Acetylpyridine Thiosemicarbazone (APTSC) to the Blood Plasma Model[a,b]

	0	1 m*M*	2 m*M* APTSC
Cu(alb) + $Cu(alb)_2$	99.7	27.0	15.4
Cu-APTSC complexes	–	72.9	84.5
Zn(alb) + Zn_2(alb)	98.8	2.8	1.2
$Zn(cysH)_2$	0.8	–	–
Zn-APTSC complexes	–	97.1	98.7

[a]Species are given as percent of total metal.
[b]Calcium and magnesium distributions not affected.

At present we have no stability constant data for methisazone-zinc complexes, but would expect a value comparable with that of the 2-acetylpyridine thiosemicarbazone-zinc complex. Alternatively, if the system is made 2 m*M* in CaDTPA, 93% of the copper and virtually all of the zinc would be computed to be bound to the DTPA (Table 5). (Computed free zinc ion concentration would fall to 6×10^{-13} *M*.)

The biological efficacy of methisazone and related compounds would seem to be related, in part, to their persistence in the body which, in turn, arises from their slow systemic absorption and their low solubility. (On the other hand, an agent must not be too insoluble: in vitro isoquinoline-1-aldehyde thiosemicarbazone is a powerful inhibitor of ribonucleoside diphosphate reductase, but it is too insoluble in water to be useful in vivo.) Because of these properties, the body tissues are continuously exposed to low concentrations of methisazone and related compounds, and their absence of charge facilitates passage across cell barriers. An important property which would be expected to improve their persistence in tissues is their lipid solubility.

Conversely, the negatively charged CaEDTA complex and the EDTA anion cannot readily cross cell membranes to chelate the free zinc ions inside cells or to react with zinc bound in the RNA polymerase. This, coupled with the rapid excretion of EDTA by the kidney, affords little opportunity to deplete the tissues of the zinc they contain (although removal of free zinc ions from blood plasma takes place as expected). To combat influenza by this method, it is necessary to deliver

TABLE 5

Effect on Metal Ion Distribution of Adding CaEDTA and CaDTPA to the Blood Plasma Model

	0	2 m*M* CaEDTA	2 m*M* CaDTPA
Cu(alb) + $Cu(alb)_2$	99.7%	94.4	6.4
Cu-EDTA	–	5.3	–
Cu–DTPA	–	–	92.6
Free Cu^{2+}	3.5×10^{-10}	2.1×10^{-10}	2.6×10^{-11} μM
Zn(alb) + Zn_2(alb)	98.8%	0.6	–
$Zn(cysH)_2$	0.8	–	–
Zn-EDTA	–	99.4	–
Zn-DTPA	–	–	99.99
Free Zn^{2+}	1.6×10^{-3}	8.5×10^{-6}	5.8×10^{-7} μM

the chelating agent to a highly selective site, with the cytoplasm of cells lining the respiratory tract. This is probably why CaEDTA, an effective inhibitor of RNA polymerase in vitro, has little effect in vivo. Similar comments apply to the disulfonates of bathocuproin and bathophenanthroline, although the parent heterocycles would probably be useful if they were not so sparingly soluble.

Finally, the wide spectrum of activities in which zinc is involved enables depletion of general levels of zinc in the body to such an extent that RNA polymerase synthesis or activity is significantly affected, and this may have serious biological disadvantages. This sets definite limits to the possible level of a chelating agent that can be administered to an animal. For example, because of its effect on body zinc ion, a dose level of only 2 m*M* calcium DTPA (diethylenetriaminepentaacetic acid) per kilogram of body weight is toxic.[59] Similarly, zinc deficiency in pregnant female rats depressed nucleic acid synthesis and caused gross congenital malformations in a high proportion of full-term fetuses.[60]

Nevertheless, there is cause for guarded optimism based on observations such as the specific inhibition by selenocystine of viral RNA polymerase but not cellular (chicken embryo) DNA-dependent RNA polymerase.[61] Against this, however, are the implications of the recent finding of an RNA-dependent RNA polymerase in rabbit reticulocytes.[62] One way to test metal-chelating agents for their suitability as virus inhibitors would be to compare the concentrations of ligands that would give 50% inhibition with the concentrations inhibiting essential cellular zinc-containing enzymes such as dehydrogenases.

The metabolic fate of administered pharmaceuticals is also important. Thus, 5-hydroxy-2-formylpyridine thiosemicarbazone is relatively inactive in vivo against ribonucleoside diphosphate reductase because of its short half-life (initially 2½ to 10 min) in the blood and the rapid formation and elimination of its glucuronide conjugate.[63] The major pathway in dogs and mice for the degradation of 1-methylisatin 3-thiosemicarbazone[64] and isoquinoline-1-aldehyde thiosemicarbazone[65] is by desulfuration to give the semicarbazone and inorganic sulfate.

Other important factors are the toxicity of the material and the therapeutic index (the ratio of therapeutic and toxic levels). Considerations such as these govern the dose rate and method of administration of drugs (Chapter 2, Volume I). For agents such as isatin 3-thiosemicarbazone, which act intracellularly against viruses, frequent and repeated dosage is necessary to maintain an adequate concentration in the cell long enough for its effects to be exerted. The low solubility of methisazone ensures its slow, continuing absorption when it is administered by mouth as an aqueous suspension.

A great deal of effort has gone into the modification of antiviral and antitumor drugs, especially the thiosemicarbazones. For example, the indiscriminate toxicity of 2-formylpyridine thiosemicarbazone has been decreased, and its solubility in water increased, by inserting a hydroxyl group in positions 3 or 5 of the pyridine ring, so that sodium salts can be formed.

VII. LIPOSOMES IN THE SELECTIVE DELIVERY OF ANTIVIRAL AGENTS

Attempts to prevent influenza, which are based on the systemic control of metal ion levels, may be unnecessarily difficult. Instead, the use of liposome-encapsulated chelating agents for the selective delivery of chelating agents to the interiors of cells of the respiratory tract is an attractive possibility.

Liposomes are finely dispersed phospholipid spherules (Figure 6), or vesicles around 1 to 10 μm in diameter, made up of concentric multiple bilayers that incorporate water and low molecular weight solutes in compartments between the

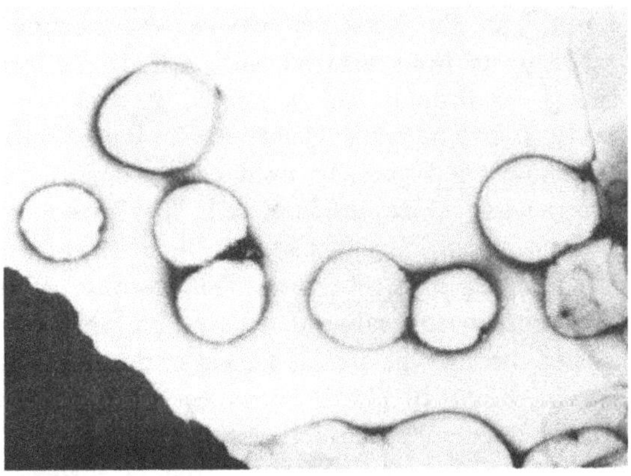

FIGURE 6. Liposome preparation with spherules approximately the diameter of influenza virus particles.

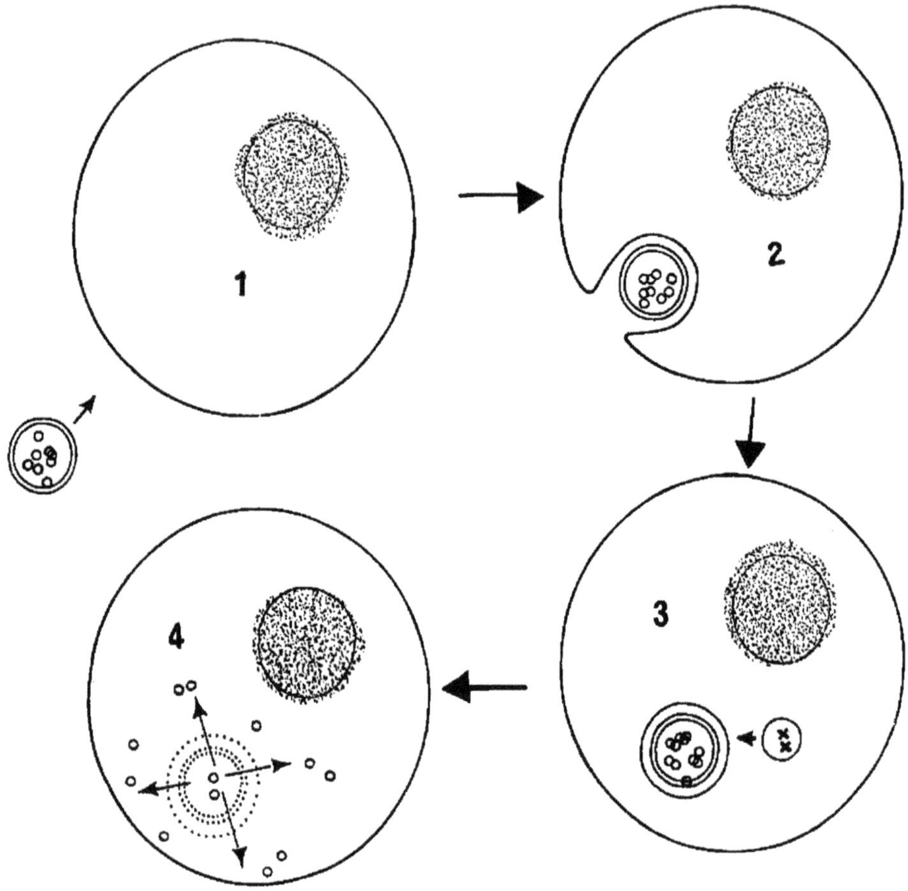

FIGURE 7. Computer-drawn representation of liposome entry into a mammalian cell. Liposome enclosing the drug approaches all (1) and penetrates the plasma membrane by pinocytosis (2). Cellular lipases disrupt the liposome coat (3) releasing drug intracellularly (4).

bimolecular lamellae.[66] They are also known as "liquid crystals" and can be formed by shaking suitable phospholipids, or their mixtures, with aqueous solutions.[67] If, for example, the solution contains CaEDTA, some of this is trapped in the interior of the liposome whereas the outside is made up of lipid material. Such a liposome looks like a fat droplet and is treated as such in the body. It can be taken up into a cell by pinocytosis (as shown in Figure 7), or it can be engulfed by phagocytes. Once inside a cell, the liposome is broken down by lysosomal lipases and the chelating agent is liberated. The function of the liposome, in concealing its true nature until it has passed through the wall, is seen as analogous to the Trojan horse of Greek antiquity. This property makes liposomes useful tools for the intracellular delivery of chelating agents or drugs. It also helps protect such drugs from metabolic modification and immunological reaction and decreases the extent of undesirable side reactions. The lipid composition may be varied considerably, giving a range of membrane structure, and charged liposomes can be formed by incorporating bases, such as stearylamine, or anionic species, such as dicetyl phosphate or phosphatidic acid.

The chelating agents, EDTA and DTPA, have been successfully encapsulated using a 3:1 mixture of phosphatidylcholine (egg lecithin) and cholesterol.[68] Electron microscopic examination following intravenous injection of such liposomes into mice showed that they were rapidly taken up by liver, spleen, and lungs. Within 15 min of injecting encapsulated [^{14}C]-EDTA, 39% of the EDTA was present in the liver, 8% in the spleen, 26% in the lungs, and 8% remained in the blood.[69] After 6 hr the figures were: liver, 41%; lungs, 10%; spleen, 11%; blood, 1%. Even after 1 day there was still 24% in the liver, 11% in spleen, 7% in lungs, and 0.5% in blood. Thus, it is possible to maintain a high intracellular level of chelating agent.

The liposomes in the liver were phagocytized by hepatocytes and Kupffer cells.[69] Initially there was close contact between the liposomal membrane and the cellular matrix. This was followed by invagination of the latter to form a distinct vesicle surrounding one or more liposomes. Such phagocytic vacuoles were found within cells of the reticuloendothelical system.[70] Morphological changes took place in hepatocytes 24 hr after injection, giving large numbers of autophagic vacuoles and, in some cells, extensive areas of focal cytoplasmic degeneration.[69] The morphology had reverted to normal 7 days after injection.[69] These lesions are similar to those seen in the kidney following intraperitoneal injection of CaEDTA solutions.[71]

By using encapsulation, the deposition and tissue-retention of EDTA is significantly increased. A high tissue concentration of EDTA can be maintained, with gradual release of chelating agent, and intracellular penetration of EDTA is achieved. Liposomes, being nonprotein, are unlikely to provoke immunological reactions. Rahman and Wright[69] conclude that "chelating agents injected in encapsulated form were not only transported across the cellular membrane, but they were released inside the cell where they successfully chelated the metal [Pb and Pu] and removed it from the cells."

In applying this approach to the chemoprophylaxis of influenza, the choice of chelating agents is reasonably wide as long as they are not inherently toxic or exert undesirable long-term effects. Results from in vitro testing become applicable, and it is now advantageous to use ionized ligands that are retained inside the target cells. CaEDTA (or better, CaDTPA) is chemically stable and would not leak out of the cell or disturb calcium metabolism; it chelates zinc and copper by selectively displacing calcium ions from CaEDTA. For a concentration of 1 m*M* CaEDTA, in a solution of 1 m*M* in free calcium and pH 7.4, free zinc ion concentration would be depressed to one millionth of what it would be in the absence of CaEDTA. Use of CaDTPA would further lower the concentration by a factor of 100. The induced zinc deficiency in the cytoplasm of the particular cell would persist until sufficient zinc ion had been transported across the cell membrane from the blood plasma. In vitro tests suggest that sodium bathophenanthroline disulfonate would also be effective, and this agent is also likely to be long persisting. By comparison, heterocyclic thiosemicarbazones would be less suitable, first because they could not be prepared in high enough concentrations inside the water-containing compartments of the liposomes and second because absence of charge would favor their rapid diffusion from the active site. The first of these disadvantages might be overcome by incorporating the thiosemicarbazone in the lipid part of the liposome.

The successful application of liposomes requires that they can be delivered to, and selectively absorbed by, a particular group of cells. Ideally, all cells that are invaded by an influenza virus should also have taken up one or more liposomes. Recent work[4] suggests that added ferritin is taken up synergistically by cells invaded by influenza virus. This may indicate an increased suceptibility of such cells to penetration by foreign materials, including liposomes. However, even a partial occupation of respiratory tract cells by chelating agents would be beneficial in slowing down the rate of propagation of virions, thereby giving the body's natural defenses more opportunity to contain the invasion and ensuring that the disease runs a milder course. With these thoughts in mind, attempts are currently being made[43] to design suitably sized liposomes and a method of administration (for example by nasal spray) for use on ferrets, (chosen as being nearest to man in the etiology of influenza).

VIII. SOME POSSIBLE DEVELOPMENTS

The suggested use of chelating agents and liposomes to combat influenza is an example of the selective manipulation of molecular architecture. Hopefully, other related possibilities will become apparent to the reader. As more becomes known about ways in which drugs could interfere with viral growth, opportunities increase for the theoretical design of improved antiviral agents. Thus, inclusion of neuraminidase in an aqueous preparation of an antiviral agent to be sprayed into the nasal cavity might be expected to facilitate penetration of the mucus layer and hence improve the drug's effectiveness.

Similarly, the administration of RNAase in liposomes to selected tissues in viral disease might break the cycle of virus replication. The rationale for this suggestion comes from work in which liposomes containing an enzyme (amyloglucosidase) have been prepared and freed from extraneous protein by column chromatography.[72] Their possible role as carriers of enzymes has been examined in relation to enzyme replacement therapy in storage diseases.[73] Following intravenous injection into rats, the liposomes are transported mainly inside the liver (parenchymal and Kupffer cells), where the proteins eventually undergo degradation by lysosomes. Free enzyme does not circulate in the blood.

Greater selectivity in the uptake of liposomes in treating influenza might be achieved by incorporating fragmented outer envelopes of influenza virions into the liposomal surface so that the liposome bears an increased resemblance to a virion. Liposome-entrapped proteins may be partially arranged at or near the surface interdigitated with the lipid.[74] The fragment containing spikes of neuraminidase and hemagglutin, by conferring greater affinity for cells likely to be invaded by influenza virus, would function as a "probe," as defined by Gregoriadis.[75] If liposomes can be made sufficiently target selective, possibilities inherent in their use would warrant a reappraisal of chemotherapeutic approaches to antiviral therapy, with results of in vitro testing having increased relevance. Already, liposomes containing inhibitors of nucleic acid synthesis, including 5-fluorouracil in dipalmitoyl lecithin-stearylamine and actinomycin B in egg lecithin-stearylamine or -phosphatidic acid, have been prepared.[76] (Actinomycin D binds specifically to double-stranded DNA.[77]) Administration of ribavirin encapsulated in liposomes would be advantageous. Ribavirin, 1-β-D-ribofuranosyl-1,2,4-triazole-3-carboxamide, is a broad-spectrum antiviral agent which inhibits the replication of influenza virus in mice[78] and ferrets[79] and is possibly a general inhibitor of RNA synthesis. It is desirable to localize such an action, rather than to affect nucleic acid synthesis throughout the entire animal. Cancer chemotherapeutic agents that inhibit nucleic acid synthesis, such as the five drugs reported by Horwitz,[80] should also be examined in this way for antiviral activity.

Neuraminidase facilitates the release of influenza virions from the cells in which they are formed.[81] There could, therefore, be advantages in using liposomes containing 2-deoxy-2,3-dehydro-*N*-trifluoroacetylneuraminic acid, which is a specific inhibitor of this enzyme.[82] Liposomes are also potentially useful as immunological adjuvants.[83]

The rapidly growing interest in liposomes leads one to expect that further useful properties will emerge. Thus, it has recently been found that when liposomes are injected intravenously into rats, there is an initial rapid rate of removal from plasma, depending on the charge carried by the liposome.[84] This is fastest for negatively charged liposomes (dicetyl phosphate and phosphatidic acid), slower for neutral liposomes, and slowest for positively charged liposomes (stearylamine). The rates, which varied about tenfold during 1 hr, paralleled the rates of uptake by liver. Such a difference might be exploited where there is competitive uptake between the liver and another organ. Again, it has been found that after an enzyme has been entrapped in a liposome, it can still be cross-linked with glutaraldehyde, and this process increases the stability of the enzyme.[84]

By appropriate choice of lipid composition, it may be possible to make liposomes in which the liposomal membrane will fuse directly with the membrane of the target cell, with simultaneous discharge of the material present in the liposome into the cytoplasm. Examples are already known.[85,86]

REFERENCES

1. **Kapikian, A. Z.,** A collaborative report: rhinoviruses – extension of the numbering system, *Virology,* 43, 524, 1971.
2. **Gross, L.,** Oncogenic viruses, Pergamon Press, Oxford, 1970.
3. **Fazekas de St. Groth, S.,** Viropexis, the mechanism of virus infection, *Nature,* 162, 294, 1948.
4. **Dourmashkin, R. R. and Tyrrell, D. A. J.,** Electron microscopic observations on the entry of influenza virus into susceptible cells, *J. Gen. Virol.,* 24, 129, 1974.
5. **Albert, A.,** *Selective Toxicity,* 5th ed., Chapman and Hall, London, 1973, 144.
6. **Oxford, J. S.,** Specific inhibitors of influenza virus replication as potential chemoprophylactic agents, *J. Antimicrob. Chemother.,* 1, 7, 1975.
7. **Hamre, D., Bernstein, J., and Donovick, R.,** Activity of *p*-aminobenzaldehyde 3-thiosemicarbazone on vaccinia virus in the chick embryo and in the mouse, *Proc. Soc. Exp. Biol. Med.,* 73, 275, 1950.
8. **Druett, H. A., Robinson, J. M., Henderson, D. W., Packman, L., and Peacock, S.,** Studies on respiratory infection. II. The influence of aerosol particle size on infection of the guinea pig with *Pasteurella pestis, J. Hyg.,* 54, 37, 1956.
9. **Thompson, R. L., Minton, S. A., Officer, J. E., and Hitchings, G. H.,** Effect of heterocyclic and other thiosemicarbazones on vaccinia infection in the mouse, *J. Immunol.,* 70, 229, 1953.
10. **Bauer, D. J. and Sadler, P. W.,** The structure-activity relationship of the antiviral chemotherapeutic activity of isatin β-thiosemicarbazone, *Br. J. Pharmacol.,* 15, 101, 1960.
11. **Bauer, D. J., St. Vincent, L., Kempe, C. H., and Downie, A. W.,** Prophylactic treatment of smallpox contacts with *N*-methylisatin β-thiosemicarbazone, *Lancet,* 2, 494, 1963.
12. **Levinson, W., Woodson, B., and Jackson, J.,** Inactivation of Rous sarcoma virus on contact with *N*-ethylisatin β-thiosemicarbazone, *Nat. New Biol.,* 232, 116, 1971.
13. **Bauer, D. J. and Apostolov, K.,** Adenovirus multiplication: inhibition by methisazone, *Science,* 154, 796, 1966.
14. **Ribiero do Valle, L. A., Raposo de Melo, P., de Salles Gomes, L. F., and Morato Proenca, L.,** Methisazone in prevention of variola minor among contacts, *Lancet,* 2, 976, 1965.
15. **Bauer, D. J.,** Clinical experience with the antiviral drug Marboran (1-methylisatin 3-thiosemicarbazone), *Ann. N.Y. Acad. Sci.,* 130, 110, 1965.
16. **Rao, A. R., McFadzean, J. A., and Squires, S.,** The laboratory and clinical assessment of an isothiazole thiosemicarbazone (M and B 7714) against pox virus, *Ann. N.Y. Acad. Sci.,* 130, 118, 1965.
17. **Tonew, M., Tonew, E., and Heinisch, L.,** Antiviral thiosemicarbazones and related compounds. II. Antiviral action of substituted isatinisothiosemicarbazones, *Acta Virol. (Engl. Ed.),* 18, 17, 1974.
18. **French, F. A. and Blanz, E. J.,** The carcinostatic activity of thiosemicarbazones of formylheteroaromatic compounds. III. Primary correlation, *J. Med. Chem.,* 9, 585, 1966.
19. **Appleyard, G., Hume, V. B. M., and Westwood, J. C. N.,** The effect of thiosemicarbazones on the growth of rabbit pox virus in tissue culture, *Ann. N.Y. Acad. Sci.,* 130, 92, 1965.
20. **Woodson, B. and Joklik, W. K.,** The inhibition of vaccinia virus multiplication by isatin β-thiosemicarbazone, *Proc. Natl. Acad. Sci. U.S.A.,* 50, 946, 1965.
21. **Haase, A. T. and Levinson, W.,** Inhibition of RNA slow viruses by thiosemicarbazones, *Biochem. Biophys. Res. Commun.,* 51, 875, 1973.
22. **Levinson, W., Faras, A., Woodson, B., Jackson, J., and Bishop, J. M.,** Inhibition of RNA-dependent DNA polymerase of Rous sarcoma virus by thiosemicarbazones and several cations, *Proc. Natl. Acad. Sci. U.S.A.,* 70, 164, 1973.
23. **Oxford, J. S. and Perrin, D. D.,** Inhibition of the particle-associated RNA-dependent RNA polymerase activity of influenza viruses by chelating agents, *J. Gen. Virol.,* 23, 59, 1974.
24. **Ho, P. P. K. and Walters, C. P.,** Inhibitors of influenza virus-induced RNA polymerase, *Ann. N. Y. Acad. Sci.,* 173, 438, 1970.
25. **Carpy, S.,** Inhibition de l'anhydrase carbonique erythrocytaire bovine B par differents agents chélateurs á pH 7.4, *Biochim. Biophys. Acta,* 151, 245, 1968.
26. **Hirose, M., Sugimoto, E., and Chiba, H.,** Studies on crystalline yeast phosphoglucomutase: the presence of intrinsic zinc, *Biochim. Biophys. Acta,* 289, 137, 1972.
27. **Valenzuela, P., Morris, R. W., Faras, A., Levinson, W., and Rutter, W. J.,** Are all nucleotidyl transferases metalloenzymes?, *Biochem. Biophys. Res. Commun.,* 53, 1036, 1973.
28. **Springgate, C. F., Mildvan, A. S., Abramson, R., Engle, J. L., and Loeb, L. A.,** *Escherischia coli* deoxyribonucleic acid polymerase. I. a zinc metalloenzyme, *J. Biol. Chem.,* 248, 5987, 1973.
29. **Slater, J. P., Mildvan, A. S., and Loeb, L. A.,** Zinc in DNA polymerases, *Biochem. Biophys. Res. Commun.,* 44, 37, 1971.
30. **Coleman, J. E.,** The role of Zn(II) in transcription by T7 RNA polymerase, *Biochem. Biophys. Res. Commun.,* 60, 641, 1974.
31. **Poiesz, B. J., Battula, N., and Loeb, L. A.,** Zinc in reverse transcriptase, *Biochem. Biophys. Res. Commun.,* 56, 959, 1974.
32. **Auld, D. S., Kawaguchi, H., Livingston, D. M., and Vallee, B. L.,** RNA-dependent DNA polymerase (reverse transcriptase) from avian myeloblastosis virus: a zinc metalloenzyme, *Proc. Natl. Acad. Sci. U.S.A.,* 71, 2091, 1974.

33. **Chang, L. M. S. and Bollum, F. J.**, Deoxynucleotide-polymerizing enzymes of calf thymus gland. IV. Inhibition of terminal deoxynucleotidyl transferase by metal ligands, *Proc. Natl. Acad. Sci. U.S.A.*, 65, 1041, 1970.
34. **Stavrianopoulos, J. G., Karkas, J. D., and Chargaff, E.**, DNA polymerases of chicken embryo: purification and properties, *Proc. Natl. Acad. Sci. U.S.A.*, 69, 1781, 1972.
35. **Scrutton, M. C., Wu, C. W., and Goldthwait, D. A.**, The presence and possible role of zinc in RNA polymerase obtained from *Escherichia coli*, *Proc. Natl. Acad. Sci. U.S.A.*, 68, 2497, 1971.
36. **Chow, N. and Simpson, R. W.**, RNA-dependent RNA polymerase activity associated with virions and subviral particles of myxoviruses, *Proc. Natl. Acad. Sci. U.S.A.*, 68, 752, 1971.
37. **Skehel, J. J.**, RNA-dependent RNA polymerase activity of the influenza virus, *Virology*, 45, 793, 1971.
38. **Moore, E. C., Zedeck, M. S., Agrawal, K. C., and Sartorelli, A. C.**, Inhibition of ribonucleoside diphosphate reductase by 1-formylisoquinoline thiosemicarbazone and related compounds, *Biochemistry*, 9, 4492, 1970.
39. **Elford, H. L., Freese, M., Passami, E., and Morris, H. P.**, Ribonucleotide reductase and cell proliferation. I. Variations of ribonucleotide reductase activity with tumour growth rate in a series of rat hepatomas, *J. Biol. Chem.*, 245, 5228, 1970.
40. **French, F. A., Blanz, E. J., Shaddix, S. C., and Brockman, R. W.**, α-(*N*)-Formylheteroaromatic thiosemicarbazones. Inhibition of tumour-derived ribonucleoside diphosphate reductase and correlation with in vivo antitumour activity. *J. Med. Chem.*, 17, 172, 1974.
41. **French, F. A. and Blanz, E. J.**, Chemotherapy studies on experimental mouse tumours. X. *Cancer Chemother. Rep.*, 2, 199, 1971.
42. **Katz, E., Margalith, E., and Winer, B.**, The effect of isatin β-thiosemicarbazone (IBT)-related compounds on IBT-resistant and on IBT-dependent mutants of vaccinia virus, *J. Gen. Virol.*, 25, 239, 1974.
43. **Oxford, J. S. and Perrin, D. D.**, in progress.
44. **Bauer, D. J., Apostolov, K., and Selway, J. W. T.**, Activity of methisazone against RNA viruses, *Ann. N.Y. Acad. Sci.*, 173, 314, 1970.
45. **Munro, T. W. 'and Sabina, L. R.**, Inhibition of infectious bovine rhinotracheitis virus multiplication by thiosemicarbazones, *J. Gen. Virol.*, 7, 55, 1970.
46. **Williams, R. J. P.**, Heavy metals in biological systems, *Endeavour*, 26, 96, 1967.
47. **Mathew, M. and Palenik, G. J.**, The crystal structure of bis(isoquinoline-1-carboxaldehyde thiosemicarbazanato) nickel(II) monohydrate, *J. Am. Chem. Soc.*, 91, 6310, 1969.
48. **Wacker, W. E. C.**, Nucleic acids and metals. III. Changes in nucleic acid, protein, and metal content as a consequence of zinc deficiency in *Euglena gracilis*, *Biochemistry*, 1, 859, 1962.
49. **Winder, F. G. and O'Hara, C.**, Effects of iron deficiency and of zinc deficiency on the composition of *Mycobacterium smegmatis*, *Biochem. J.*, 82, 98, 1962.
50. **Weser, U., Seeber, S., and Warnecke, P.**, Reactivity of Zn^{2+} on nuclear DNA and RNA biosynthesis of regenerating rat liver, *Biochim. Biophys. Acta*, 179, 422, 1969.
51. **Fujioka, M. and Lieberman, I.**, A Zn^{2+} requirement for synthesis of deoxyribonucleic acid by rat liver, *J. Biol. Chem.*, 239, 1164, 1964.
52. **Terhune, M. W. and Sandstead, H. H.**, Decreased RNA polymerase activity in mammalian zinc deficiency, *Science*, 177, 68, 1972.
53. **McQuitty, J. T., DeWys, W. D., Monaco, L., Strain, W. H., Job, C. C., Apgar, J., and Pories, W. J.**, Inhibition of tumour growth by dietary zinc deficiency, *Cancer Res.*, 30, 1387, 1970.
54. **Ferguson, L. N.**, Cancer and chemicals, *Chem. Soc. Rev.*, 4, 289, 1975.
55. **Oxford, J. S.**, An inhibitor of the particle-associated RNA-dependent RNA polymerase of influenza A and B viruses, *J. Gen. Virol.*, 18, 11, 1973.
56. **Perrin, D. D. and Sayce, I. G.**, Computer calculation of equilibrium concentrations in mixtures of metal ions and complexing species, *Talanta*, 14, 833, 1967.
57. **Hallman, P. S., Perrin, D. D., and Watt, A. E.**, The computed distribution of copper(II) and zinc(II) ions among seventeen amino acids present in human blood plasma, *Biochem. J.*, 121, 549, 1971.
58. **Agarwal, R. P. and Perrin, D. D.**, unpublished.
59. **Auth, U.**, Metabolismus und Toxizität therapeutischer Chelatbildner. XIII Mitteilung: Einfluss von Ca-DTPA auf die Zn-Konzentration der Organe, *Strahlentherapie*, 146, 490, 1973.
60. **Hurley, L. S.**, Zinc deficiency in the developing rat, *Am. J. Clin. Nutr.*, 22, 1332, 1969.
61. **Ho, P. P. K., Walters, C. P., Streightoff, F., Baker, L. A., and DeLong, D. C.**, In vivo activity of an inhibitor of influenza virus-induced ribonucleic acid polymerase, *Antimicrob. Agents Chemother.*, p. 636, 1967.
62. **Downey, K. M., Byrnes, J. J., Jurmark, B. S., and So, A. G.**, The effect of 6-mercaptopurine and its derivatives on mammalian RNA-dependent RNA synthesis, *Biochem. Biophys. Res. Commun.*, 56, 227, 1974.
63. **DeConti, R. C., Toftness, B. R., Agrawal, K. C., Tomchick, R., Mead, J. A. R., Bertino, J. R., Sartorelli, A. C., and Creasey, W. A.**, Clinical and pharmacological studies with 5-hydroxy-2-formylpyridine thiosemicarbazone, *Cancer Res.*, 32, 1455, 1972.
64. **Nutting, L. A., Weber, E. M., and Tyron, J. L.**, Metabolic removal of sulphur from 1-methylisatin 3-thiosemicarbazone, *J. Virol.*, 1, 650, 1967.

65. **Creasey, W. A., Agrawal, K. C., Capizzi, R. L., Stinson, K. K., and Sartorelli, A. C.**, Studies of the antineoplastic activity and metabolism of α-(*N*)-heterocyclic carboxaldehyde thiosemicarbazones in dogs and mice, *Cancer Res.*, 32, 565, 1972.
66. **Bangham, A. D., Standish, M. M., and Watkins, J. C.**, Diffusion of univaient ions across the lamellae of swollen phospholipids, *J. Mol. Biol.*, 13, 238, 1965.
67. **Sessa, G. and Weissmann, G.**, Phospholipid spherules (liposomes) as a model for biological membranes, *J. Lipid Res.*, 9, 310, 1968.
68. **Rahman, Y. E., Rosenthal, M. W., Cerny, E. A., and Moretti, E. S.**, Preparation and prolonged tissue retention of liposome-encapsulated chelating agents, *J. Lab. Clin. Med.*, 83, 640, 1974.
69. **Rahman, Y. E. and Wright, B. J.**, Liposomes containing chelating agents, *J. Cell. Biol.*, 65, 112, 1975.
70. **Segal, A. W., Wills, E. J., Richmond, J. E., Slavin, G., Black, C. D. V., and Gregoriadis, G.**, Morphological observations on the cellular and subcellular destination of intravenously administered liposomes, *Br. J. Exp. Pathol.*, 55, 320, 1974.
71. **Schwartz, S. L., Johnson, C. B., and Doolan, P. D.**, Study of the mechanism of renal vacuologenesis induced in the rat by ethylenediaminetetraacetate, *Mol. Pharmacol.*, 6, 54, 1970.
72. **Gregoriadis, G., Leathwood, P. D., and Ryman, B. E.**, Enzyme entrapment in liposomes, *FEBS Lett.*, 14, 95, 1971.
73. **Ryman, B. E.**, Possible use of liposomes in enzyme replacement therapy in storage diseases, in *Enzyme Therapy in Lysosomal Storage Diseases*, Tager, J. M., Hooghwinkel, G. J. M., and Daems, W. Th., Eds., North-Holland, Amsterdam, 1974, 149.
74. **Magee, W. E. and Miller, O. V.**, Liposomes containing antiviral antibody can protect cells from virus infection, *Nature*, 235, 339, 1972.
75. **Gregoriadis, G.**, Structural requirements for the specific uptake of macromolecules and liposomes by target tissues, in *Enzyme Therapy in Lysosomal Storage Diseases*, Tager, J. M., Hooghwinkel, G. J. M., and Daems, W. Th., Eds., North-Holland, Amsterdam, 1974, 131.
76. **Gregoriadis, G.**, Drug entrapment in liposomes: possibilities for chemotherapy, *Biochem. Soc. Trans.*, 2, 117, 1974.
77. **Reich, E.**, Actinomycin: correlation of structure and function of its complexes with purines and DNA, *Science*, 143, 684, 1964.
78. **Sidwell, R. W., Huffman, J. H., Khare, G. P., Allen, L. B., Witkowski, J. T., and Robbins, R. K.**, Broad spectrum antiviral activity of virazole: 1-β-D-ribofuranosyl-1,2,4-triazole-3-carboxamide, *Science*, 177, 705, 1972.
79. **Potter, C. W. and Oxford, J. S.**, unpublished.
80. **Horwitz, S. B.**, Novel inhibitors of RNA synthesis, *Fed. Proc.*, 33, 2281, 1974.
81. **Seto, J. T. and Rott, R.**, Functional significance of sialidase during influenza virus multiplication, *Virology*, 30, 731, 1966.
82. **Palese, P., Schulman, J. L., Bodo, G., and Meindl, P.**, Inhibition of influenza and parainfluenza virus replication in tissue culture by 2-deoxy-2,4-dehydro-*N*-trifluoroacetylneuraminic acid (FANA), *Virology*, 59, 490, 1974.
83. **Allison, A. C. and Gregoriadis, G.**, Liposomes as immunological adjuvants, *Nature*, 252, 252, 1974.
84. **Gregoriadis, G. and Neerunjun, D. E.**, Control of the rate of hepatic uptake and catabolism of liposome-entrapped proteins injected into rats, *Eur. J. Biochem.*, 47, 179, 1974.
85. **Grant, C. W. and McConnell, H. M.**, Fusion of phospholipid vesicles with viable *Acholeplasma laidlawii*, *Proc. Natl. Acad. Sci. U.S.A.*, 70, 1238, 1973.
86. **Papahadjopoulos, D., Poste, G., and Schaeffer, B. E.**, Fusion of mammalian cells by unilamellar lipid vesicles: influence of lipid surface charge, fluidity and cholesterol, *Biochim. Biophys. Acta*, 323, 23, 1973.

Guide to Index

Page numbers followed by **I** indicate Volume I.
Page numbers followed by **II** indicate Volume II.

INDEX

A

B

C

D

E

F

G

H

I

N

O

P

Q

R

S

T

TITLES OF INTEREST

CRC PRESS HANDBOOKS

CRC HANDBOOK OF CHEMISTRY AND PHYSICS, 57th Edition
Edited by **Robert C. Weast, Ph.D.**, Consolidated Natural Gas Co., Inc.

CRC HANDBOOK OF LABORATORY ANIMAL SCIENCE
Edited by **Edward C. Melby, Jr., D.V.M.**, The Johns Hopkins University School of Medicine, and **Norman H. Altman, V.M.D.**, Papanicolaou Cancer Research Institute.

CRC HANDBOOK OF MICROBIOLOGY
Edited by **Allen I. Laskin, Ph.D.**, Exxon Research and Engineering Co., and **Hubert Lechevalier, Ph.D.**, Institute of Microbiology, Rutgers University.

CRC HANDBOOK SERIES IN CLINICAL LABORATORY SCIENCE
Chief Executive Editor, **David Seligson, Sc.D., M.D.**, Yale University School of Medicine.

CRC PRESS "UNISCIENCE"™ TITLES

CHEMOTHERAPY OF INFECTIOUS DISEASE
By **Hans H. Gadebusch, Ph.D.**, Squibb Institute for Medical Research.

METHODS OF DETECTION AND IDENTIFICATION OF BACTERIA
By **Brij M. Mitruka, M.S., Ph.D.**, assisted by **Mary J. Bonner, M.S., Ph.D.**, both with the University of Pennsylvania Medical Center.

SELECTIVE INHIBITORS OF VIRAL FUNCTIONS
Edited by **William A. Carter, M.D.**, State University of New York at Buffalo.

CRC 'MONOTOPIC"™ TITLES

AUSTRALIA ANTIGEN AND HEPATITIS
By **Baruch S. Blumberg, M.D., Ph.D., Alton I. Sutnick, M.D., W. Thomas London, M.D.**, and **Irving Millman, M.S., Ph.D.**, all with the Institute for Cancer Research.

CRC CRITICAL REVIEWS™ JOURNALS

CRC CRITICAL REVIEWS™ IN MICROBIOLOGY
Edited by **Allen Laskin, Ph.D.**, Exxon Research and Engineering Co., and **Hubert Lechevalier, Ph.D.**, Institute of Microbiology, Rutgers University.

Direct all inquiries to CRC Press, Inc.